Handbook of

Single-Molecule Electronics

Handbook of SINGLE-MOLECULE ELECTRONICS

edited by

Kasper Moth-Poulsen

CRC Press
Taylor & Francis Group
Boca Raton London New York

CRC Press is an imprint of the
Taylor & Francis Group, an **informa** business

Handbook of Single-Molecule Electronics

First published in 2016 by Pan Stanford Publishing Pte. Ltd

Published 2022 by CRC Press
Taylor & Francis Group
6000 Broken Sound Parkway NW, Suite 300
Boca Raton, FL 33487-2742

ISBN-13: 978-981-4463-38-6 (hbk)

British Library Cataloguing-in-Publication Data
A catalogue record for this book is available from the British Library.

Visit the Taylor & Francis Web site at
http://www.taylorandfrancis.com

and the CRC Press Web site at
http://www.crcpress.com

Contents

Preface

Since the first visionary publications about the concepts of single-molecule electronics came out about 40 years ago, the research field has evolved into a truly interdisciplinary effort attracting contributors from various disciplines such as synthetic chemistry, theoretical chemistry and physics, experimental physics, and electronic engineering. Although great progress in both experimental realizations and theoretical understanding of single-molecule devices has been made, making it possible (to some extent) to predict and design molecular electronic components with tailored properties is still a challenge and still many steps are to be taken to overcome them before we will see widespread use of computing devices based on single molecules.

The aim of this book is to introduce a new generation of scientists to the fascinating research field of single-molecule electronics. The book consists of an introduction plus 12 chapters seeking to give a balanced view of the research field as it is seen through the eyes of experts coming from different areas of research—from synthetic chemistry, through modeling to experimental approaches and systems view. Needles to say, this book could not have come about without the contributions of the authors, and I would like to take this opportunity to thank specifically Anders Bergvall, Andrey Danilov, Mikael Fogelström, Thomas Frederiksen, Tina A. Gschneidtner, Joshua Hihath, Jaap Hoekstra, Cecilia Holmqvist, Won H. Jeong, Tomas Löfwander, Sergey Kubatkin, Shubhaditya Majumdar, Jonathan A. Malen, Mogens B. Nielsen, Kasper Nørgaard, Pramod S. Reddy, Jos S. Seldenthuis, Gemma C. Solomon, Jos M. Thijssen, and Herre S. J. van der Zant. I would also like to thank Pan Stanford Publishing for this opportunity, my collaborators and co-workers at Chalmers University of Technology and abroad, and all scientists who have and are currently contributing to bring single-molecule electronics research forward by creating a stimulating and creative world community. Finally, I would like to thank my family for their love and support.

This is the beginning!

Kasper Moth-Poulsen
Chalmers University of Technology
Sweden
Summer 2015

Chapter 1

Introduction

Kasper Moth-Poulsen
Department of Chemistry and Chemical Engineering,
Chalmers University of Technology, Gothenburg, 41296 Sweden
kasper.moth-poulsen@chalmers.se

Microfabrication and field-effect transistors are two key enabling technologies for today's information society. It is hard to imagine superfast and omnipresent electronic devices, information technology, the Internet, and mobile communication technologies without the access to continuously cheaper and miniaturized microprocessors. The giant leap in performance of microprocessors from the first personal computing machines to today's mobile devices is to a large extent realized via development of new materials and fabrication methods leading to miniaturization of the active components.

The ultimate limit of miniaturization of electronic components is the realization of components based on single molecules or even single atoms. Today there is a vibrant academic community that seeks to develop the basic understanding of and experimental realization of single-molecule components. The research field is very interdisciplinary, spanning from organic synthesis for the design and preparation of active molecular components, to experimental

Handbook of Single-Molecule Electronics
Edited by Kasper Moth-Poulsen

ISBN 978-981-4463-38-6 (Hardcover), 978-981-4463-39-3 (eBook)
www.panstanford.com

physics for the investigation and evaluation of the systems, to theoretical efforts seeking to understand electron transport and new physical phenomenon occurring at the molecular length scale (10^{-10} m to 10^{-9} m).

The scope of this book is to introduce a new generation of scientists to the exciting field of single-molecule electronics. The chapters are written with a nonspecialist audience in mind, yet the extensive nature of the chapters has allowed the authors to introduce and explain their research at a detailed level typically not found in the primary literature. Therefore, I hope that also the specialized reader will find useful references to the most recent literature. The ambition has been to cover all aspects of single-molecule electronics, from the theoretical aspects, through the experimental realizations and chemical synthesis of molecular components to how molecular components are thought to be implemented in future integrated circuits.

The book is organized as follows: Chapter 2 by *K. Moth-Poulsen* gives an introduction to the most commonly employed experimental techniques and discusses the challenges associated with trying to contact single molecules by two or three electrodes located a few nanometers from each other. Chapter 3 by *A. Bergvall, M. Fogelström, C. Holmqvist,* and *T. Löfwander* introduces a theoretical framework to describe electron transport through single-molecule systems. The chapter sets out from single-level systems and extends to descriptions of more complicated structures, including molecular systems coupled to graphene electrodes. Chapter 4 by *T. Frederiksen* extends on the theoretical treatment by employing first-principles simulations to describe electron transport in atomic-scale systems. The theoretical framework in this chapter is based on the density-functional theory (DFT) for the electronic structure in combination with nonequilibrium Green's functions (NEGF) for the transport. Recent developments based on the DFT–NEGF approach are used to describe electron-vibration interactions in molecular junctions, local heating effects, and inelastic signatures in device current–voltage and shot noise characteristics. Chapter 5 by *J. Hihath* focuses on the importance of controlling the molecule–electrode interface in single-molecule devices.

A simple transport model is used to demonstrate how contact affects transport properties, explore the use of various molecule–electrode linker chemistries, and also discuss controlling the molecule–electrode contact mechanically to control the device's charge transport properties. Chapter 6 by *J. S. Seldenthuis, H. S. J. van der Zant,* and *J. M. Thijssen* describes vibrational excitations in single-molecule junctions. The effect of vibrational excitations on transport strongly depends on the coupling between the molecule and the electrodes. In the weak-coupling regime, transport is incoherent and the current is dominated by sequential tunneling processes. In the opposite limit, the strong-coupling regime, transport is coherent. Electrons are delocalized over the molecule and the electrodes. The current is generally dominated by elastic tunneling processes, which do not excite vibrational modes. It is, however, possible for electrons to tunnel inelastically by emitting or absorbing a phonon. Such a process opens a new transmission channel, which can lead to either an increase or a reduction in conductance. The chapter discusses experimental observations of both regimes of transport related to vibrational effects. Chapter 7 by *T. Gschneidtner* and *K. Moth-Poulsen* introduces basic concepts of self-assembly at interfaces with a special emphasis on molecules at metal surfaces. New electrode materials such as graphene and carbon nanotubes are also discussed. Chapter 8 by *M. B. Nielsen* presents chemical approaches for the design and synthesis of molecular switches. This chapter describes how a variety of redox- and photoactive molecules have been used for conductance switching. Synthetic protocols for some of the switches will also be presented. Molecular switches and switching mechanisms are further discussed in Chapter 9 by *A. Danilov* and *S. Kubatkin.* In this chapter, basic switching mechanisms in molecular switches—thermal fluctuations, current-induced excitations, and quantum tunneling—are reviewed. It is demonstrated how experimental observations can be used to distinguish between different switching mechanisms in single-molecule devices. It is also discussed how the intrinsic switching properties may be affected when the molecule is bridged to electrodes, and how to distinguish whether the switching happens in the molecular kernel or at the molecule-to-electrode interface. Chapter 10 by *S. Majumdar,*

W. H. Jeong, J. A. Malen, and *P. S. Reddy* introduces a theoretical framework needed to describe thermoelectricity in molecular junctions and discusses experimental observations. Chapter 11 by *G. C. Solomon* introduces several models to describe and evaluate molecular systems capable of demonstrating quantum interference effects. Chapter 12 by *K. Nørgaard* and *T. Jain* introduces possible fabrication schemes for molecular junctions through self-assembly processes as well as strategies for incorporating these molecular junctions into functional devices. The scope for up-scaling these concepts into devices consisting of multiple molecular junctions will be discussed. Chapter 13 by *J. Hoekstra* discusses circuit design in future single-molecule electronics systems where the active components are based on single molecules.

Chapter 2

Experimental Techniques

Kasper Moth-Poulsen
Department of Chemistry and Chemical Engineering,
Chalmers University of Technology, Gothenburg, 41296 Sweden
kasper.moth-poulsen@chalmers.se

Bridging the gap between the molecular and mesoscopic length scales is one of the most fundamental challenges in nanoscience. A particular challenge is to develop methods for selective functionalization of nanostructures with single-molecule spatial resolution and stoichiometry. In this context, the research field of single-molecule electronics has seen impressive progress and creativity in the design of single-molecule devices. This chapter introduces the main concepts and experimental techniques employed for the study of electron transport through single molecules. Methods such as scanning probe techniques, mechanically controlled break junctions (MCBJs), and electromigration techniques have been introduced.

2.1 Introduction

The realization of single-molecule electronic devices is challenging in several ways. First, the typical length of molecules used in the research field is in the order of 1–2 nm. Fabrication of electrodes,

Handbook of Single-Molecule Electronics
Edited by Kasper Moth-Poulsen

ISBN 978-981-4463-38-6 (Hardcover), 978-981-4463-39-3 (eBook)
www.panstanford.com

typically made of noble metals, separated by 1–2 nm is beyond the limits of classical top-down lithographic techniques [16]. Second, due to the tiny dimensions of the molecule, it is typically impractical to place the molecule in the nanogap by direct manipulation. Instead, chemical interaction between the molecule and the electrode is needed for positioning of a molecule in the gap between the electrodes. Third, since the electrodes are typically much larger than the molecules, it is an additional challenge to make sure that only a single molecule is placed in each functional device. In addition to these three basic challenges, other challenges such as device stability, uniformity, yield, and scalability are equally important. This chapter gives an overview of the most common experimental techniques employed toward addressing these challenges. The chapter is divided into two subsections: The first gives a general introduction to different experimental techniques. The second part discusses single-molecule signatures, statistics, and data selection. The details of self-assembly at surfaces in particular are discussed further in Chapter 7, "Self-Assembly at Interfaces"; how the interface between the molecule and the electrode is formed and how molecules can be placed in nanogaps by chemical means. Chapter 12, "Parallel Self-Assembly Strategies toward Multiple Single-Molecule Devices," will go into more details with self-assembled parallel fabrication of multiple single-molecule devices.

2.2 Experimental Techniques

The fabrication of nanogap electrodes for single-molecule electronics can be roughly divided into three general classes: (1) mechanical methods, (2) nanofabrication methods, and (3) self-fabrication methods. In addition to these three general types of devices, the experiments can be categorized as either two-terminal or three-terminal devices, depending on the number of electrodes in proximity to the molecular entity. Two-terminal devices allow for basic transport measurements, while three-terminal devices, employing a third, so-called gate electrode, allow for more thorough manipulation of the energy levels of the molecule with respect to the source and drain electrodes. The strong interaction between the

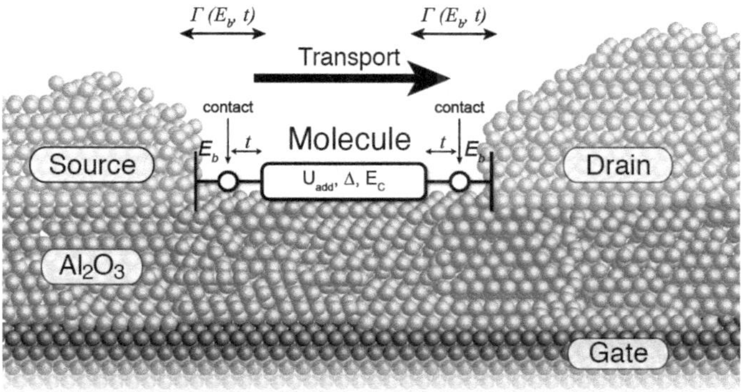

Figure 2.1 Sketch of a molecule placed in a nanogap, highlighting some of the important features: the coupling between the molecule and the electrode, and the energy levels in the molecule. Reprinted with permission from Ref. [17]. Copyright (2009) Nature Publishing Group.

molecule and its surroundings means that an honest description of a molecule in the device needs to take all these interactions into account [17]. Figure 2.1 outlines the basic forces acting on a molecule placed between three electrodes, where the electronic coupling between the molecule and the electrodes, together with the gate dielectric and electrode material, defines the properties of the single-molecule device.

The following sections will briefly review the basic features of these three methods.

2.2.1 *Mechanical Techniques*

Mechanical techniques involve a direct mechanical formation of a nanogap between two electrodes. These two electrodes can be formed from the breaking of a thin wire (MCBJ technique) or formed via close contact between a movable atomic force microscope (AFM) or a scanning tunneling microscope (STM) tip and a substrate. Molecules are typically introduced to the devices via self-assembly from solution or evaporation via the gas phase [18].

A strength of mechanical methods is the rate with which single-molecule experiments can be performed; in ideal cases, thousands of single-molecule devices can be formed and evaluated in a single day [19]. The challenge with the mechanical methods is typically long-term device stability and that the methods are restricted to the realization of a single device at a time.

2.2.1.1 Scanning tunneling microscope

The STM was the first technique to directly image single atoms and molecules [20]. The method is based on precise positioning of a sharp conducting tip over a conducting surface. Since the tip is placed close to the surface, a tunneling current can be measured when a voltage bias i applied between the tip and the surface. By operating the STM in either constant current mode or constant height mode, basic information about the electronic transparency of the molecules can be observed [21–23]. A simple expression for the tunneling current through single molecules is given by Eq. 2.1–2.3, which describe the tunneling current through the system as a so-called double-layer model [24]. Since the tip is placed at some distance d from the molecule, tunneling through the surrounding media (typically vacuum or air) needs to be taken into account. Equations 2.1 and 2.2 describe the tunneling through air and through the molecule as simple tunneling barriers. Due to the geometry of the experiment, the two tunneling barriers can be combined to yield Eq. 2.3 as a simple expression to describe the tunneling through the whole system.

$$G = G_{\mathrm{air}} \exp(-\alpha_i d_i) \tag{2.1}$$

$$G = G_i \exp(-\beta_i h_i) \tag{2.2}$$

$$G = G_i \exp(-\beta_i h_i) G_{\mathrm{air}} \exp(-\alpha_i d_i) \tag{2.3}$$

G is the transconductance of the system under study. G_i denotes the contact conductance defined by the molecule metal coupling, β is the tunneling decay constant within the molecule, and h is the height of the molecule. G_{air}, d, and α denote the corresponding constants for the surrounding media. Tunneling mechanisms and more refined models for tunneling than the double-layer model are discussed in Chapter 5, "Controlling the Molecule–Electrode Contact in Single-Molecule Devices."

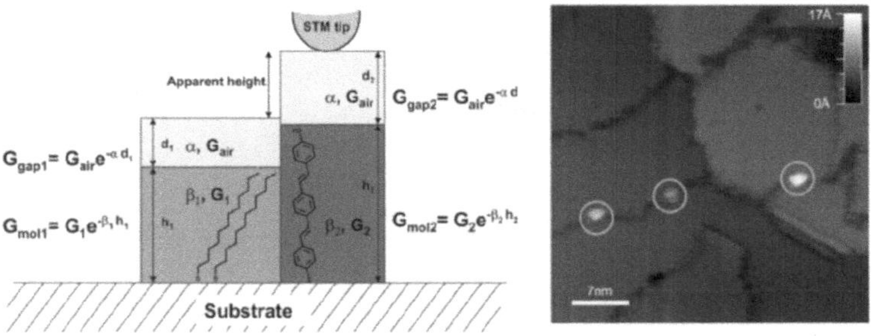

Figure 2.2 Sketch of two-layer model for a conducting molecule embedded in a matrix of dodecane thiols (left). STM picture of dodecane thiols on a gold substrate (right). The blue rings highlight inserted conducting molecules (thiol end-capped oligo phenylenes). Adapted with permission from Ref. [23]. Copyright (2005) American Chemical Society.

If the studied molecules are of similar type, matrix molecules can be used as an internal reference point to estimate the tunneling decay parameter (β) of the molecule under study [23, 24]. Figure 2.2 shows (left) a double-layer model illustrating tunneling barriers of dodecane thiol and a π-conjugated OPV3 molecule and (right) an STM image of a self-assembled monolayer of dodecane thiols on gold with three OPV3 molecules inserted (blue circles).

The so-called in situ STM method is a further enhancement of the STM method in the context of single-molecule electronics. In situ STM is carried out in condensed media such as water [5, 6, 18, 25] or ionic liquids [26]. These more advanced experiments allow for the use of additional electrodes for electrochemical control of the redox levels of the molecule under study. This makes it possible to carry out scanning tunneling spectroscopy on single molecules and to demonstrate diode- and transistor-like function at room temperature [5, 6, 25]. Figure 2.3 shows examples of redox-active osmium compounds used to demonstrate transistor-like behavior.

2.2.1.2 Atomic force microscope

The AFM differs from the STM in the way the feedback for the detailed determination of the position of the tip is constructed. While the STM relies on the exponential decay of the tunneling current with distance, the AFM uses a flexible cantilever, which

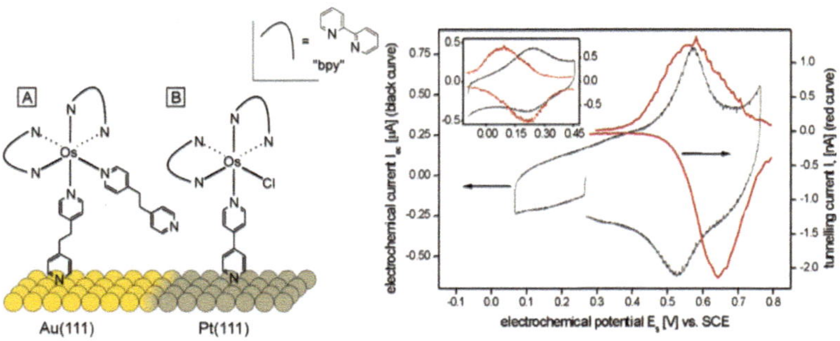

Figure 2.3 Transistor-like function in single molecules studied by in situ single-molecule STM: (left) studied osmium compounds; (right) in situ scanning tunneling spectroscopy of single molecules (red curves) and bulk measurements (black curves); large graph is molecule A, and insert is molecule B. Reprinted with permission from Ref. [6]. Copyright (2005) American Chemical Society.

bends when put in contact with the surface [27]. The position of the cantilever is then determined by a laser reflected from the rear of the cantilever onto an array of photodiodes. This concept has been refined during the last decades, and today it is possible not only to study molecular or atomic structures with AFM, but also to retrieve detailed information of forces and friction between the AFM tip and the surface. If the AFM tip is conductive, conductivity measurements can be performed concurrently with measurements of molecular forces [28–32].

For both AFM and STM, several ways of isolating and contacting single molecules have been developed. Weiss and coworkers used embedded π-conjugated molecular wires in a matrix of dodecane thiols to secure that they were probing a single molecule at a time [21]. Gold nanoparticles have been assembled on top of molecular wires to facilitate contact with the single molecules (Fig. 2.4) [33, 34].

Another widely applied method is the so-called STM break junction method [12]. Here the tip is crashed into a surface on which the molecules under study are assembled. The tip is then retracted from the surface, and the conductivity through the molecule is studied until it finally breaks. The advantage of this method is that many such measurements can be performed after

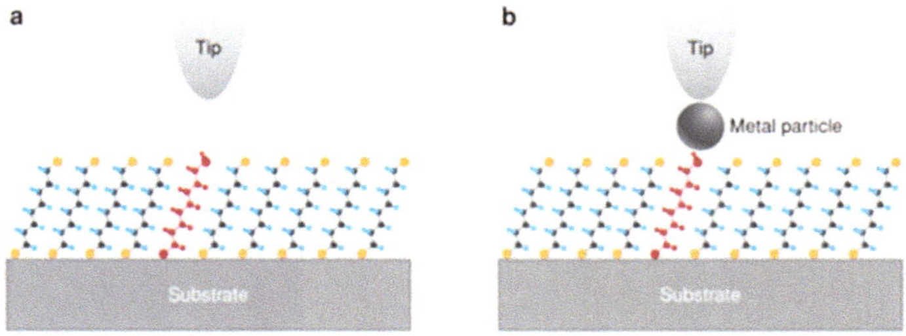

Figure 2.4 Schematic illustration of molecular wires embedded in a matrix of dodecane thiols: (A) molecules in matrix, and (B) molecules in matrix with nanoparticle on top. Reprinted with permission from Ref. [5]. Copyright (2005) American Chemical Society.

each other, making useful statistics, and the possibility to form histograms of the conductivity. Also, just before the junction breaks, the probability of having only a single molecule in the junction is relatively high. Figure 2.5 illustrates conductance traces, possible device geometries, and conductance histograms from STM break junction measurements. The conductance traces (Fig. 2.5A,C,E, left) correspond to the conductance as a function of electrode separation. The stepwise decline in conductivity seen in the conductance traces corresponds to integer variations in the available conductance channels, e.g., number of bridging gold atoms (Fig. 2.5A) or bridging molecules (Fig. 2.5C).

2.2.1.3 Mechanically controlled break junction

In MCBJs, a thin wire is placed on top of a flexible substrate. By applying force from a pushing rod, the flexible substrate can be bent. The device is designed so that a large change in the positioning of the pushing rod results in a minor change in the gap between the two electrodes. This makes it possible to break the wire to form nanogap junctions with well-controlled distance between the electrodes (Fig. 2.6).

Molecules can be introduced to the wire surface, either before or after the breaking. The measurements can be carried out in a solution of molecules or in vacuum where small molecules such as H_2 can be introduced via the gas phase [35–37]. The

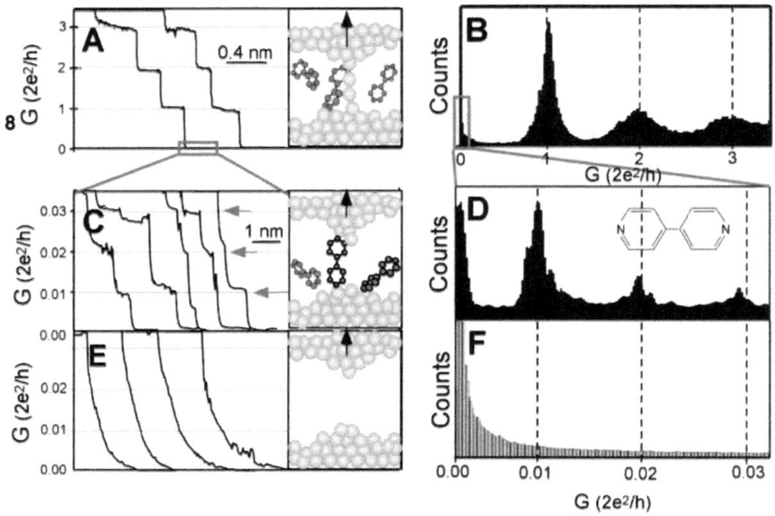

Figure 2.5 STM break junction method. Conductance traces and possible device geometries (A—gold wires, C—molecular conductance, E—without molecules) corresponding conductance histograms (B, D, F). Reprinted with permission from Ref. [12]. Copyright (2003) American Association for the Advancement of Science.

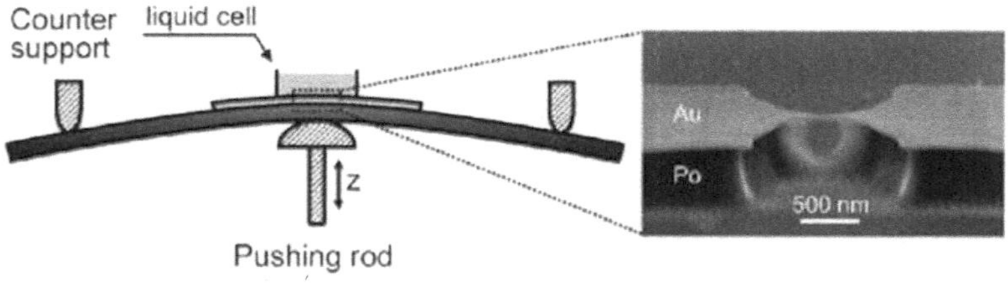

Figure 2.6 Lithographically prepared MCBJ setup: (left) schematic, and (right) SEM picture. Reprinted with permission from Ref. [14]. Copyright (2008) American Chemical Society.

first examples of molecular break junction devices were indeed based on breaking thin metal wires glued to a flexible substrate [38–40]. Today lithographically prepared break junction devices are widely applied [37]. The breaking of the electrodes is often repeated in a way similar to STM break junction method, so-called "crash and withdraw," the motivation being to expose new molecules and clean electrodes to each "crash and withdraw" cycle. A strength of the mechanical break junction method is that it can be refined to "high throughput" measurements, meaning many device realizations in a single day [41]. While the first mechanical break junction setups operated at low temperatures at ultra-high vacuum conditions, today examples of measurements performed at ambient conditions are more frequent. The integration of mechanical break junction experiments in electrochemical environment has been demonstrated [42, 43] as well as the incorporation of a third gate electrode extending the range of possible experiments [44–46].

2.2.2 *Nanofabrication Methods*

Nanofabrication methods seek to fabricate nanogap electrodes on substrates via advanced nanofabrication schemes. Since a 1–2 nm gap between two electrodes is beyond the capabilities of today's top-down lithographic techniques, some unconventional "tricks" are required. An advantage of the nanofabrication of molecular devices on a solid support is the typical longer stability of devices. This allows for electronic characterization of the devices to be performed over extended periods of time. Another advantage is the possibility to integrate a third so-called gate electrode that enables a more detailed electronic characterization of the single molecules in the devices.

2.2.2.1 Electromigration

The electromigration method is based on the breaking of a thin wire by the momentum of electrons transported through the wire [47–52]. The detailed structure of the formed nanogap is determined by several factors, including initial crystallinity of the electrode material and the voltage drop and current across the electromigrated wire [53]. The breaking of the wire is often carried out with a variable resistor placed in the circuit to minimize the voltage drop over the nanogap when it is breaking [54]. A too high voltage drop will lead to larger gaps between the nanosized electrodes and will increase the chance of forming gold islands

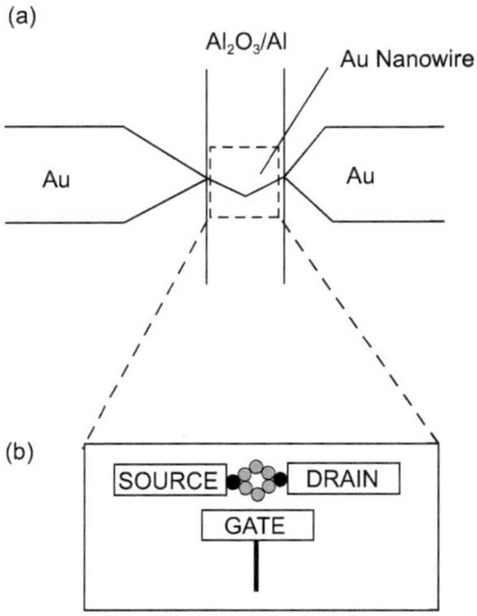

Figure 2.7 (A) Illustration of electromigration device [1]. A thin gold nanowire is placed between two larger electrodes. An oxidized aluminum back gate is placed under the gold nanowire. The tiny dimensions of the central gold nanowire ensure that the nanogap formation occurs in this central part of the device. (B) Illustration of a molecule in the nanogap (not drawn to scale).

in the nanogap that can display transport properties somewhat resembling that of molecular systems [54–59].

An advantage of the electromigration method is that it can be carried out on semiconductor surfaces. This makes it possible to place a third gate electrode under the nanogap electrodes. The illustration of an electromigration device is shown in Fig. 2.7. Integration of this third gate electrode allows for a more detailed electronic investigation of the studied molecule, since it can be used to manipulate the energy levels of the molecule independently with respect to the source and drain bias voltages. Molecules are typically introduced to the device after the formation of the nanogap electrodes, typically via solution-based self-assembly.

2.2.2.2 Evaporation methods

Another way to fabricate nanogap electrodes is to use the evaporation of metal electrode material at an oblique angle at a position

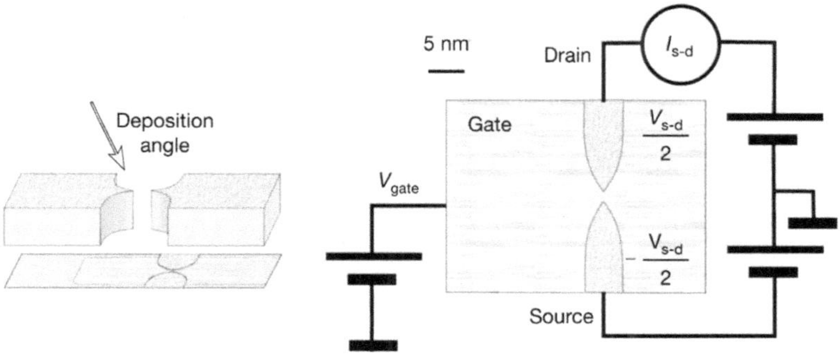

Figure 2.8 Fabrication of nanogap electrodes via evaporation of metal at an oblique angle. The molecules are introduced to the device by the evaporation of sub-monolayers followed by electrostatic capture of the molecules in the nanogap at cryogenic temperatures. Reprinted with permission from Ref. [8]. Copyright 2011, AIP Publishing LLC.

defined by a shadow mask (Fig. 2.8) [60–62]. By carefully controlling the evaporation angles, electrodes with nanometer separation can be fabricated. The distance between the electrodes is monitored by measuring the tunneling current between the source and drain electrodes [60–62]. When very narrow electrode gaps are achieved, a tunneling current in the nanoampere regime can be measured. Kubatkin, Danilov, and coworkers have then introduced molecules into the nanogaps via the evaporation of the molecules onto the surface during the same vacuum cycle that the electrodes were formed [60]. This ensures very clean surfaces and electrodes.

2.2.2.3 Direct manipulation methods

One of the strengths of STM and AFM experiments is the possibility to manipulate atomic and molecular structures directly at the surface. Eigler originally demonstrated the direct manipulation of atoms at surfaces [63]. An attractive strategy in the context of single-molecule electronics is the use of direct manipulation of atoms and molecules via STM to build devices in situ [64]. The advantage of this approach is that the combination of electronic measurements and STM imaging provides high-level structural information about the device geometry at the same time as the electronic measurements

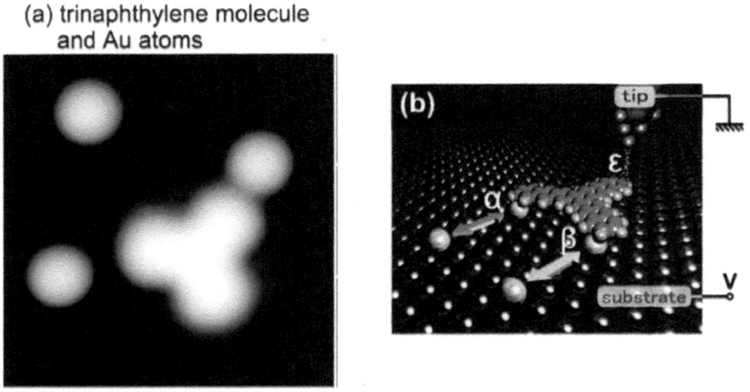

Figure 2.9 (a) STM picture of trinaphtyl molecule and three gold atoms. Logic function was demonstrated by manipulating the position of the gold atoms. (b) Artist's impression of the device geometry. Reprinted with permission from Ref. [3]. Copyright (2011) American Chemical Society.

are performed [65, 66]. Joachim and coworkers used this approach to demonstrate basic logic function in a device consisting of a trinaphtyl molecule and isolated gold atoms (Fig. 2.9). The device properties were changed by manipulating the position of the gold atoms and thereby the local coupling between the molecule and the substrate [3].

2.2.2.4 Low-dimensional electrode materials

Most of the methods described above are based on noble metals as the electrode material. Noble metals such as gold are three-dimensional materials, meaning that a reduction of dimension in three directions is needed if one wishes to construct atomic point contacts to single molecules. A way to reduce this challenge is to use two-dimensional materials such as graphene, which only extends one carbon atom in the third dimension, thus making it a very interesting material for single-molecule devices. Another advantage of the reduced dimensionality is that it might lead to reduced ambiguity in determining the molecule–electrode interface structure [67, 68]. Some groups have demonstrated that it is possible to fabricate nanometer electrode gaps in grapheme [69–

72], and devices of molecules attached to graphene electrodes have been demonstrated [69].

The use of nanorods [73–75] or carbon nanotubes reduces dimensionality of the electrodes further, since there is only space for a single or a few molecules in the nanogap between such small electrode structures [2, 76–79]. Nuckolls and coworkers have demonstrated that it is possible to use e-beam followed by plasma treatment to form nanometer gaps between two carbon nanotube electrodes [2, 80, 81].

The oxidized carbon nanotube contacts contain carboxyl groups, which make it possible to functionalize the electrodes with functional molecules via covalent chemical bonds. Figure 2.10 illustrates the fabrication scheme for single-wall carbon nanotube (SWCNT) nanogap electrodes. An SWCNT is placed between two macroscopic gold electrodes. The SWCNT is covered with PMMA resist, and e-beam lithography is used to expose a small section of the nanotube. Plasma etch is used to remove the exposed SWCNT and to oxidize the ends of the formed electrodes to form chemically addressable carboxyl groups (Fig. 2.10B).

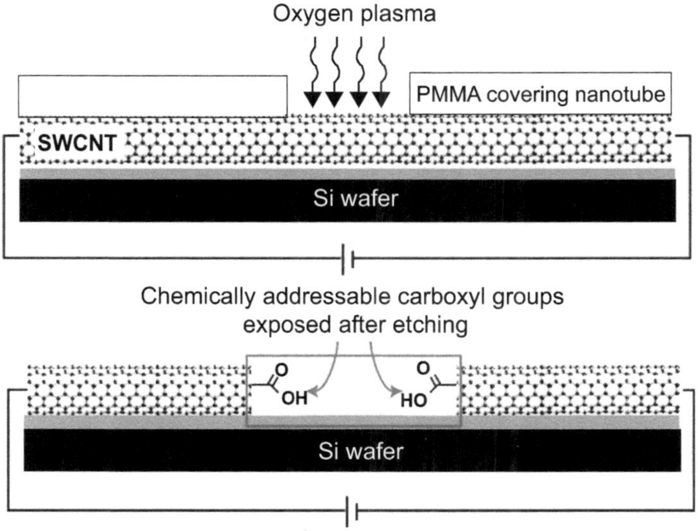

Figure 2.10 Carbon nanotube electrodes: Device fabrication scheme.

2.2.3 *Self-Assembled Devices*

Self-assembly is the assembly of supramolecular structures without direct human intervention [82]. Self-assembled single-molecule electronic devices can be seen as one step toward the self-fabrication of electronic circuits. Nobel laureate J.-M. Lehn introduced the term self-fabrication relating to the "controlled assembly of ordered, fully integrated, and connected operational systems by hierarchical growth" [83].

While most single-molecule electronic experiments rely on self-assembly of molecules at metal surfaces, some attempts are made to assemble larger parts of the single-molecule devices. The idea is to use self-assembly starting from the single molecule and forming larger structures around these using self-assembly on multiple length scales [8, 11, 73–75]. This approach should be seen in contrast to the top-down based mechanical and nanofabrication methods described above where nanogap electrodes are formed and molecules are subsequently placed in these nanogaps.

Israel and coworkers assembled gold nanoparticle dimers around thiol end-capped molecular wires [11]. The nanoparticle dimers were fabricated in a way that, statistically, only a single molecule would bridge the gap in each gold nanoparticle dimer. The nanoparticle dimers are later assembled between microfabricated electrodes via AC-field trapping (Fig. 2.11).

One of the strengths of this method is that the size of the nanogap is defined by the molecule that has to fit into the nanogap. Also, molecules are used to direct the assembly of the nanoscale features that are too small, or at least very challenging to fabricate via top-down fabrication methods [8, 11].

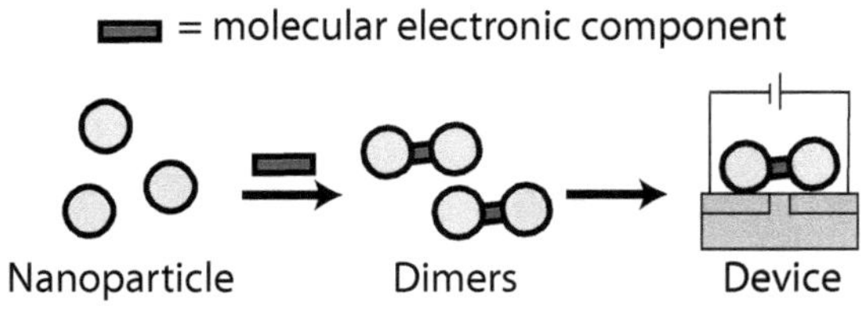

Figure 2.11 Assembly of nanoparticle dimers from molecular electronic components and subsequent assembly onto functional device.

Self-assembly of molecules at surfaces is further discussed in Chapter 7, "Self-Assembly at Interfaces," while self-assembly of full devices is further discussed in Chapter 12, "Parallel Self-Assembly strategies toward Multiple Single-Molecule Devices."

2.3 Identifying Single Molecules in Devices

A general challenge in single-molecule electronics is that it is typically not possible to directly image the molecular structures inside the devices. Therefore, one has to rely on indirect techniques, such as detailed analysis of the conductance properties, or statistics, to substantiate the claim that single-molecule devices are produced.

2.3.1 *Statistics*

In mechanical break junctions and in STM break junction measurements where large data sets of molecular measurements can be recorded in reasonable time, statistical methods are often applied (Fig. 2.12). The experimental data is typically represented as selected conductance traces, meaning the conductivity measured as a function of electrode separation (Fig. 2.12, left). Stepwise decrease in conductivity is taken as the indication that atomic or molecular junctions are formed. The last step before complete breaking of the device is the single-atom or single-molecule conductivity and used as input in the conductance histograms (Fig. 2.12, right). Not all devices break at the low molecular conductance; many break at a conductance quantum of 1 corresponding to the breaking of single-atom metal (gold) contacts.

While conductance histograms reveal some information about molecular devices, a more detailed two-dimensional histogram analysis of the experimental data might reveal more [4, 84–86]. An example is shown in Fig. 2.13; here Lörtscher and coworkers have plotted 500 conductance traces on top of each other. The color code gives an impression of the statistics and the general trends of typical devices. The plot reveals details about differences of the conductivity of the naked break junction device (a), with solvent (dichlorobenzene) (b), with C_{60} molecules (c), and with a

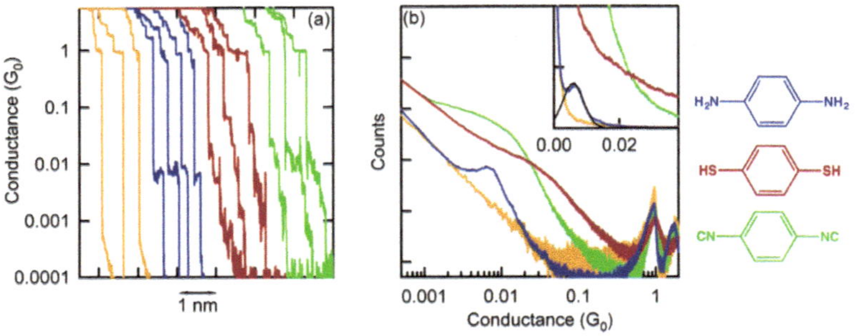

Figure 2.12 Conductance traces (left), conductance histograms (middle), and molecular structures (right) of p-diaminobenzene (blue), p-benzenedithiol (red), and p-diisocyanobenzene (green). Reprinted with permission from Ref. [13]. Copyright (2006) American Chemical Society.

functionalized C_{60} "tadpole" molecule (d). The rapid breaking of the naked device (a) showing conductivity drop from 1 G_0 to the 10^{-4} to 10^{-6} G_0 regime within 0.2 nm is assumed to be due to the release of stress from the breaking wire. The corresponding traces from the device dipped in dichlorobenzene (Fig. 2.13b) show similar features, but a bigger spread of conductances is observed, possibly due to the remaining physisorbed solvent molecules. The traces with C_{60} molecules in the gap show a high conductance (0.83 G_0) plateau extending from 0 nm to 0.5 nm electrode separations, in the same size range as the diameter of a single C_{60} molecule (approximately 1.1 nm). The trend is even more pronounced when analyzing the conductance trace histograms of the C_{60} tadpole molecule (Fig. 2.13d) where the "high conductance" plateau extends up to 0.75 nm electrode separation, possibly due to the added length of the molecule compared to C_{60} and due to possible interactions between gold and the nitrogen on the pyrrolidine part of the molecule [4].

2.3.2 *Single-Molecule Signatures*

Single-molecule conductance at variable bias can be measured by the break junction method incorporating feedback loops to control electrode separation, once the single-molecule junction is formed. By using molecular switches, showing reversible and controllable

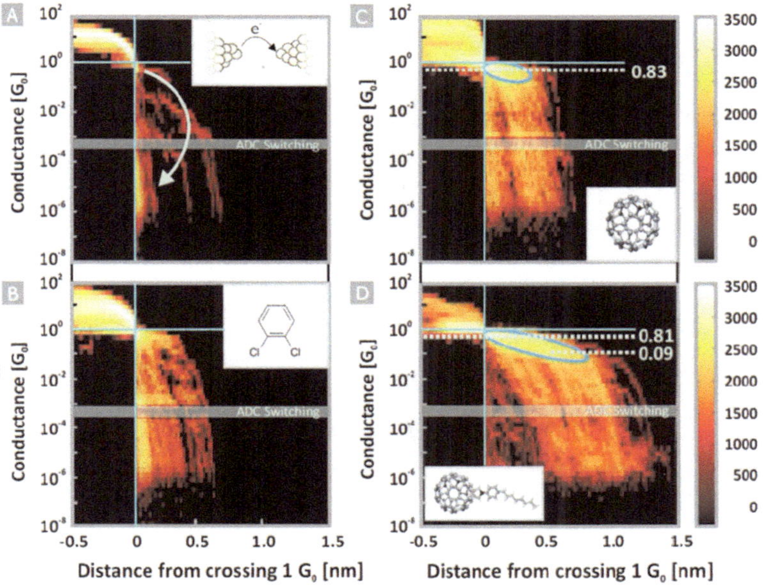

Figure 2.13 Histograms of conductance traces, plotted from 500 conductance traces: (a) naked electrodes, (b) electrodes exposed to dichlorobenzene, (c) electrodes exposed to dichlorobenzene and C_{60}, and (d) electrodes exposed to dichlorobenzene and "C_{60} tadpole." The color code indicates the number of data points. Reprinted with permission from Ref. [4], Copyright © 2013, John Wiley and Sons.

bistability, a molecular memory was demonstrated by Lörtscher and coworkers [15]. By applying a "write" pulse to the molecular device, the molecule is switched to an "on" state, while an "erase" pulse switched the molecule to an "off" state. The reversibility, and the fact that bistability was only observed in devices where the molecular switch was used, proves that single-molecule species are indeed present in the single-molecule devices (Fig. 2.14).

For three-terminal devices, the "proof" of having a single molecule in the device relies on a detailed analysis of the experimental observation of multiple conductance traces with variations in source/drain bias and gate voltage. A plot of such variations is called a stability diagram, and molecular features such as charging energies, vibrational features, Kondo effect, spin transitions, and different types of molecular switches are taken as

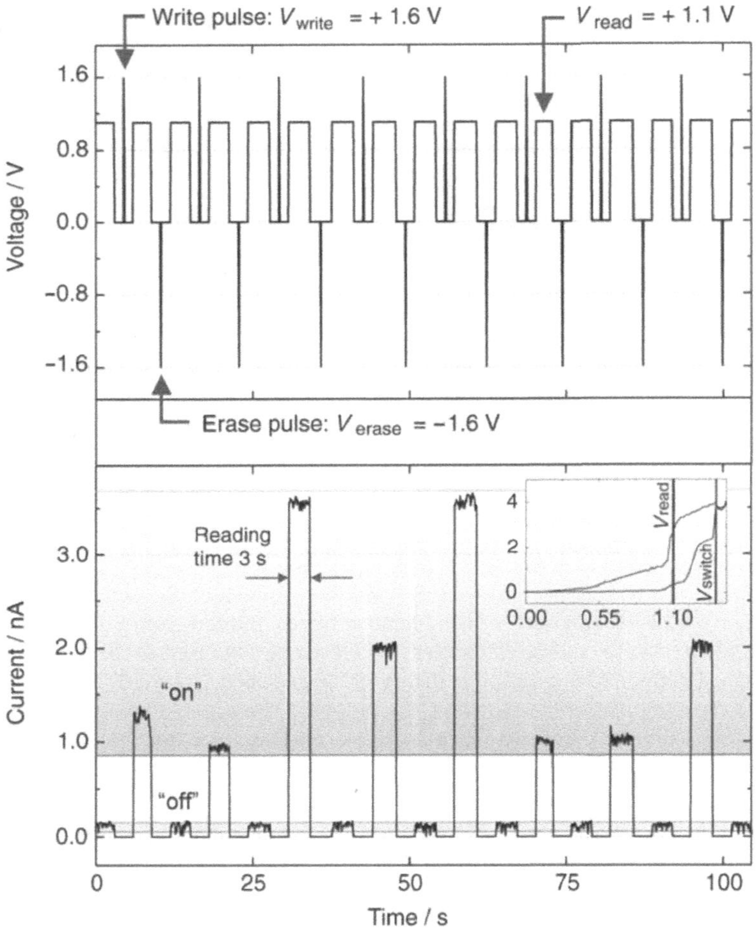

Figure 2.14 Molecular memory based on a molecular switch. By using a write pulse or an erase pulse, the device could be manipulated to "on" or "off" states. Reprinted with permission from Ref. [15]. Copyright © 2006, John Wiley and Sons.

evidence that a single molecule is present in the studied device. Figure 2.15 illustrates a stability diagram where both so-called Coulomb diamonds and molecular vibrations are visible in the measurement [9]. Excitation energies (E_{exc}), as well as addition energies (E_{add}), are directly observed from such experiments.

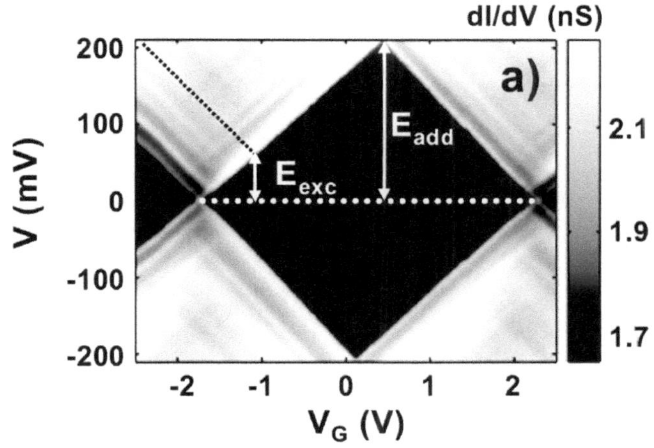

Figure 2.15 Stability diagram of OPV5 molecule in an electromigrated device. Reprinted with permission from Ref. [9]. Copyright © 2007, John Wiley and Sons.

Vibrational effects are further discussed in Chapter 6, "Vibrational Excitations in Single-Molecule Junctions."

2.3.3 *Possible Artifacts*

When considering single-molecule transport measurements, it is important to be vigilant toward possible artifacts and misinterpretations. An example of nonmolecular systems that might be present from device fabrications is small metal islands. Such metal islands can have quantitized charging energies leading to Coulomb blockade type transport signatures, somewhat similar to that observed in molecular devices [10]. Figure 2.16 shows a stability diagram of electronic transport through a metal island featuring Coulomb-charging phenomena. The electronic measurement is complemented by TEM inspection of the device (Fig. 2.16, right) showing the detailed structure of the device and the position of the metal island.

The molecular transport measurements can be differentiated from metal artefacts by difference in charging energies. The metal islands typically have much lower charging energies compared to single molecules.

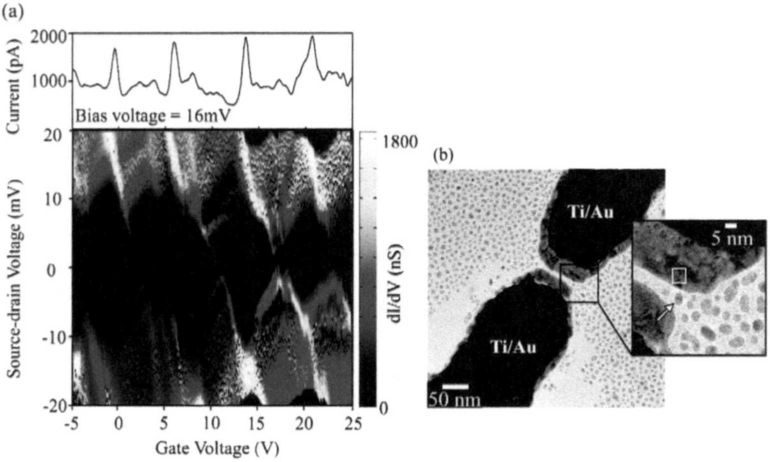

Figure 2.16 Electron transport through metal island (left) and TEM inspection of Ti/Au nanogap electrodes (right). Reproduced with permission from Ref. [10], copyright IOP Publishing. All rights reserved.

2.4 Summary and Conclusion

This chapter has given an introduction to the most widely used experimental techniques in single-molecule electronics experiments. The methods are grouped in three general classes: (i) mechanical methods including STM, AFM, and mechanical break junction methods; (ii) lithographic methods, including electromigration techniques, evaporation techniques, and devices with reduced dimensions such as devices with graphene and carbon nanotube as electrode materials; and (iii) self-assembled devices. In addition, general experimental challenges such as statistics, data selection, and possible artifacts have been discussed.

Problems

1. Give examples of the three classes of experimental techniques developed for single-molecule electronics measurements.
2. Describe the general challenges associated with realizing single-molecule electronic devices.

3. The electromigration technique is used to form nanogap electrodes: Sketch an electromigration device and describe how the nanogap is formed.
4. Equation 2.1 gives an expression for the tunneling current through a molecule. Use the double-layer model as described in Fig. 2.2 to make an expression for the decay function β of molecule *b* given by known parameters for molecule *a* and the experimental observed apparent height.

References

1. Song, H., et al., Observation of molecular orbital gating. *Nature*, 2009. **462**(7276): pp. 1039–1043.
2. Guo, X.F., et al., Covalently bridging gaps in single-walled carbon nanotubes with conducting molecules. *Science*, 2006. **311**(5759): pp. 356–359.
3. Soe, W.H., et al., Manipulating molecular quantum states with classical metal atom inputs: demonstration of a single molecule NOR logic gate. *ACS Nano*, 2011. **5**(2): pp. 1436–1440.
4. Lortscher, E., et al., Bonding and electronic transport properties of fullerene and fullerene derivatives in break-junction geometries. *Small*, 2013. **9**(2): pp. 209–214.
5. Xiao, X.Y., et al., Electrochemical gate-controlled conductance of single oligo(phenylene ethynylene)s. *Journal of the American Chemical Society*, 2005. **127**(25): pp. 9235–9240.
6. Albrecht, T., et al., Transistor-like behavior of transition metal complexes. *Nano Letters*, 2005. **5**(7): pp. 1451–1455.
7. Kushmerick, J., Nanotechnology: Molecular transistors scrutinized. *Nature*, 2009. **462**(7276): pp. 994–995.
8. Guttman, A., et al., Self-assembly of metallic double-dot single-electron device. *Applied Physics Letters*, 2011. **99**(6): p. 3.
9. Osorio, E.A., et al., Addition energies and vibrational fine structure measured in electromigrated single-molecule junctions based on an oligophenylenevinylene derivative. *Advanced Materials*, 2007. **19**(2): p. 281.
10. Gao, B., et al., Three-terminal electric transport measurements on gold nanoparticles combined with ex situ TEM inspection. *Nanotechnology*, 2009. **20**(41). doi:10.1088/0957-4484/20/41/415207.

11. Dadosh, T., et al., Measurement of the conductance of single conjugated molecules. *Nature*, 2005. **436**(7051): pp. 677–680.
12. Xu, B.Q. and N.J.J. Tao, Measurement of single-molecule resistance by repeated formation of molecular junctions. *Science*, 2003. **301**(5637): pp. 1221–1223.
13. Venkataraman, L., et al., Single-molecule circuits with well-defined molecular conductance. *Nano Letters*, 2006. **6**(3): pp. 458–462.
14. Huber, R., et al., Electrical conductance of conjugated oligomers at the single molecule level. *Journal of the American Chemical Society*, 2008. **130**(3): pp. 1080–1084.
15. Lortscher, E., et al., Reversible and controllable switching of a single-molecule junction. *Small*, 2006. **2**(8–9): pp. 973–977.
16. Li, T., W.P. Hu, and D.B. Zhu, Nanogap electrodes. *Advanced Materials*, 2010. **22**(2): pp. 286–300.
17. Moth-Poulsen, K. and T. Bjornholm, Molecular electronics with single molecules in solid-state devices. *Nature Nanotechnology*, 2009. **4**(9): pp. 551–556.
18. Chen, F., et al., Measurement of single-molecule conductance. *Annual Review of Physical Chemistry*, 2007. **58**: pp. 535–564.
19. van Ruitenbeek, J., E. Scheer, and H.B. Weber, Contacting individual molecules using mechanically controllable break junctions. *Lecture Notes in Physics*, 2005. **680**: pp. 253–275.
20. Binnig, G. and H. Rohrer, Scanning tunneling microscopy. *Helvetica Physica Acta*, 1982. **55**(6): pp. 726–735.
21. Bumm, L.A., et al., Are single molecular wires conducting? *Science*, 1996. **271**(5256): pp. 1705–1707.
22. Patrone, L., et al., Direct comparison of the electronic coupling efficiency of sulfur and selenium anchoring groups for molecules adsorbed onto gold electrodes. *Chemical Physics*, 2002. **281**(2–3): pp. 325–332.
23. Moth-Poulsen, K., et al., Probing the effects of conjugation path on the electronic transmission through single molecules using scanning tunneling microscopy. *Nano Letters*, 2005. **5**(4): pp. 783–785.
24. Bumm, L.A., et al., Electron transfer through organic molecules. *Journal of Physical Chemistry B*, 1999. **103**(38): pp. 8122–8127.
25. Albrecht, T., et al., In situ scanning tunnelling spectroscopy of inorganic transition metal complexes. *Faraday Discussions*, 2006. **131**: pp. 265–279.
26. Albrecht, T., et al., Scanning tunneling spectroscopy in an ionic liquid. *Journal of the American Chemical Society*, 2006. **128**(20): pp. 6574–6575.

27. Binnig, G., C.F. Quate, and C. Gerber, Atomic force microscope. *Physical Review Letters*, 1986. **56**(9): pp. 930–933.

28. Frei, M., et al., Linker dependent bond rupture force measurements in single-molecule junctions. *Journal of the American Chemical Society*, 2012. **134**(9): pp. 4003–4006.

29. Hauptmann, N., et al., Force and conductance during contact formation to a C-60 molecule. *New Journal of Physics*, 2012. **14**: p. 14.

30. Hong, W.J., et al., Single molecular conductance of tolanes: experimental and theoretical study on the junction evolution dependent on the anchoring group. *Journal of the American Chemical Society*, 2012. **134**(4): pp. 2292–2304.

31. Nef, C., et al., Force-conductance correlation in individual molecular junctions. *Nanotechnology*, 2012. **23**(36): p. 8.

32. Xu, B., X. Xiao, and N.J. Tao, Measurements of single-molecule electromechanical properties. *Journal of the American Chemical Society*, 2003. **125**(52): pp. 16164–16165.

33. Andres, R.P., et al., "Coulomb staircase" at room temperature in a self-assembled molecular nanostructure. *Science*, 1996. **272**(5266): pp. 1323–1325.

34. Cui, X.D., et al., Reproducible measurement of single-molecule conductivity. *Science*, 2001. **294**(5542): pp. 571–574.

35. Smit, R.H.M., et al., Measurement of the conductance of a hydrogen molecule. *Nature*, 2002. **419**(6910): pp. 906–909.

36. Kergueris, C., et al., Electron transport through a metal-molecule-metal junction. *Physical Review B*, 1999. **59**(19): pp. 12505–12513.

37. Martin, C.A., et al., Lithographic mechanical break junctions for single-molecule measurements in vacuum: possibilities and limitations. *New Journal of Physics*, 2008. **10**: p. 065008.

38. Moreland, J. and J.W. Ekin, Electron-tunneling experiments using Nb-Sn break junctions. *Journal of Applied Physics*, 1985. **58**(10): pp. 3888–3895.

39. Reed, M.A., et al., Conductance of a molecular junction. *Science*, 1997. **278**(5336): pp. 252–254.

40. vanRuitenbeek, J.M., et al., Adjustable nanofabricated atomic size contacts. *Review of Scientific Instruments*, 1996. **67**(1): pp. 108–111.

41. Martin, C.A., et al., A versatile low-temperature setup for the electrical characterization of single-molecule junctions. *Review of Scientific Instruments*, 2011. **82**(5): p. 8.

42. Yang, Y., et al., An electrochemically assisted mechanically controllable break junction approach for single molecule junction conductance measurements. *Nano Research*, 2011. **4**(12): pp. 1199–1207.
43. Yi, Z.W., et al., Molecular junctions based on intermolecular electrostatic coupling. *Chemical Communications*, 2010. **46**(42): pp. 8014–8016.
44. Ballmann, S. and H.B. Weber, An electrostatic gate for mechanically controlled single-molecule junctions. *New Journal of Physics*, 2012. **14**: p. 9.
45. Champagne, A.R., A.N. Pasupathy, and D.C. Ralph, Mechanically adjustable and electrically gated single-molecule transistors. *Nano Letters*, 2005. **5**(2): pp. 305–308.
46. Martin, C.A., J.M. van Ruitenbeek, and H.S.J. van der Zant, Sandwich-type gated mechanical break junctions. *Nanotechnology*, 2010. **21**(26).
47. Park, H., et al., Fabrication of metallic electrodes with nanometer separation by electromigration. *Applied Physics Letters*, 1999. **75**(2): pp. 301–303.
48. Liang, W.J., et al., Kondo resonance in a single-molecule transistor. *Nature*, 2002. **417**(6890): pp. 725–729.
49. Park, J., et al., Coulomb blockade and the Kondo effect in single-atom transistors. *Nature*, 2002. **417**(6890): pp. 722–725.
50. Prins, F., et al., Room-temperature stability of Pt nanogaps formed by self-breaking. *Applied Physics Letters*, 2009. **94**(12): p. 3.
51. Strachan, D.R., et al., Controlled fabrication of nanogaps in ambient environment for molecular electronics. *Applied Physics Letters*, 2005. **86**(4): p. 3.
52. van der Zant, H.S.J., et al., Electromigrated molecular junctions. *Physica Status Solidi B-Basic Solid State Physics*, 2006. **243**(13): pp. 3408–3412.
53. Girod, S., et al., Real time atomic force microscopy imaging during nanogap formation by electromigration. *Nanotechnology*, 2012. **23**(36): p. 4.
54. Taychatanapat, T., et al., Imaging electromigration during the formation of break junctions. *Nano Letters*, 2007. **7**(3): pp. 652–656.
55. Heersche, H.B., et al., Kondo effect in the presence of magnetic impurities. *Physical Review Letters*, 2006. **96**(1).
56. Houck, A.A., et al., Kondo effect in electromigrated gold break junctions. *Nano Letters*, 2005. **5**(9): pp. 1685–1688.
57. Strachan, D.R., et al., Clean electromigrated nanogaps imaged by transmission electron microscopy. *Nano Letters*, 2006. **6**(3): pp. 441–444.

58. Natelson, D., et al., Single-molecule transistors: Electron transfer in the solid state. *Chemical Physics*, 2006. **324**(1): pp. 267–275.

59. Sordan, R., et al., Coulomb blockade phenomena in electromigration break junctions. *Applied Physics Letters*, 2005. **87**(1).

60. Kubatkin, S., et al., Single-electron transistor of a single organic molecule with access to several redox states. *Nature*, 2003. **425**(6959): pp. 698–701.

61. Kubatkin, S.E., et al., Coulomb blockade effects at room temperature in thin-film nanoconstrictions fabricated by a novel technique. *Applied Physics Letters*, 1998. **73**(24): pp. 3604–3606.

62. Kubatkin, S.E., et al., Tunneling through a single quench-condensed cluster. *Journal of Low Temperature Physics*, 2000. **118**(5–6): pp. 307–316.

63. Eigler, D.M. and E.K. Schweizer, Positioning single atoms with a scanning tunneling microscope. *Nature*, 1990. **344**(6266): pp. 524–526.

64. Huang, Y.L., et al., Reversible single-molecule switching in an ordered monolayer molecular dipole array. *Small*, 2012. **8**(9): pp. 1423–1428.

65. Gross, L., et al., Measuring the charge state of an adatom with noncontact atomic force microscopy. *Science*, 2009. **324**(5933): pp. 1428–1431.

66. Liljeroth, P., J. Repp, and G. Meyer, Current-induced hydrogen tautomerization and conductance switching of naphthalocyanine molecules. *Science*, 2007. **317**(5842): pp. 1203–1206.

67. Bergvall, A., et al., Graphene nanogap for gate-tunable quantum-coherent single-molecule electronics. *Physical Review B*, 2011. **84**(15).

68. Peterfalvi, C.G. and C.J. Lambert, Suppression of single-molecule conductance fluctuations using extended anchor groups on graphene and carbon-nanotube electrodes. *Physical Review B*, 2012. **86**(8).

69. Prins, F., et al., Room-temperature gating of molecular junctions using few-layer graphene nanogap electrodes. *Nano Letters*, 2011. **11**(11): pp. 4607–4611.

70. Jin, C.H., et al., Deriving carbon atomic chains from graphene. *Physical Review Letters*, 2009. **102**(20).

71. He, Y.D., et al., Graphene and graphene oxide nanogap electrodes fabricated by atomic force microscopy nanolithography. *Applied Physics Letters*, 2010. **97**(13).

72. Standley, B., et al., Graphene-based atomic-scale switches. *Nano Letters*, 2008. **8**(10): pp. 3345–3349.

73. Jain, T., et al., Aligned growth of gold nanorods in PMMA channels: parallel preparation of nanogaps. *ACS Nano*, 2012. **6**(5): pp. 3861–3867.

74. Jain, T., et al., Self-assembled nanogaps via seed-mediated growth of end-to-end linked gold nanorods. *ACS Nano*, 2009. **3**(4): pp. 828–834.

75. Tang, Q., et al., Self-assembled nanogaps for molecular electronics. *Nanotechnology*, 2009. **20**(24).

76. Guo, X.F. and C. Nuckolls, Functional single-molecule devices based on SWNTs as point contacts. *Journal of Materials Chemistry*, 2009. **19**(31): pp. 5470–5473.

77. Grunder, S., et al., Synthesis and optical properties of molecular rods comprising a central core-substituted naphthalenediimide chromophore for carbon nanotube junctions. *European Journal of Organic Chemistry*, 2011. (3): pp. 478–496.

78. Marquardt, C.W., et al., Electroluminescence from a single nanotube-molecule-nanotube junction. *Nature Nanotechnology*, 2010. **5**(12): pp. 863–867.

79. Qi, P.F., et al., Miniature organic transistors with carbon nanotubes as quasi-one-dimensional electrodes. *Journal of the American Chemical Society*, 2004. **126**(38): pp. 11774–11775.

80. Whalley, A.C., et al., Reversible switching in molecular electronic devices. *Journal of the American Chemical Society*, 2007. **129**(42): p. 12590.

81. Guo, X.F., et al., Chemoresponsive monolayer transistors. *Proceedings of the National Academy of Sciences of the United States of America*, 2006. **103**(31): pp. 11452–11456.

82. Lehn, J.-M., Supramolecular chemistry: Scope and perspectives molecules, supermolecules, and molecular devices (Nobel Lecture). *Angewandte Chemie International Edition in English*, 1988. **27**(1): pp. 89–112.

83. Lehn, J.M., Toward complex matter: Supramolecular chemistry and self-organization. *Proceedings of the National Academy of Sciences of the United States of America*, 2002. **99**(8): pp. 4763–4768.

84. Makk, P., et al., Correlation analysis of atomic and single-molecule junction conductance. *ACS Nano*, 2012. **6**(4): pp. 3411–3423.

85. Martin, C.A., et al., Fullerene-based anchoring groups for molecular electronics. *Journal of the American Chemical Society*, 2008. **130**(40): pp. 13198–13199.

86. Fock, J., et al., A statistical approach to inelastic electron tunneling spectroscopy on fullerene-terminated molecules. *Physical Chemistry Chemical Physics*, 2011. **13**(32): pp. 14325–14332.

Chapter 3

Basic Theory of Electron Transport Through Molecular Contacts

Anders Bergvall,[a] **Mikael Fogelström,**[a] **Cecilia Holmqvist,**[b] **and Tomas Löfwander**[a]

[a] *Department of Microtechnology and Nanoscience, Chalmers University of Technology, SE-412 96 Göteborg, Sweden*

[b] *Fachbereich Physik, Universität Konstanz, D-78457 Konstanz, Germany*

tomas.lofwander@chalmers.se

3.1 Introduction

In this chapter, we will introduce a basic theoretical description of coherent electron transport through low-dimensional junctions and molecular devices. The description introduced is based on quantum transport theory using a tight-binding description of molecules and lead materials. We apply this theory in a few worked examples on junctions based on graphene and carbon nanotubes and on molecular-superconducting hybrid junctions.

In Fig. 3.1, we show a pictorial view of a generic two-terminal molecular electronics device. Clearly, a number of complications are involved in modeling electron transport through such a device. We need to know about the properties of the leads, such as their

Handbook of Single-Molecule Electronics
Edited by Kasper Moth-Poulsen

ISBN 978-981-4463-38-6 (Hardcover), 978-981-4463-39-3 (eBook)
www.panstanford.com

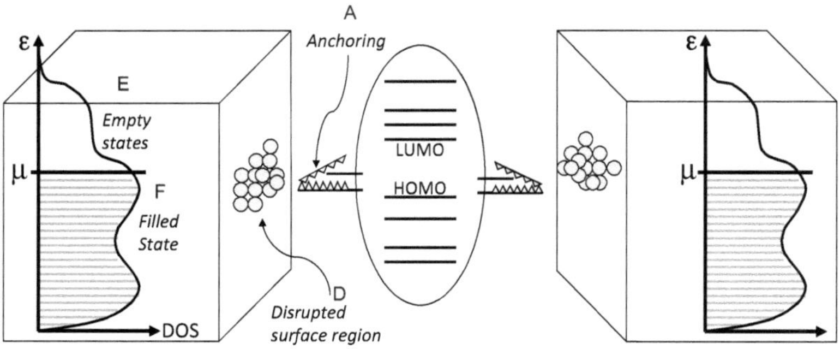

Figure 3.1 Illustration of a molecular-scale contact, including leads with disrupted surface regions and a molecule with anchoring to the lead surfaces via alligator clips.

bandstructure and the reconstruction at the surfaces. The leads may also display broken symmetry states with macroscopic ordering, such as ferromagnetism or superconductivity. We will discuss the latter case in this review. For the molecule, we need to know about the molecular orbitals and how the anchoring to the leads will affect them. This includes broadening and level shifts and should incorporate the effects of the contact potentials between the leads and the molecule. All these complications mean that molecular electronics is a truly interdisciplinary field of research, where a range of subtopics traditionally categorized as Physics, Chemistry, or Electrical Engineering come together. The modeling strategies of such devices can be grouped into categories: either one tries a bottom-up approach of varying complexity (for instance density functional theory), or one utilizes a heuristic technique based on simplified Hamiltonians. In this review, we will focus mainly on heuristic techniques based on tight-binding theory. Depending on the system, one method or the other may lead to a better understanding of experiments.

Traditionally, two transport regimes are considered in molecular electronics: (1) the quantum-coherent transport regime, or (2) the Coulomb blockade regime. The two transport regimes describe two opposite physical limits and are characterized by opposite relations between two energy scales: on the one hand, the tunneling rate $\gamma = \hbar/\tau_{\mathrm{D}}$, where τ_{D} is the dwell time in the weak link (the molecule), and on the other hand, the charging energy U, which characterizes the strength of electron–electron interactions in the weak link. If the charging energy is large, $U \gg \gamma$, then interactions play an important role and limit transport. We are in the Coulomb blockade regime. The molecule is described in terms of many-body states involving a certain number of electrons, and transport is described in terms of transitions between such states [1, 2]. On the other hand, if $U \ll \gamma$, the tunneling rate is big, interactions are less important, and transport is elastic and fully quantum coherent (in the absence of scattering by phonons or other degrees of freedom). The molecular orbitals are then broadened by the coupling to the leads, and we have to consider a broadened level spectrum and how it is populated. The occupation numbers are not integers; rather, we have a fractional charge transfer between molecule and leads, which we have to compute. In this review, we will focus on the small U limit and neglect electron–electron interactions and take transport to be fully quantum coherent. We will also touch upon electron–phonon scattering in the contact.

The outline of this review is as follows. In the first section, we discuss the simplest example imaginable, namely, a single-level quantum dot with good coupling to leads. This allows us to introduce basic concepts of quantum transport and start develop the mathematical framework. In the following section, we discuss in detail the recursive Green's function technique, which is a very powerful and useful algorithm that can be used to study nanoscale devices numerically. To illustrate the technique, we focus on the benzene ring, which serves as an example that can easily be worked out with pen and paper. In the following section, we discuss transport through the benzene ring with metallic leads and graphene leads. The electron transport through the benzene ring can be used to illustrate in the most straightforward way quantum interference effects in molecular electronics. The graphene leads neatly illustrate the crucial influence of contacts on transport in molecular electronics. At the same time, graphene leads allow us to reach an interesting gate-tunable quantum-coherent transport regime [3], which will be discussed in detail. In the last section, we come back to the single-level quantum dot, but coupled to superconducting leads. In this case, a supercurrent can flow through the dot. In addition, we discuss the influence of coupling of vibrations to electron flow by adding a single phonon mode coupled to the electrons flowing through the dot. At the end of the chapter, we present a few problems that students can use to learn the basic concepts of coherent quantum transport through molecular-size contacts, including the effects of graphene leads or superconducting leads. There are already a number of excellent review articles, collections, and books that treat quantum-coherent transport in molecular-size contacts [4–9]. We encourage students to read those in addition to this review, since some parts skipped over here are treated in detail in those. The idea of this review is to cover a few topics usually not discussed in detail, such as graphene contacts or superconducting contacts.

3.2 Electron Transport Through a Single-Level Quantum Dot

An idealized molecular transistor consists of two reservoirs of electrons, usually referred to as leads, which are electrically connected over a molecule or molecular structure. The molecule has a set of energy levels or molecular levels with corresponding wave functions or orbitals. These orbitals may be localized and thus do not contribute to the direct charge transfer between the two reservoirs, or they may be extended and have a finite overlap with the states of the reservoirs. Only the extended orbitals provide the necessary quantum channels for charge transport to take place over the molecular contact. For a charge current to flow, we need to establish an externally controllable voltage bias V between the two leads so that one or more molecular levels lay in the energy window eV. In the ideal case, we will assume that the molecule may be electrostatically gated so that the molecular levels may be moved in and out of, or shifted through, the bias window.

Before we formulate the theory of charge transport based on nonequilibrium Green's functions, let us sketch how transport between two reservoirs of electrons over a single level happens. The current is defined as the average change of charge over time, i.e.,

$$e\langle\dot{\mathcal{N}}\rangle \approx e\frac{\Delta\mathcal{N}}{\Delta t} \tag{3.1}$$

Here e is the electron charge, $e = -|e|$, and $\Delta\mathcal{N}$ is the change in the number of charge carriers. In equilibrium, the number of charge carriers, $\mathcal{N}$, of a reservoir is determined by the Fermi–Dirac distribution $f(\varepsilon; T, \mu)$ at energy ε, temperature T, and chemical potential μ,

$$\mathcal{N} = 2f(\varepsilon; T, \mu) = 2 \times \frac{1}{\exp\left[(\varepsilon - \mu)/k_{\mathrm{B}}T\right] + 1} \tag{3.2}$$

The factor 2 accounts for the electron spin-degeneracy. The number of charge carriers, $\mathcal{N}_d$, of the single level is unknown and needs to be determined. Δt in Eq. 3.1 can be estimated by a rate of change $\hbar/\gamma_{L,R}$ that quantifies the electrons' ability to hop between the left/right (L/R) reservoir and the single level or dot (d). $\hbar$ is Plank's constant divided by 2π. The energy of the level is ε, and it is assumed now to be sharply defined. Inserting the number of electrons on the level $\mathcal{N}_d$ and in the left, $\mathcal{N}_L$, and the right, $\mathcal{N}_R$, reservoirs, Eq. 3.1 reads

$$I_L = e\frac{\gamma_L}{\hbar}(\mathcal{N}_L - \mathcal{N}_d) \quad \text{and} \quad I_R = e\frac{\gamma_R}{\hbar}(\mathcal{N}_d - \mathcal{N}_R) \tag{3.3}$$

The number of electrons on the island can now be determined by posing the condition of a conservation of current, i.e., $I_L = I_R$, or

$$\mathcal{N}_d = \frac{\gamma_L\mathcal{N}_L + \gamma_R\mathcal{N}_R}{\gamma_L + \gamma_R} = 2\frac{\gamma_L f(\varepsilon; T_L, \mu_L) + \gamma_R f(\varepsilon; T_R, \mu_R)}{\gamma_L + \gamma_R} \tag{3.4}$$

Knowing $\mathcal{N}_d$, we can determine the current that flows over the single level:

$$I = I_L = I_R = \frac{2e}{\hbar}\frac{\gamma_L\gamma_R}{\gamma_L + \gamma_R}\left[f(\varepsilon; T_L, \mu_L) - f(\varepsilon; T_R, \mu_R)\right] \tag{3.5}$$

We have assumed that both temperature ($T_{L/R}$) and chemical potential ($\mu_{L/R}$) can be accessed and varied separately in each reservoir. This simple model has been introduced in many introductory texts on quantum transport (see the book by Datta [7]) Despite its simplicity, it emphasizes several key features. The electrical current over the single level will flow only when the energy of the level lies inside the energy window $eV = \mu_R - \mu_L$. The maximum current a single level can support is a given fraction $(= 2\pi \frac{\gamma_L \gamma_R}{\gamma_L + \gamma_R})$ of $2e/h$.

There is a problem though with this simple treatment; the current will grow as $\sim \gamma$ with increasing tunneling rate (assuming for simplicity $\gamma_L = \gamma_R \equiv \gamma$ for the moment). This can be remedied by introducing a so-called density of state function for molecular level that depends on the coupling to the two reservoirs:

$$\mathcal{D}(E) = \frac{1}{2\pi} \frac{\gamma_L + \gamma_R}{(E - \varepsilon)^2 + (\frac{\gamma_L + \gamma_R}{2})^2} \tag{3.6}$$

which integrated over energy gives back the single state, i.e.,

$$\int_{-\infty}^{\infty} \mathcal{D}(E) dE = 1 \tag{3.7}$$

We will justify this by solving the quantum-mechanical tunneling problem below, but for now we assume this broadening of the single level by coupling to the leads as given. If we now insert $\mathcal{D}(E)$ into Eq. 3.5 and write the current as an integral over energy as

$$I = \frac{2e}{\hbar} \int_{-\infty}^{\infty} \mathcal{D}(E) \frac{\gamma_L \gamma_R}{\gamma_L + \gamma_R} \left[f(E; T_L, \mu_L) - f(E; T_R, \mu_R) \right] dE \tag{3.8}$$

and take the zero temperature limit, we get

$$I = \frac{2e}{\hbar} \frac{\gamma_L \gamma_R}{\gamma_L + \gamma_R} \int_{\mu_L}^{\mu_R} \mathcal{D}(E) dE \tag{3.9}$$

To calculate the conductance of the level, $G = \partial I / \partial V$, we assume a small enough bias voltage so that $\mathcal{D}(E)$ can be considered a constant. We then obtain

$$G = \frac{2e^2}{h} \frac{\gamma_L \gamma_R}{\left(E_f - \varepsilon\right)^2 + \left(\frac{\gamma_L + \gamma_R}{2}\right)^2} \tag{3.10}$$

We can now identify the quantum of conductance per spin as

$$G_q = \frac{e^2}{h} \tag{3.11}$$

We see that the maximum conductance one can get over a single molecular level is equal to $2G_q$ when the coupling to the two leads are symmetric, i.e., $\gamma_L = \gamma_R$, and the level is on resonance, i.e., aligned with the Fermi energy of the leads, $\varepsilon = E_f$.

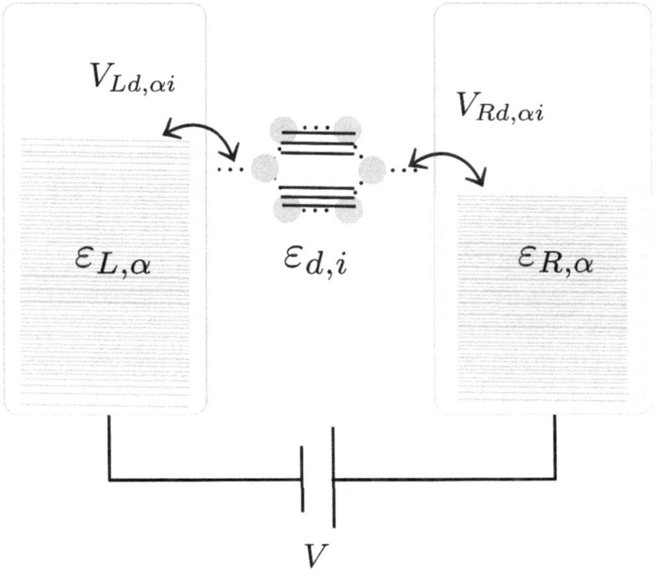

Figure 3.2 Schematic setup of a molecular transistor.

The behavior of the current over single molecular levels is in stark contrast to the macroscopic Ohm's law for which the current increases linearly with the applied voltage. What this simple model could not capture was how the coupling between the molecule and the leads modifies the energy structure of the molecule (and the leads). This we had to put in by hand. We now refine the treatment using quantum mechanical tools. In what follows, we will consider coherent transport across a single molecular level.

The next step is to formulate a quantum mechanical theory to calculate charge transport over a molecule that is coupled to two large leads as schematically shown in Fig. 3.2 The derivation we sketch here has been discussed by several authors, and this technique is described in more detail in Refs. [9–15]. The leads will be considered to be reservoirs with a dense set of electron states described by their spectra, $\varepsilon_{L,\alpha}$ and $\varepsilon_{R,\alpha}$, and their densities of states, $\mathcal{D}(\varepsilon_{L,\alpha})$ and $\mathcal{D}(\varepsilon_{R,\alpha})$. The index α numerates the possible set of quantum numbers, such as wavenumber, spin, site index, etc., that describe the reservoir eigenstates. The two corresponding

Hamiltonians for the reservoirs read

$$\hat{\mathcal{H}}_L = \sum_{\alpha} \varepsilon_{L,\alpha} \hat{c}^{\dagger}_{L,\alpha} \hat{c}_{L,\alpha} \quad \text{and} \quad \hat{\mathcal{H}}_R = \sum_{\alpha} \varepsilon_{R,\alpha} \hat{c}^{\dagger}_{R,\alpha} \hat{c}_{R,\alpha} \tag{3.12}$$

Here $\hat{c}^{\dagger}_{L,\alpha}$, $\hat{c}_{L,\alpha}$ and $\hat{c}^{\dagger}_{R,\alpha}$, $\hat{c}_{R,\alpha}$ are creation and annihilation operators that add ($\hat{c}^{\dagger}_{L,\alpha}$, $\hat{c}^{\dagger}_{R,\alpha}$) or remove ($\hat{c}_{L,\alpha}$, $\hat{c}_{R,\alpha}$) an electron in state α in either lead. The diagonal form of the Hamiltonians in Eq. 3.12 indicates that we know all the properties of the lead material.

Next we describe the quantum state of the *isolated* molecule, which we continue to denote by a subscript d for dot. It is described by its spectrum $\varepsilon_{d,i}$, and creation and annihilation operators $\hat{d}^{\dagger}_i$, $\hat{d}_i$. The dot-Hamiltonian reads

$$\hat{\mathcal{H}}_d = \sum_{i} \varepsilon_{d,i} \hat{d}^{\dagger}_i \hat{d}_i \tag{3.13}$$

The index i enumerates the energy levels of the molecule. We will use a compact notation; the operators $\hat{d}^{\dagger}_i$, $\hat{d}_i$ denote the field operator in second quantization so that for extended quantum systems, such as a benzene ring, $\hat{d}^{(\dagger)}_i \equiv \hat{d}^{(\dagger)}_i(x)$ (it gives the amplitude of the wave function on carbon site x in state i).

To couple these three isolated quantum systems, we introduce tunneling. This is done in a perturbative way by considering the overlap between the reservoir states and the extended states of the molecule. This overlap is parametrized by tunneling amplitudes $V_{Ld,\alpha i}$, $V_{Rd,\alpha i}$ that enter in the tunneling Hamiltonian coupling the levels α of the left lead and the levels i of the dot

$$\hat{\mathcal{H}}_{T,Ld} = \sum_{\alpha,i} \left(V_{Ld,\alpha i} \hat{c}^{\dagger}_{L,\alpha} \hat{d}_i + V_{dL,i\alpha} \hat{d}^{\dagger}_i \hat{c}_{L,\alpha} \right) \tag{3.14}$$

and likewise for dot to right lead

$$\hat{\mathcal{H}}_{T,Rd} = \sum_{\alpha,i} \left(V_{Rd,\alpha i} \hat{c}^{\dagger}_{R,\alpha} \hat{d}_i + V_{dR,i\alpha} \hat{d}^{\dagger}_i \hat{c}_{R,\alpha} \right) \tag{3.15}$$

The tunneling amplitudes obey the relation $V_{L(R)d,\alpha i} = V^{*}_{dL(R),i\alpha}$. We will assume that the tunneling amplitudes are free parameters that we can tune as we wish, but in principle one can compute them from overlap integrals between orbital wave functions of the lead and the molecule.

The total Hamiltonian describing the coupled system is the sum of the Hamiltonians above:

$$\hat{\mathcal{H}}_{Tot} = \hat{\mathcal{H}}_L + \hat{\mathcal{H}}_R + \hat{\mathcal{H}}_d + \hat{\mathcal{H}}_{T,Ld} + \hat{\mathcal{H}}_{T,Rd} \quad (3.16)$$

Our task is to compute the current that flows over the molecule bridging the gap when a voltage is applied as shown in Fig. 3.2, and the average current into either lead (L/R) is determined by the equation of motion for the corresponding number operator:

$$\mathcal{J}_{L/R} = e\langle\dot{\hat{\mathcal{N}}}_{L/R}\rangle = \frac{ie}{\hbar}\langle[\hat{\mathcal{H}}_{Tot}, \hat{\mathcal{N}}_{L/R}]\rangle \quad (3.17)$$

where $\langle\hat{\mathcal{O}}\rangle = \mathrm{Tr}[\hat{\rho}\hat{\mathcal{O}}]$ denotes the quantum mechanical average with the density matrix $\rho = \exp[-\beta(\hat{\mathcal{H}} - \mu\hat{\mathcal{N}})]/\mathcal{Z}$, the partition function $\mathcal{Z} = \mathrm{Tr}[\hat{\rho}]$, and the inverse temperature $\beta = 1/k_B T$. The commutators $[\hat{\mathcal{H}}_{Tot}, \hat{\mathcal{N}}_{L/R}]$ in the expression for the current Eq. 3.17 are straightforward to evaluate. We get for the current over either link:

$$\mathcal{J}_L = -\frac{ie}{\hbar}\sum_{\alpha,i}\left\{V_{Ld,\alpha i}\langle\hat{c}^{\dagger}_{L,\alpha}\hat{d}_i\rangle - V_{dL,i\alpha}\langle\hat{d}^{\dagger}_i\hat{c}_{L,\alpha}\rangle\right\} \quad (3.18)$$

$$\mathcal{J}_R = -\frac{ie}{\hbar}\sum_{\alpha,i}\left\{V_{Rd,\alpha i}\langle\hat{c}^{\dagger}_{R,\alpha}\hat{d}_i\rangle - V_{dR,i\alpha}\langle\hat{d}^{\dagger}_i\hat{c}_{R,\alpha}\rangle\right\} \quad (3.19)$$

We introduce Green's functions that are defined as $\mathcal{G}^{<}_{dL,i\alpha} = i\langle\hat{c}^{\dagger}_{L,\alpha}\hat{d}_i\rangle$ and $\mathcal{G}^{<}_{Ld,\alpha i} = i\langle\hat{d}^{\dagger}_i\hat{c}_{L,\alpha}\rangle$. The current formula can then be rewritten as

$$\mathcal{J}_L = -\frac{e}{\hbar}\sum_{\alpha,i}\left\{V_{Ld,\alpha i}\mathcal{G}^{<}_{dL,i\alpha} - V_{dL,i\alpha}\mathcal{G}^{<}_{Ld,\alpha i}\right\} \quad (3.20)$$

$$= -\frac{e}{\hbar}\mathrm{Tr}\left\{V_{Ld}\circ\mathcal{G}^{<}_{dL} - V_{dL}\circ\mathcal{G}^{<}_{Ld}\right\} \quad (3.21)$$

In the second line, we introduce the noncommutative $\circ$-product, which is shorthand for the convolution over the internal degrees of freedom, and we can suppress the internal indices (α, i). The operation of a trace $\equiv$ Tr introduced above is short for a sum over all discrete and an integral over all continuous quantum degrees of freedom.

The advantage of introducing nonequilibrium Green's functions $\mathcal{G}^{<,>}$ is that we, in a self-consistent way, can describe coherent transport, including inelastic processes, i.e., coupling to external modes, etc. For an excellent introduction to nonequilibrium Green's function methods applied to molecular electronics, we refer to the book by Cuevas and Scheer [9]. The theory of tunneling that we develop here can be applied for extended systems involving detailed structure of the molecule and nanostructured contact regions. This theory is known as the recursive Green's function theory and has been widely applied to transport through small finite-sized regions. That will be exemplified in the subsequent sections of this chapter. Here we will lay down the basis by first explicitly applying it to transport through a single resonant level.

The goal is to calculate the current over the molecular constriction. To do this, we need to find the Green's functions $\mathcal{G}^<_{Ld/dL}$ and the same over the right contact, $\mathcal{G}^<_{Rd/dR}$. We now define an extended notation with a system Green's function $\check{\mathcal{G}}$ as

$$\check{\mathcal{G}} = \begin{pmatrix} \check{G}_L & \check{G}_{Ld} & \check{G}_{LR} \\ \check{G}_{dL} & \check{G}_d & \check{G}_{dR} \\ \check{G}_{RL} & \check{G}_{Rd} & \check{G}_R \end{pmatrix}, \quad \check{\mathcal{G}}_0 = \begin{pmatrix} \check{G}^0_L & 0 & 0 \\ 0 & \check{G}^0_d & 0 \\ 0 & 0 & \check{G}^0_R \end{pmatrix}, \tag{3.22}$$

$$\text{and} \quad \check{\mathcal{V}} = \begin{pmatrix} 0 & \check{V}_{Ld} & 0 \\ \check{V}_{dL} & 0 & \check{V}_{dR} \\ 0 & \check{V}_{Rd} & 0 \end{pmatrix} \tag{3.23}$$

and the "check" on each component denotes that they are Keldysh matrices of the form

$$\check{X} = \begin{pmatrix} X^R & X^K \\ 0 & X^A \end{pmatrix} \tag{3.24}$$

The greater/lesser functions in Eq. 3.21 are given as $X^{>,<} = \frac{1}{2}(X^K \pm (X^R - X^A))$. The Green's function is determined by a Dyson equation

$$\begin{aligned} \check{\mathcal{G}} &= \check{\mathcal{G}}_0 + \check{\mathcal{G}}_0 \circ (\check{\mathcal{V}} + \check{\Sigma}) \circ \check{\mathcal{G}} \\ &= \check{\mathcal{G}}_0 + \check{\mathcal{G}} \circ (\check{\mathcal{V}} + \check{\Sigma}) \circ \check{\mathcal{G}}_0. \end{aligned} \tag{3.25}$$

The Dyson equation is a perturbation series that sums a series of tunneling on and off the molecule. In its form above, we assume that the tunneling elements are large and we neglect charging effects. The self-energy $\check{\Sigma} \equiv \Sigma_d\{\check{\mathcal{G}}\}$ in Eq. 3.25 will initially be set to zero but will later be used to incorporate interactions on the molecular level with internal dynamics or with some oscillator representing a coupling between the tunneling charge and vibrational degrees of freedom. The reservoir Green's functions $\check{G}^0_L$ and $\check{G}^0_R$ are assumed to be given accounting for interactions in the reservoirs.

Working out the component equations of Eq. 3.25 by straightforward matrix multiplication, we can express the two Green's functions $\check{G}_{dL}$ and $\check{G}_{Ld}$ with the help of the level Green's function $\check{G}_d$ as

$$\check{G}_{Ld} = \check{G}^0_L \circ \check{V}_{Ld} \circ \check{G}_d \tag{3.26}$$
$$\check{G}_{dL} = \check{G}_d \circ \check{V}_{dL} \circ \check{G}^0_L \tag{3.27}$$

It is natural to introduce the dressed tunneling elements or reservoir self-energies $\check{\Gamma}_{L/R} = \check{V}_{dL/R} \circ \check{G}^0_{L/R} \circ \check{V}_{L/Rd}$. With the help of these we can write the current formula as

$$j^K_L(t, t') = \frac{e}{\hbar}\mathrm{Tr}[\check{\Gamma}_L \circ \check{G}_d - \check{G}_d \circ \check{\Gamma}_L]^K(t, t') \tag{3.28}$$

$\check{G}_d$ entering in Eq. 3.28 is given by

$$\check{G}_d = \check{G}^0_d + \check{G}^0_d \circ (\check{\Gamma}_L + \check{\Sigma}_d + \check{\Gamma}_R) \circ \check{G}_d \tag{3.29}$$

This equation follows directly from Eq. 3.25. For equilibrium situations, the $\circ$-product reduces to simple multiplication, and Eq. 3.29 is easily solved by matrix inversion, after making a Fourier transform over the time-variable to frequency, and we get

$$\check{G}_d = [[\check{G}^0_d]^{-1} - \check{\Gamma}_L - \check{\Gamma}_R - \check{\Sigma}_d]^{-1} = [[\check{G}^0_d]^{-1} - \check{\Gamma}_\Sigma - \check{\Sigma}_d]^{-1} \tag{3.30}$$

The Green's function of a free molecule is given by its discrete spectrum $\{E_i\}$, and for each level, we get

$$\check{G}^0_{d,i}(\varepsilon) = \begin{pmatrix} G^{0,R}_{d,i}(\varepsilon) & G^{0,K}_{d,i}(\varepsilon) \\ 0 & G^{0,A}_{d,i}(\varepsilon) \end{pmatrix} = \begin{pmatrix} \dfrac{1}{\varepsilon + i\eta - E_i} & -\dfrac{i\eta}{(\varepsilon - E_i)^2 + \eta^2}\phi_0(\varepsilon) \\ 0 & \dfrac{1}{\varepsilon - i\eta - E_i} \end{pmatrix} \tag{3.31}$$

where $\eta \to 0^+$ and $\phi_0(\varepsilon)$ is the distribution function of the level that gives the probability of an energy level being occupied at energy ε for the uncoupled molecule. $\phi_0(\varepsilon)$ is in principle unknown. For the reservoir self-energies, we write in a similar way

$$\check{\Gamma}_{L/R/\Sigma} = \begin{pmatrix} \Gamma^R_{L/R/\Sigma} & \Gamma^K_{L/R/\Sigma} \\ 0 & \Gamma^A_{L/R/\Sigma} \end{pmatrix} \tag{3.32}$$

In equilibrium, the Keldysh components are given as

$$\begin{aligned}\Gamma^K_{L/R} &= (\Gamma^R_{L/R} - \Gamma^A_{L/R})\tanh[(\varepsilon - \mu_{L/R})/2T_{L/R}] \\ &= (\Gamma^R_{L/R} - \Gamma^A_{L/R})\phi_{L/R}(\varepsilon) \\ \Gamma^K_\Sigma &= \Gamma^K_L + \Gamma^K_R\end{aligned}$$

Evaluating Eq. 3.30 for a single level, $E_i \equiv E$, we get the Green's functions of the level, $G^A(\varepsilon) = [G^R(\varepsilon)]^\dagger$,

$$G^R(\varepsilon) = \frac{1}{\varepsilon + i\eta - E - \Gamma^R_\Sigma - \Sigma^R_d} \tag{3.33}$$

$$G^K(\varepsilon) = \frac{2i\{\mathrm{Im}[\Gamma^R_R]\phi_R(\varepsilon) + \mathrm{Im}[\Gamma^R_L]\phi_L(\varepsilon)\} + \Sigma^K_d}{(\varepsilon - E - \mathrm{Re}[\Gamma^R_\Sigma + \Sigma^R_d])^2 + (\mathrm{Im}[\Gamma^R_\Sigma + \Sigma^R_d])^2} \tag{3.34}$$

Turning to the current across the single molecular level, assuming a stationary situation, neglecting for the moment any intrinsic self-energies $\check{\Sigma}_d$, and setting $\tilde{E} = E + \mathrm{Re}[\Gamma^R_\Sigma]$, we can evaluate the commutator and get

$$\mathcal{J} = 2 \times \frac{e}{\hbar}\int_{-\infty}^{\infty} \frac{d\varepsilon}{2\pi} \frac{4\mathrm{Im}[\Gamma^R_R]\mathrm{Im}[\Gamma^R_L]}{(\varepsilon - \tilde{E})^2 + (\mathrm{Im}[\Gamma^R_\Sigma])^2}(\phi_L(\varepsilon) - \phi_R(\varepsilon)) \tag{3.35}$$

with $\mathcal{J} = \mathcal{J}_L = -\mathcal{J}_R$ and where the prefactor 2 accounts for the spin degeneracy. The functions $\gamma_{L/R}$ that we introduced in the beginning of this section, which describe coupling to the leads, now appear naturally in the formula as $\gamma_{L/R} = -2\mathrm{Im}[\Gamma^R_{L/R}]$.

In contrast to the result of our initial simple consideration of charge transport across a single level dot in Eq. 3.5, the full quantum treatment of coherent transport also includes, in a natural way, how the energy spectrum of the dot is modified by the coupling to the leads. The energy of the level is shifted, $E \rightarrow E + \mathrm{Re}[\Gamma^R_\Sigma + \Sigma^R_d]$, by an amount that depends both on the level structure of the leads and the overlap between the level and lead orbitals. The level also becomes broadened by an amount given by $\mathrm{Im}[\Gamma^R_\Sigma + \Sigma^R_d]$ giving the tunneling charge a dwell time on the molecule. The level-broadening due to the coupling to the reservoirs can be viewed as a dwell time $\sim\hbar/\mathrm{Im}[-\Gamma^R_\Sigma]$ of the tunneling charge on the molecule. If there are internal degrees of freedom on the molecule, then these may couple to transport. This is captured by the self-energy $\check{\Sigma}_d$. This will introduce some decoherence to the transport as energy is exchanged between the tunneling electrons and the internal degrees of freedom. The form of $\check{\Sigma}_d$ depends on what the internal degrees of freedom are, and we will exemplify it in a section below as an oscillator degree of freedom. Finally, this treatment allows to determine the occupation of the single level from $G^K(\varepsilon)$.

3.3 Recursive Green's Function Technique

In the last section, we introduced a general formalism for transport starting from a Hamiltonian with terms describing the uncoupled molecule, the leads, and coupling terms (tunneling). We showed that the strategy of computation is to eliminate the leads in favor of a molecule Green's function that includes the effect of tunneling to the leads in terms of self-energies. In this section, we will look into the uncoupled molecule in more detail and introduce a technique that can be used to treat what is sometimes called an extended molecule (or system): what in the last section was called the uncoupled dot will now include the molecule itself but also portions of the leads. The idea is to treat the disrupted surface regions in more detail. Then the surface regions are coupled to leads that connect the device to the rest of the circuit. The advantage of this strategy is that the leads can be taken as perfect waveguides that connect the system to reservoirs in equilibrium with well-defined chemical potentials and temperatures in the spirit of Landauer–Büttiker scattering theory.

The technique we will discuss here is called the recursive Green's function technique [16, 17]. To understand the underlying mechanics of this algorithm, we start out by looking at the simple, yet nontrivial, atomic system shown in Fig. 3.3a. In the next section, we discuss graphene leads, in which case one can include large pieces of the graphene in the system together with the molecule and then have perfect graphene ribbon waveguides coupled to the outside circuit. One may also use this method to describe general nanostructures, as for instance a graphene Hall bar with multiple terminals [18]. But here we focus on the benzene ring to keep the discussion simple. In the benzene ring, every atom is assumed to have only a single π-orbital, and only orbitals of directly adjacent (nearest-neighbor) atoms have a nonvanishing overlap (as indicated by the solid lines connecting the dots). If we denote the hopping

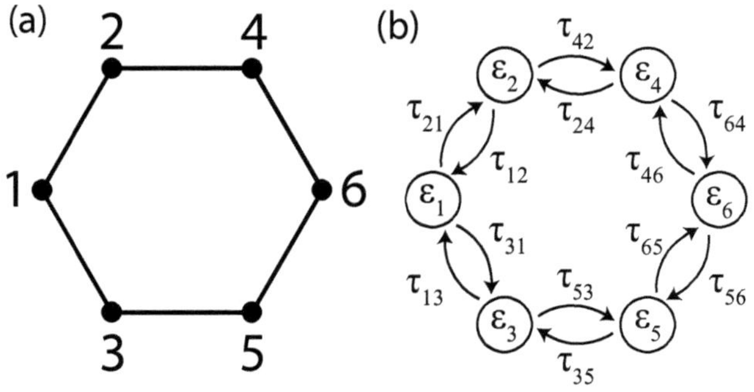

Figure 3.3 System.

energy from atom i to a neighboring atom j (as illustrated in Fig. 3.3b) as τ_{ji}, we may go ahead and write down the system's Hamiltonian matrix as

$$H = \begin{pmatrix} \epsilon_1 & \tau_{12} & \tau_{13} & 0 & 0 & 0 \\ \tau_{21} & \epsilon_2 & 0 & \tau_{24} & 0 & 0 \\ \tau_{31} & 0 & \epsilon_3 & 0 & \tau_{35} & 0 \\ 0 & \tau_{42} & 0 & \epsilon_4 & 0 & \tau_{46} \\ 0 & 0 & \tau_{53} & 0 & \epsilon_5 & \tau_{56} \\ 0 & 0 & 0 & \tau_{64} & \tau_{65} & \epsilon_6 \end{pmatrix} \tag{3.36}$$

where ϵ_i, the onsite energy, is the eigenvalue of the Hamiltonian H_i for an isolated atom i.

Theoretically, we could now find the (retarded) Green's function of the system by using the definition

$$G = [(E + i\eta)1 - H]^{-1} \tag{3.37}$$

where 1 is the identity matrix and η a small positive number. We note, however, that doing so would require us to invert a matrix of size $N \times N$, an operation whose computational cost scales as $\mathcal{O}(N^3)$. In the example above, where $N = 6$, and for other small systems (a couple of hundred atoms), this would pose no problem, but if we were to increase the system size even further, we would soon run into trouble since the time required for the matrix inversion would grow too large. Thus, we require a more efficient way of calculating the Green's functions than the "brute force" method based on direct inversion of the full system's Hamiltonian. This is where the recursive Green's function method comes into play.

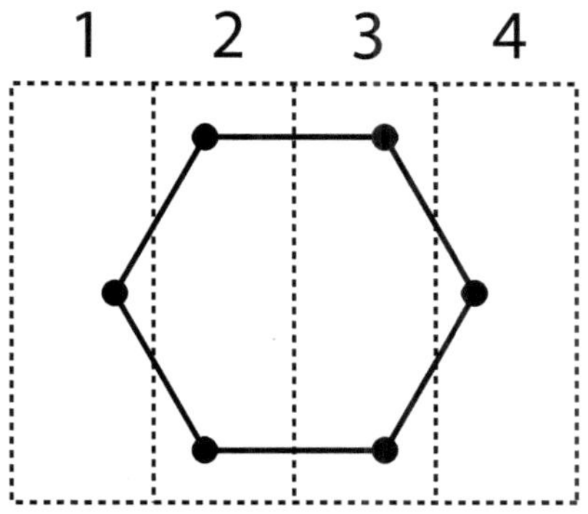

Figure 3.4 Partitions.

Before we start explaining how the method works, we notice two things. First, each atom in the system has only a few nearest neighbors. This makes the full system Hamiltonian H rather sparse, and we can group atoms together such that one group of atoms has only nonvanishing hopping energies to atoms located in the same or in a directly adjacent group. One such partitioning into smaller groups, or slices, is shown in Fig. 3.4 and the corresponding partitioned Hamiltonian becomes

$$H = \begin{pmatrix} H_1 & V_{12} & 0 & 0 \\ V_{21} & H_2 & V_{23} & 0 \\ 0 & V_{32} & H_3 & V_{34} \\ 0 & 0 & V_{43} & H_4 \end{pmatrix} \tag{3.38}$$

where $H_1 = \epsilon_1$, $H_4 = \epsilon_4$,

$$H_2 = \begin{pmatrix} \epsilon_2 & 0 \\ 0 & \epsilon_3 \end{pmatrix}, \quad H_3 = \begin{pmatrix} \epsilon_4 & 0 \\ 0 & \epsilon_5 \end{pmatrix}, \tag{3.39}$$

and

$$\begin{aligned} V_{12} &= \begin{pmatrix} \tau_{12} & \tau_{13} \end{pmatrix}, \quad V_{21} = \begin{pmatrix} \tau_{21} \\ \tau_{31} \end{pmatrix}, \quad V_{23} = \begin{pmatrix} \tau_{24} & 0 \\ 0 & \tau_{35} \end{pmatrix}, \\ V_{32} &= \begin{pmatrix} \tau_{42} & 0 \\ 0 & \tau_{53} \end{pmatrix}, \quad V_{34} = \begin{pmatrix} \tau_{46} \\ \tau_{56} \end{pmatrix}, \quad V_{43} = \begin{pmatrix} \tau_{64} & \tau_{65} \end{pmatrix}. \end{aligned} \tag{3.40}$$

After partitioning the system, we can redraw our benzene ring as shown in Fig. 3.5. The transformed system now looks like a simple one-dimensional linear chain, except that we have matrices instead of scalar quantities to deal with.

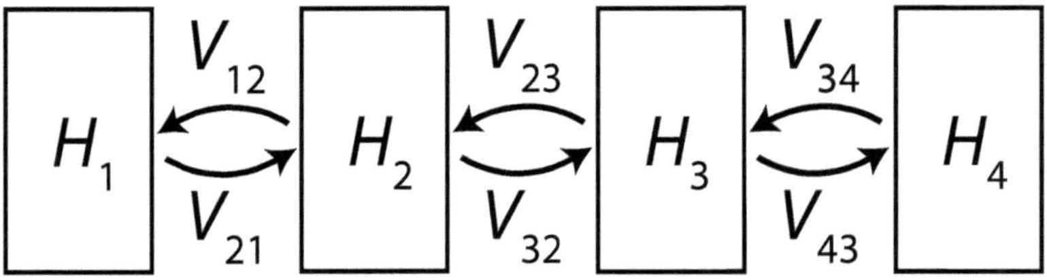

Figure 3.5 Blocks.

Second, for most of our calculations we will not need to calculate all the matrix elements of G. For example, if we were to connect leads to our benzene ring at atoms 1 and 6, a calculation of the transmission through the ring would require knowledge about the elements G_{16} and G_{61}, but not about, e.g., G_{25} and G_{52}. In our partitioned system, finding the transmission through the ring would require us to find the Green's functions (or propagators) between slices 1 and 4 (since these are the slices in which the atoms labeled 1 and 6 are contained). We will now show how to do this in a recursive manner, by starting with a single slice and then adding the other slices one by one, re-calculating only the necessary inter- and intra-slice propagators as we go along.

Let us denote by g_{mm} the Green's function of the isolated slice m, given by

$$g_{mm} = [(E + i\eta)1 - H_m]^{-1} \tag{3.41}$$

and by G^A_{nm} the propagator between slices m and n after having added A of the total S slices. After adding the first slice, we simply find that $G^1_{11} = g_{11}$. By using the Dyson equation $G = g + gVG = g + GVg$, we see that we can write

$$\begin{cases} G^2_{22} = g_{22} + g_{22}V_{21}G^2_{12} \\ G^2_{12} = g_{11}V_{12}G^2_{22} = G^1_{11}V_{12}G^2_{22}. \end{cases} \tag{3.42}$$

These equations form a closed system, and we may (by inserting one into the other) extract that

$$G^2_{22} = \left(1 - g_{22}V_{21}G^1_{11}V_{12}\right)^{-1} g_{22} \tag{3.43}$$

$$G^2_{12} = G^1_{11}V_{12}G^2_{22}, \tag{3.44}$$

$$G^2_{21} = G^2_{22}V_{21}G^1_{11} \tag{3.45}$$

and

$$G_{11}^{2} = G_{11}^{1} + G_{11}^{1} V_{12} G_{21}^{2} \tag{3.46}$$

More generally, when adding slice A, we use the recurrence equations

$$G_{AA}^{A} = \left(1 - g_{AA} V_{A,A-1} G_{A-1,A-1}^{A-1} V_{A-1,A}\right)^{-1} g_{AA} \tag{3.47}$$

$$G_{1A}^{A} = G_{1,A-1}^{A-1} V_{A-1,A} G_{AA}^{A} \tag{3.48}$$

$$G_{A1}^{A} = G_{AA}^{A} V_{A,A-1} G_{A-1,1}^{A-1} \tag{3.49}$$

and

$$G_{11}^{A} = G_{11}^{A-1} + G_{1,A-1}^{A-1} V_{A-1,A} G_{A1}^{A} \tag{3.50}$$

until we finally find G_{11}^{S}, G_{1S}^{S}, G_{S1}^{S}, and G_{SS}^{S} (where $S = 4$ since we have a total of four slices). These will contain the elements given by G_{11}, G_{16}, G_{61}, and G_{66}, and we may continue calculating the transmission through the system. Instead of using $\mathcal{O}(N^3)$ operations to invert the full $N \times N$ Hamiltonian H, dividing the system into S slices (each containing a maximum of M atoms) requires only $\mathcal{O}(SM^3)$ operations (S inversions and limited number of multiplications of $M \times M$ matrices). For our particular example, when $S = 4$ and $M = 2$, the difference is not that big, but as N grows large, we see that a cost of $\mathcal{O}(SM^3) = \mathcal{O}(NM^2)$ will be much more beneficial than $\mathcal{O}(N^3)$ as long as we can partition the system so that $M \ll N$. The price we pay is that we must use a more complex method than direct inversion, and that we no longer have access to all the matrix elements of G. As we will show in the next subsection, we can, however, regain knowledge about some of the matrix elements G_{ji} by a slight increase in computational cost and required memory.

3.3.1 *Local Properties*

When calculating only the transmission through the benzene ring, we did not need any other matrix elements of G than those between the atoms where we have put the leads. If we, on the other hand, were interested in finding the (local) density of states, or the current between two neighboring atoms i and j, we would need other matrix elements that we did not calculate while going in the forward direction. Knowledge about these elements can be achieved by applying the recursive method once again, but this time going in the other direction. After finishing the first run, where we ended up finding G^4_{11}, G^4_{14}, G^4_{41}, and G^4_{44}, we now start by trying to calculate

$$G^4_{34} = G^3_{33} V_{34} G^4_{44} \tag{3.51}$$

$$G^4_{43} = G^4_{44} V_{43} G^3_{33} \tag{3.52}$$

and

$$G^4_{33} = G^3_{33} + G^3_{33} V_{34} G^4_{43} \tag{3.53}$$

In the above equations, we see that we would need G^3_{33}. We calculated this propagator while applying the method in the forward direction, but we threw it away after using it to find G^4_{44}. If we had chosen to save that element for later use, we would have been easily able to solve for G^4_{34} and G^4_{43} by using the equations above. The same applies to the elements G^3_{13} and G^3_{31}, which would be needed to find

$$G^4_{13} = G^3_{13} + G^3_{13} V_{34} G^4_{43} \tag{3.54}$$

and

$$G^4_{31} = G^3_{31} + G^4_{34} V_{43} G^3_{31} \tag{3.55}$$

More generally, if we always make sure that we save G^A_{1A}, G^A_{A1}, and G^A_{AA} for later, the backward recurrence equations would read

$$G^S_{A,A+1} = G^A_{AA} V_{A,A+1} G^S_{A+1,A+1} \tag{3.56}$$

$$G^S_{AS} = G^A_{AA} V_{A,A+1} G^S_{A+1,S} \tag{3.57}$$

$$G^S_{A+1,A} = G^{A+1}_{A+1,A+1} V_{A+1,A} G^A_{AA} \tag{3.58}$$

$$G^S_{SA} = G^S_{S,A+1} V_{A+1,A} G^A_{AA} \tag{3.59}$$

$$G^S_{AA} = G^A_{AA} + G^A_{AA} V_{A,A+1} G^S_{A+1,A} \tag{3.60}$$

$$G^S_{1A} = G^A_{1A} + G^A_{1A} V_{A,A+1} G^S_{A+1,A} \tag{3.61}$$

and

$$G^S_{A1} = G^A_{A1} + G^S_{A,A+1} V_{A+1,A} G^A_{A1} \tag{3.62}$$

Knowing G^S_{AA} would allow us to calculate the local density of states (LDOS) on atom A, and by using G^S_{Ac}, G^S_{cA}, $G^S_{A+1,c}$, and $G^S_{c,A+1}$, where $c = \{1, S\}$, would let us calculate $G^{<S}_{A,A+1}$ and $G^{<S}_{A+1,A}$, which are needed to find the currents going between slices A and $A+1$.

Applying the method backwards thus makes it possible to extract also local properties of the system. The computational cost will increase a bit since we are required to perform a lot of extra matrix multiplications. With growing N, the scaling is, however, still $\mathcal{O}(NM^2)$, and the performance of the extended method is thus equal to the one we used earlier. What will increase is the consumed memory. Earlier, we were only required to save $\mathcal{O}(M^2)$ data elements to perform the algorithm. Now, with the demand that we save some elements after each forward step, the required memory will be of order $\mathcal{O}(SM^2) = \mathcal{O}(NM)$. For a square atomic system (where $M \approx \sqrt{N}$ and $S \approx M$), this limits the system size to a couple of hundred thousand atoms on a normal PC. If we, however, study systems where $M \ll N$, the total amount of atoms may be much larger than that before we run out of memory.

3.3.2 *Further Optimizations*

Since the computational cost of the recursive Green's function method scales as $\mathcal{O}(NM^2)$, it appears that reducing the number of atoms per slice, M, as much as possible would be beneficial. The ideal case would be a system consisting of a one-dimensional linear chain of atoms where only nearest-neighbor hopping is considered. For such a system, each slice would have only a single atom and both the computational cost and the memory consumption would thus scale as $\mathcal{O}(N)$. For other less trivial systems, the situation is different. For our benzene ring, partitioning the system into slices each having a single atom would not reduce the computational cost to $\mathcal{O}(N)$, the reason being that we could no longer just add a slice and forget about it as we did earlier. As an example, earlier we could forget about slice two the moment we started to add slice three. All we had to do was to calculate and save the elements G^2_{12}, G^2_{21}, and G^2_{22}. Now if we instead put only one atom in each slice (thus increasing the number of slices from four to six), we realize that to add atom four we would need to know the propagators G^3_{12}, G^3_{21}, and G^3_{22}. Before we did not need to calculate or save these, but now we have to. Thus, reducing the number of atoms per slice to one allows us to work with only scalar quantities, but we are instead forced to do $\mathcal{O}(K^2)$ more operations when adding each slice. Here, K will be the average number of atoms that, after each round, still do not have all their neighbors already added to the system. Usually, $K \approx M$, and we do not reduce the computational time by reducing the number of atoms per slice. We can, as first shown in Ref. [19], however, reduce the amount of needed memory from $\mathcal{O}(NM)$ to $\mathcal{O}(\max(N, M^2))$ by this method at the cost of making our algorithm more complicated. One advantage of the method of adding atoms one by one is that more complicated geometries with multiple terminals can be treated.

3.4 Graphene Leads: Gate-Tunable Quantum Coherent Transport

In the previous section, we explained how to compute the uncoupled system Green's function within a tight-binding approximation by utilizing a recursive technique. We illustrated the technique by computing the Green's function of the benzene ring. In this section, we will couple the benzene ring to metallic leads and study transport through it. In doing so, we can illustrate how quantum mechanical coherence influences transport through interference effects inside the molecule. In addition, we will couple the benzene ring to graphene leads. We will then see the crucial influence of the disrupted lead surface region in a spectacular way. The zigzag graphene edges that the benzene ring will be coupled to are known to harbor surface states at zero energy [20]. If the surface states and a molecular orbital are overlapping (in space and energy), they hybridize, which influences greatly the transport through the molecular level.

In addition, we will discuss a setup where the graphene leads and the molecule can be gated and discuss how the transport can be tuned even when we are in the quantum coherent transport regime. In the Coulomb blockade regime, three-terminal molecular devices have been fabricated [21]. The third terminal acts as a gate, tuning the molecular levels into and out of resonance with the leads, which greatly affects the source–drain current. Typically, first-order tunneling processes dominate in this regime and transport is incoherent. Higher-order coherent corrections (co-tunneling) can, of course, be considered when necessary. In contrast, in the coherent regime, two-terminal devices are typically envisaged since the gate is not efficient because of screening by the electrons in the massive metallic leads.

In a recent publication, we proposed that by utilizing graphene leads, a gate can effectively tune the transport properties of the device even if we are in the quantum coherent transport regime. The idea is that the electron density of the graphene leads can be tuned electrostatically by applying a voltage between a gate capacitively coupled to the graphene sheet. This means that the gate might not affect the molecule too much, but rather the leads. In pristine charge-neutral graphene, the Fermi level is located exactly at the so-called Dirac point in the band structure. The density of states goes to zero at this point. As the charge density is tuned by the gate, the Fermi energy can be located either in the hole band or in the electron band. The density of states at the Fermi level is then finite and grows linearly as the Fermi energy is shifted away from the Dirac point. The gate would also affect the molecular levels. But if the coupling of the gate to the molecular levels is sufficiently different from the coupling to the large-area graphene leads, the device will display a transistor effect, although the coupling between leads and molecule is sufficiently strong that we are in the quantum coherent transport regime. The transistor effect is optimal when the gate can tune the Fermi level between the hole and electron bands of graphene, through the Dirac point, while at the same time the molecular level remains close to the Fermi energy, as illustrated in Fig. 3.6.

As an illustration of this concept, we study the simple and naive model system shown in Fig. 3.7. This model system also serves as an illustration of the recursive technique presented in the preceding section. A planar molecular junction is created by first cutting a graphene nanoribbon perpendicular to its edges, and then bridging the gap with the benzene molecule discussed earlier. The two ribbons are connected to reservoirs and thus serve as source and drain leads. We will study the two cases where the right lead is

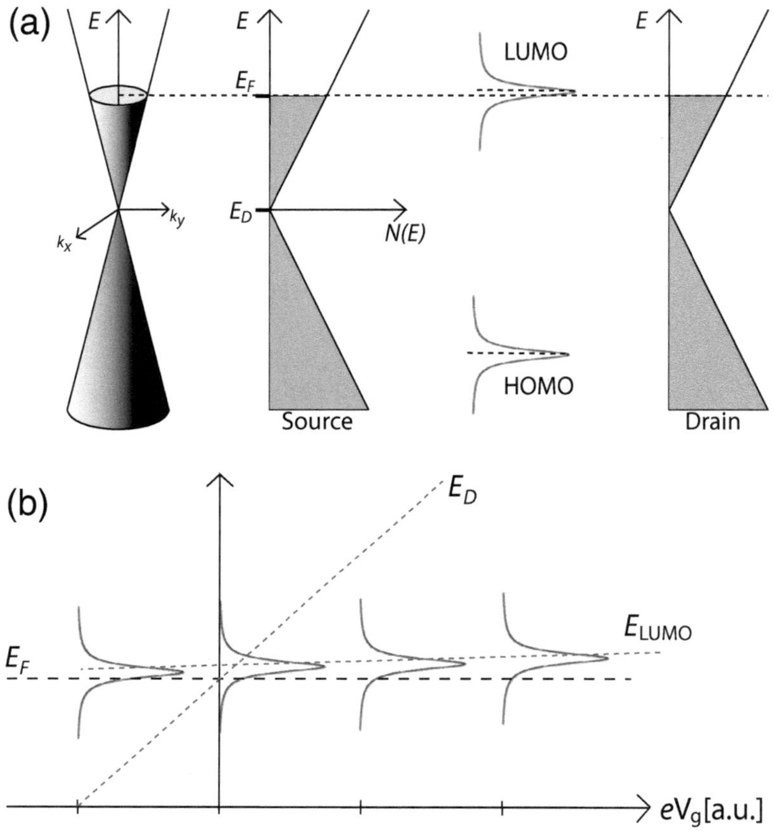

Figure 3.6 (a) Energy level diagram at zero source–drain voltage of a molecule with the LUMO aligned close to the Fermi energy of the graphene leads. The work function of the leads can be changed by a few hundred meV via a back gate so that the Dirac point E_D in the graphene density of states $N(E)$ is above or below (depicted) the Fermi level E_F. (b) A sketch of the movement of the Dirac point through the molecular level as the device is gated. Here we illustrate a case when the gate is well coupled to the leads but badly coupled to the molecule (by construction), which leads to different velocities for the movement of the graphene band structure (illustrated by the Dirac point) and the molecular levels (illustrated by the LUMO).

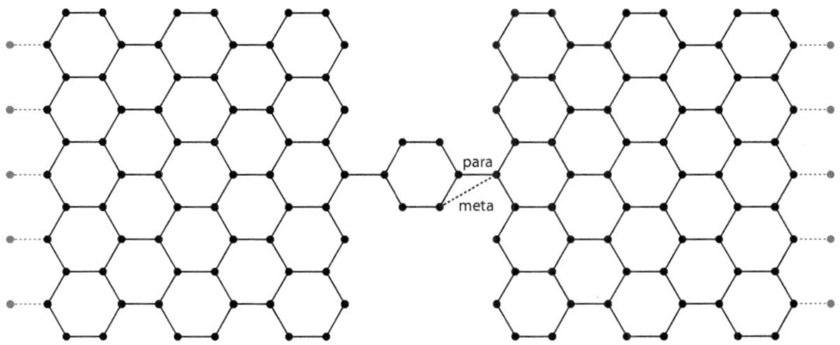

Figure 3.7 Extended system consisting of a benzene ring coupled to graphene ribbons. Here we take the nanogap edge to be of the zigzag type. The graphene is coupled to perfect ribbon leads at the red points to the left and right. In the calculations in this section, the graphene ribbons are much wider (25 nm) than showed in this illustration. The benzene ring can be coupled in the *para*- or the *meta*-orientation. Here the *para*-configuration is illustrated, while the *meta*-configuration is only indicated by a dashed line since the actual geometry would be different in that case.

connected to either atom 5 or atom 6 of the benzene molecule, referred to as the *meta*- and *para*-configurations, respectively.

Using the recursive Green's function technique (or, for this small system, direct inversion would suffice), we can calculate the propagators $G(E)$ for the benzene ring. The leads are included in the recursive technique in the same way as described in Section 3.2 for the transport through the single molecular level. After elimination of the leads in favor of the molecule Green's function, the influence of the leads is contained in the lead self-energies. In the spirit of the previous section, we write it down explicitly for the left lead connecting site number one of the benzene ring:

$$\Gamma_L(E) = \begin{pmatrix} \Gamma_L(E) & 0 & 0 & 0 & 0 & 0 \\ 0 & 0 & 0 & 0 & 0 & 0 \\ 0 & 0 & 0 & 0 & 0 & 0 \\ 0 & 0 & 0 & 0 & 0 & 0 \\ 0 & 0 & 0 & 0 & 0 & 0 \\ 0 & 0 & 0 & 0 & 0 & 0 \end{pmatrix} \tag{3.63}$$

In this example, the molecule is only connected to one atom of the graphene leads and the self-energy is $\Gamma_L = \tilde{\tau} g_s(E)\tilde{\tau}^*$, where $g_s(E)$ is the surface Green's function of the graphene lead at the site connecting the molecule. The self-energy for the right lead is, in an analogous way, a 6 by 6 matrix, but with the self-energy $\Gamma_R(E)$ entering on the fifth or sixth diagonal element depending on if we connect atom 5 or atom 6 of the benzene ring to the right lead. The propagators $G(E)$ are computed as $G(E) = [E - H - \Gamma_L(E) - \Gamma_R(E)]^{-1}$. In the recursive technique, we therefore replace H used in the previous section by $H + \Gamma_L(E) + \Gamma_R(E)$.

For transport, the lead-molecule coupling functions $\gamma_L(E) = i\left[\Gamma_L(E) - \Gamma_L(E)^\dagger\right]$ and $\gamma_R(E) = i\left[\Gamma_R(E) - \Gamma_R(E)^\dagger\right]$ are also needed. The Landauer–Büttiker transmission function is defined as

$$T(E) = \mathrm{Tr}\left\{\gamma_L(E)G(E)\gamma_R(E)G(E)^\dagger\right\} \tag{3.64}$$

For our benzene molecule, it reduces to

$$T(E) = \gamma_R(E)|G_{61}(E)|^2\gamma_L(E) \tag{3.65}$$

for the *para*-configuration. For the *meta*-configuration, we instead obtain $T(E) = \gamma_R(E)|G_{51}(E)|^2\gamma_L(E)$. The transmission has a very intuitive form with three factors: jumping into the molecule quantified by $\gamma_L(E)$, propagation through the molecule with an amplitude $G_{61}(E)$ [the probability $|G_{61}(E)|^2$ enters $T(E)$], and finally jumping out to the second lead as quantified by $\gamma_R(E)$. The propagation through the molecule includes all quantum interference effects. The fact that the electrons can take different paths through the molecule is, in practice, included through the matrix inversion when $G(E)$ is computed.

For the simulation in this section, we set the different hopping energies to $\tau_l = \tau$ within the leads, $\tau_b = 0.3\tau$ within the benzene molecule, and $\tilde{\tau} = 0.2\tau$ between the leads and the molecule. The onsite energies, both in the leads and in the molecule, are all initially set to zero, $\epsilon_l = \epsilon_b = 0$.

It is instructive to first consider normal metallic leads, as considered for instance in Ref. [22]. We may, for simplicity, consider the wide-band limit when we compute the surface Green's function. It is then energy independent, given by $g_s(E) = -i/W$, where W is some energy scale here put to 2τ. The density of states (DOS), $\mathcal{N}(E) = -(1/\pi)\mathrm{Tr}[\mathrm{Im}G(E)]$, on the benzene ring is shown in Fig. 3.8a. The molecular levels of the benzene ring coupled to the

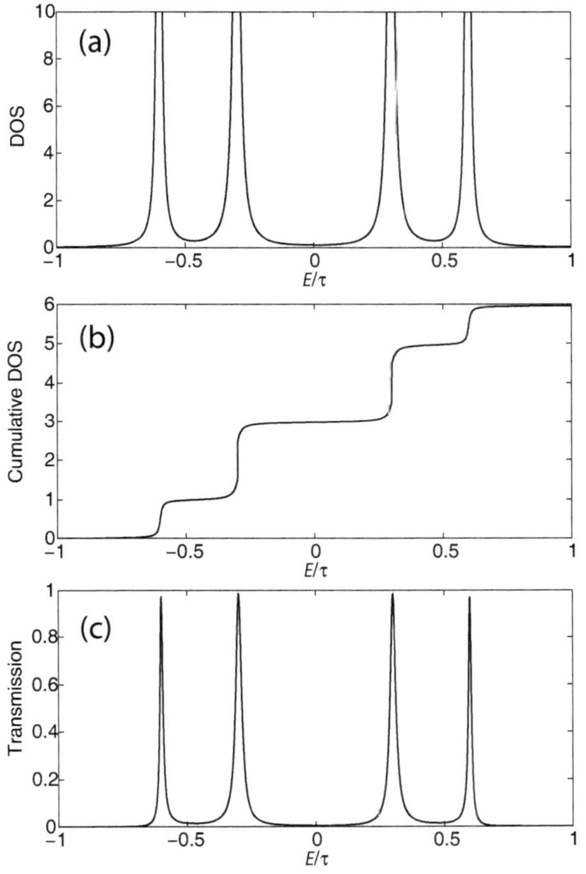

Figure 3.8 (a) Density of states, (b) cumulated DOS (see text), and (c) transmission function through a benzene molecule with metallic leads modeled by a wide-band surface Green's function. This figure is for the *para*-configuration of anchoring the benzene molecule.

leads (indicated by peaks in the DOS) are located at the energies of the isolated molecule, which are $\pm|\tau_b|$ and $\pm 2|\tau_b|$. The DOS peaks are broadened by the coupling to the leads. We shall see that when we go beyond the wide-band limit below, the surface Green's function acquires a real part and the levels are also shifted. But here we see only level broadening. The molecular levels at $\pm|\tau_b|$ are doubly degenerate. This can also be seen in Fig. 3.8b, where we display the integrated DOS, $n(E) = \int_{-\infty}^{E} \mathcal{N}(\tilde{E})d\tilde{E}$. The steps at $\pm|\tau_b|$ have double height as compared with the steps at $\pm 2|\tau_b|$.

We plot the transmission versus energy for the wide-band limit of the leads in Fig. 3.8c. The transmission shows a resonant behavior around the molecular levels. For the *meta*-configuration, the lower molecular levels at $E = \pm|\tau_b|$ split into two because of a destructive quantum interference effect (see Fig. 3.9).

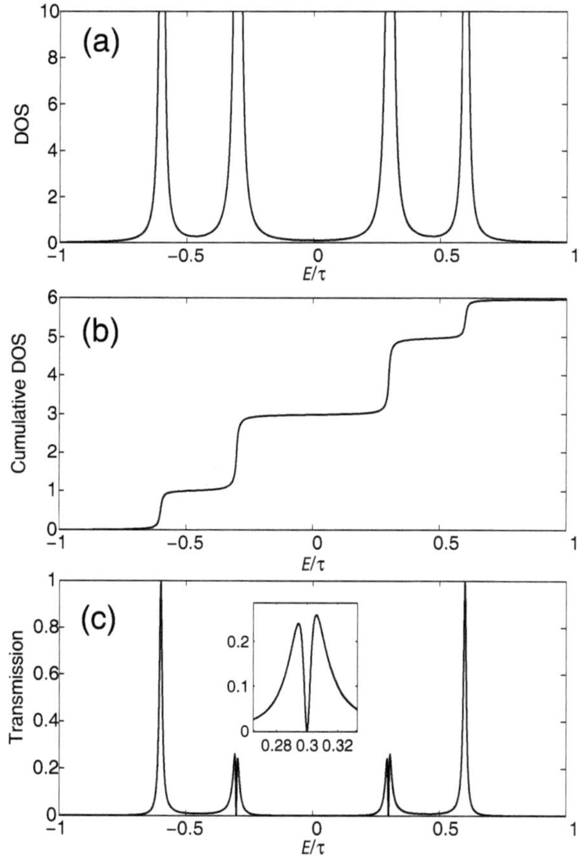

Figure 3.9 (a) Density of states, (b) cumulated DOS (see text), and (c) transmission function through a benzene molecule with metallic leads modeled by a wide-band surface Green's function. This figure is for the *meta*-configuration of anchoring the benzene molecule.

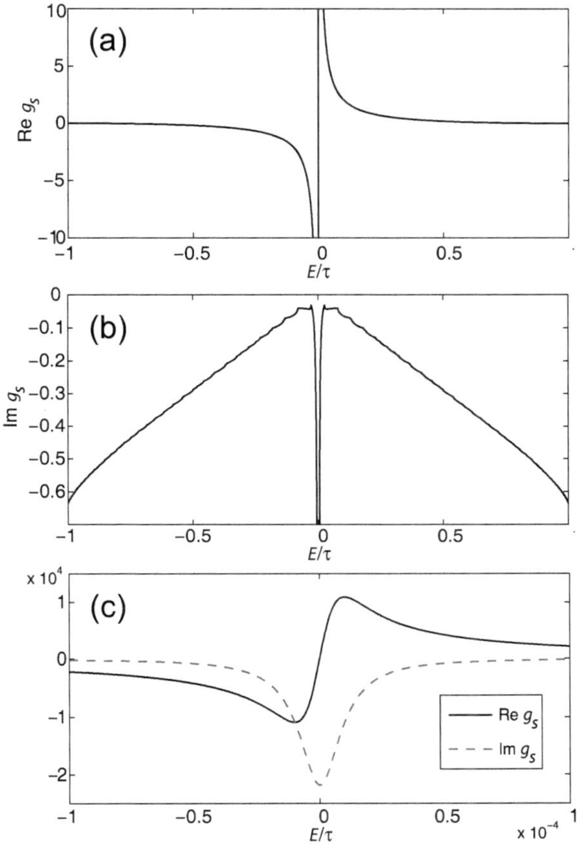

Figure 3.10 Surface Green's function for a graphene lead with a zigzag termination at the nanogap. Panel (c) is a zoom in near the Dirac point, where the zigzag edge state is reflected as a pole in the Green's function. The wiggly behavior seen in panel (b) is due to the finite width of the graphene lead (about 25 nm), which induces a finite number of transverse modes through quantum confinement. This effect is not of importance to the discussion in this chapter.

We are now in a position to expand the discussion to include the graphene leads. The surface Green's function $g_S(E)$ now has a nontrivial energy dependence as shown in Fig. 3.10 for the case of nanoribbon leads with zigzag edges at the nanogap, where the molecule is anchored. The imaginary part contains a term linear in energy, which reflects the linear DOS of graphene. We can contrast that with typical metallic leads, which have more or less constant energy-independent DOS near the Fermi energy (as in the wide-band approximation). Near the Dirac point, the surface Green's function also has a pole due to the zigzag edge state. This is reflected as a sharp peak at the Dirac point E_{D} in the imaginary part of $g_s(E)$ and the typical principal part behavior $1/(E - E_{\mathrm{D}})$ of the real part. We can, therefore, say by inspection that the surface Green's function to a reasonably good approximation has the form

$$g_s(E) \approx -ic_1|E - E_{\mathrm{D}}| + c_2\frac{1}{E + i\delta - E_{\mathrm{D}}} \tag{3.66}$$

where the imaginary part δ is small because the zigzag edge state turns out to be quite weakly coupled to the bulk of the leads. In the figure, we have $E_{\mathrm{D}} = 0$ since the onsite energies in the leads are $\epsilon_l = 0$. The weights of the two terms were here arbitrarily set to constants c_1 and c_2. In the following, we do not use Eq. 3.66 for the simulations, but instead use the full surface Green's function of graphene nanoribbon leads of width 25 nm computed numerically.

The transmission for the case of graphene leads is shown in Fig. 3.11. The energy-dependent surface Green's function, which has both real and imaginary parts, leads to both broadening and shifts of the energy levels. This is seen for the *para*-configuration in Fig. 3.11a. For the *meta*-configuration, it is even more apparent: the quantum interference is now considerably changed as compared with the metallic wide-band leads and complete destructive interference is not present anymore, although the double-peak structure remains (see Fig. 3.11b).

The influence of a back gate of the device can be included by shifting the benzene onsite energies $\epsilon_b \rightarrow \epsilon_b + \alpha_b eV_g$ and the graphene leads onsite energies by $\epsilon_l \rightarrow \epsilon_l + \alpha_l eV_g$. A nice gate effect can be achieved [3] if the gate couples differently to the leads and the molecule, i.e., $\alpha_b \neq \alpha_l$. In the following, we consider the extreme case when the molecule is unaffected, i.e., $\alpha_b = 0$. In Fig. 3.12, we display the resulting gate effect. As the gate shifts the Dirac point from $E_{\mathrm{D}} = 0$ to $E_{\mathrm{D}} = \alpha_l eV_g = |\tau_b|$, it coincides with the molecular level. Ignoring for a moment the zigzag edge state (this would in fact correspond to armchair nanogap termination), the DOS in the leads goes to zero at the Dirac point and the molecule is increasingly

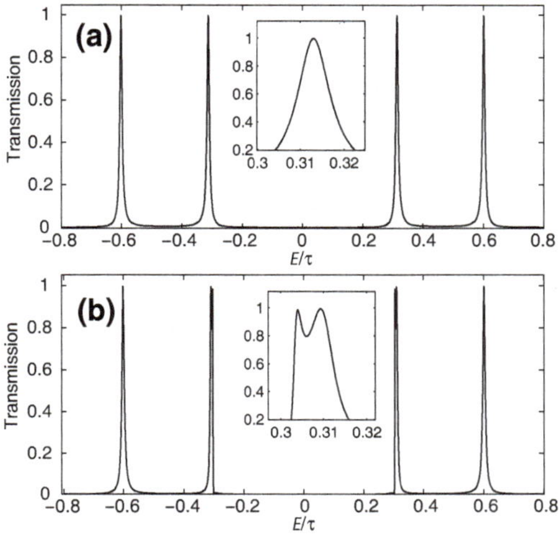

Figure 3.11 Transmission through the benzene molecule with graphene leads for (a) the *para*-configuration, and (b) the *meta*-configuration.

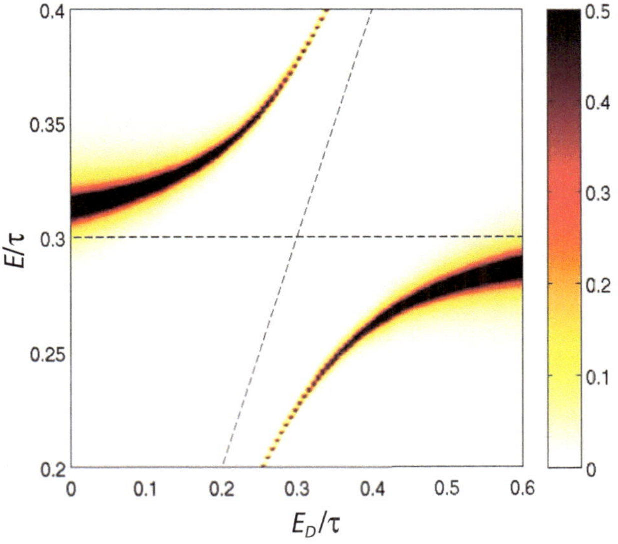

Figure 3.12 Map of the transmission function for different gate voltages (horizontal axis). The dashed lines are a guide to the eye of the unperturbed molecular level (horizontal line) and the zigzag edge state, which follows the Dirac point along the diagonal line. The molecule is here anchored in the *para*-configuration.

decoupled, and the transmission peaks near the molecular level become increasingly sharp as $E_{\rm D}$ approaches the molecular level at $|\tau_b|$. This is the main transistor effect. Eventually, of course, the charging energy becomes important as the coupling to the leads becomes too small, $\gamma \ll U$. This appears in a small window around the gate voltage in the plot where the Dirac point coincides with the molecular level and we neglect this effect here. For the special case of zigzag nanogap edges used in this example, the edge states hybridize with the molecular levels and an anticrossing effect is achieved on top of the increasing sharpness of the transmission resonances. The anticrossing is very much apparent in the plot in Fig. 3.12. The corresponding conductance map for the *meta*-configuration is displayed in Fig. 3.13. Here the double-peak interference in the molecule is also affected by the hybridization of the molecular levels with the zigzag edge states. The zigzag edge state-molecule level

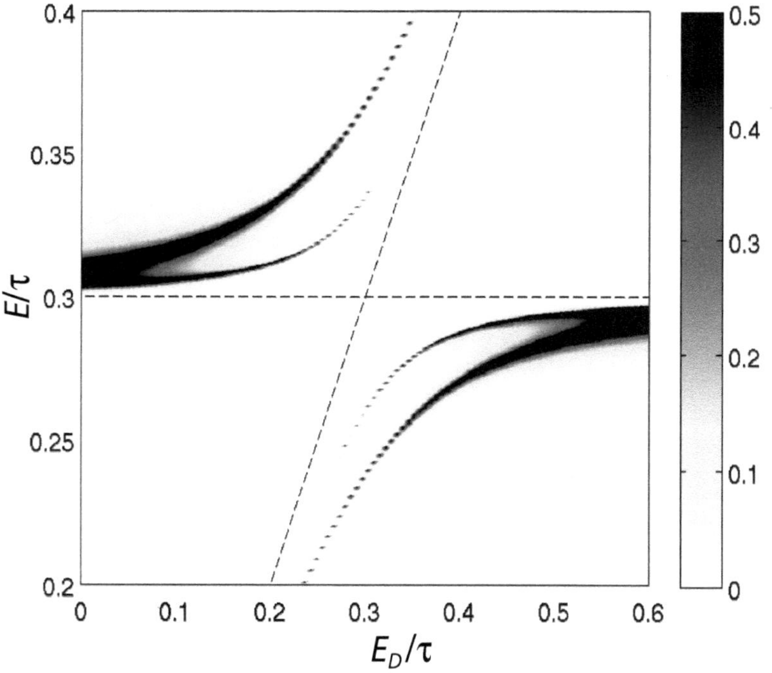

Figure 3.13 Same as in Fig. 3.12 but for the *meta*-configuration.

hybridization has been studied in several theory papers recently [3, 23–26].

3.5 Molecular Contact between Superconducting Leads

As another application of the theory introduced in Section 3.2, we will consider a simplified model of a single-wall carbon nanotube (swCNT) embedded between two superconducting metal leads. Such a system has been shown to support a supercurrent [27–34]. The electronic energy levels of the swCNT, formed through size quantization, can be tuned in and out of resonance with the lead Fermi levels by gating the swCNT. Such superconducting swCNT transistors can further be switched from a Coulomb blockade regime, to a Kondo regime, to a weakly interacting Fabry–Pérot regime by changing a back gate voltage [31]. The potential for applications of such swCNT quantum-level junctions was demonstrated through the fabrication and detailed functional control of a nano-SQUID, involving two gated swCNT junctions with controlled on-off states as well as controlled 0-π SQUID states [32]. Similar control has been demonstrated using semiconducting nanowire junctions [35–37].

Supercurrents through mesoscopic or nanosized Josephson junctions are mainly carried by spectroscopically sharp subgap bound states, so-called Andreev bound states (ABS) [38], as predicted theoretically in Refs. [39, 40]. When the weak link is a quantum dot with a single or a few discrete levels, as for the swCNT junction, supercurrents are also predicted to flow mainly through ABS [41]. ABS come in pairs, one state above and one below the Fermi level. The two ABS of the pair have opposite dispersions with the superconducting phase difference over the junction and carry supercurrent in opposite directions across the junction. These two states form a well-defined two-level system that has been suggested as a qubit [42, 43]. Before such a qubit can be realized, experimental detection and characterization of this engineered two-level system should be carried out.

We have proposed a method for direct ABS spectroscopy, based on dispersive measurement of a polaritonic state formed

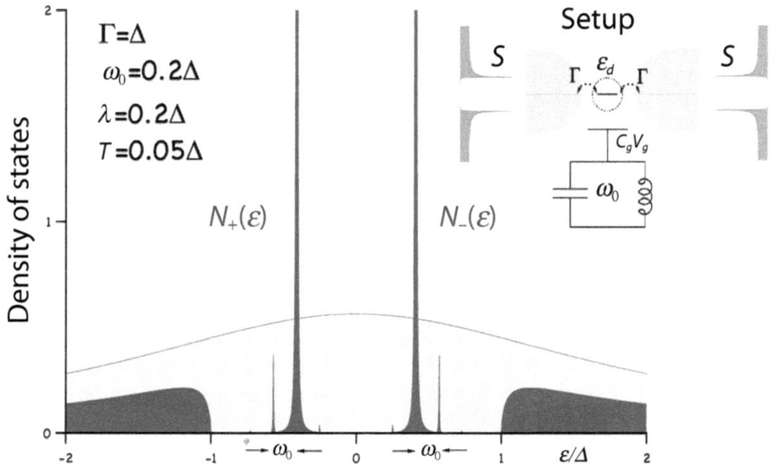

Figure 3.14 Andreev bound state spectrum and schematic setup of the system. Superconducting density of states for $\varphi = 0.6\pi$ resolved into right, $N_+(\varepsilon)$, and left, $N_-(\varepsilon)$, current-carrying branches. The lightly shaded background is the density of states of the resonant level when the contacts are in the normal state. The electron–oscillator coupling leads to dressed Andreev bound states with sidebands separated in energy by the oscillator frequency ω_0. The inset shows a sketch of the considered system.

by the ABS strongly coupled to a QED cavity mode [44]. Consider two superconducting leads connected by a molecule with one resonant level, as shown in Fig. 3.14. The superconducting proximity effect leads to a split of the resonant level into a pair of ABS, where the ABS level splitting can be controlled by tuning the superconducting phase difference in a SQUID setup. When the gate electrode is coupled to an LC-oscillator, induced quantum fluctuations of the gate potential lead to dressed Andreev bound states. By tuning the ABS into resonance with the cavity and measuring the shift of the cavity base frequency, the ABS energy can be observed. The complete ABS energy dispersion with respect to the superconducting phase can be extracted by using a cavity with a variable base frequency [45, 46]. The suggested dispersive method does not involve real inter-level transitions and could be simpler to realize than methods based on microwave absorption [47]. Another advantage of the suggested spectroscopy is that it makes it possible to directly access and characterize individual ABS without measuring dc contact current–voltage characteristics [48]. Similar dispersive measurement method has earlier been applied to detect quantum states of superconducting charge qubits [49].

3.5.1 *General Methods*

In the previous section, we derived the Dyson equation (3.25) for $\check{\mathcal{G}}$ reading

$$\check{\mathcal{G}} = \check{\mathcal{G}}_o + \check{\mathcal{G}}_o \circ (\check{\mathcal{V}} + \check{\Sigma}) \circ \check{\mathcal{G}} \tag{3.67}$$

The coupling between the molecular level and the leads is treated to infinite order in the hopping, which enters in the Dyson equation through the potential $\check{\mathcal{V}}$, as described above. The uncoupled subsystems are described by Green's functions with subscript o. Effects of coupled electron and oscillator degrees of freedom may be included via $\check{\Sigma}$, the local self-energy on the dot describing the coupling to the oscillator. Here we treat this coupling perturbatively within the self-consistent Born approximation [50–54]. We thus assume that the tunneling rate Γ is large compared with the electron–oscillator coupling λ.

The solution strategy is to first eliminate the leads in favor of the renormalized molecular level Green's function as written in Eq. 3.30:

$$\check{G}(\varepsilon) = [\check{G}_0^{-1}(\varepsilon) - \check{\Gamma}(\varepsilon) - \check{\Sigma}(\varepsilon)]^{-1} \tag{3.68}$$

The bare level retarded Green's function for $\epsilon_d = 0$ is simply $[\hat{G}_0^R(\varepsilon)]^{-1} = \varepsilon^R \hat{\tau}_3$, where $\varepsilon^R = \varepsilon + i0^+$, and $\hat{\tau}_i$, $(i = 1, 2, 3)$ are the Pauli matrices in electron–hole space that spans the solution space of a superconducting system. The superconducting leads enter, after summing over all tunneling k-states, via the self-energy $\check{\Gamma}(\varepsilon) = \check{\Gamma}_L(\varepsilon) + \check{\Gamma}_R(\varepsilon)$ (omitting the subscript Σ in Eq. 3.30) with components

$$\hat{\Gamma}^R(\varepsilon) = \left[\tilde{\varepsilon}^R \hat{\tau}_3 - \Delta \cos(\phi/2) i \hat{\tau}_2\right] / \bar{\Omega}^R(\varepsilon) \tag{3.69}$$

$$\hat{\Gamma}^A(\epsilon) = \hat{\tau}_3 [\hat{\Gamma}^R(\epsilon)]^\dagger \hat{\tau}_3 \tag{3.70}$$

$$\hat{\Gamma}^K(\varepsilon) = \left[\hat{\Gamma}^R(\varepsilon) - \hat{\Gamma}^A(\varepsilon)\right] n_e(\varepsilon) \tag{3.71}$$

where Δ is the superconducting order-parameter and $\tilde{\varepsilon}^R = \varepsilon^R + i\gamma$, $\bar{\Omega}^R(\varepsilon) = \sqrt{|\Delta|^2 - (\tilde{\varepsilon}^R)^2}/2\Gamma$, and $n_e(\varepsilon) = \tanh \frac{\varepsilon}{2T}$. ϕ is the phase difference between the two superconducting leads.

We have assumed for simplicity that all the tunneling amplitudes are the same, i.e., $V_{di,Lk} = v$ so that

$$\hat{\Gamma}^R_\alpha(k,\varepsilon) = |v|^2 \sum_k G^R_\alpha(k,\varepsilon) = \frac{|v|^2}{2\pi}\int d\xi_k \mathcal{D}(\xi_k)\hat{G}^R_\alpha(\xi_k,\varepsilon)$$
$$\approx -i\frac{|v|^2}{\pi}\mathcal{D}(E_{F,\alpha})\hat{g}^R_\alpha(\varepsilon) \tag{3.72}$$

Here $\mathcal{D}(E_{F,\alpha})$ is the density of states at the Fermi level of the lead α, and we have introduced the *quasi classical* Green's function $\hat{g}^R_\alpha(\varepsilon)$ of that lead. For a superconducting state, $\hat{g}^R_\alpha(\varepsilon)$ is given as

$$\hat{g}^R_\alpha(\varepsilon) = -\frac{\pi}{\sqrt{|\Delta|^2 - (\varepsilon^R)^R}}\begin{pmatrix} \epsilon^R & -\Delta_\alpha \\ \Delta^*_\alpha & -\epsilon^R \end{pmatrix} \tag{3.73}$$

We set $\Delta_\alpha = |\Delta(T)|\exp(\pm\phi/2)$.

We now need to consider the coupling to the oscillator to compute $\check{\Sigma}$. Corrections to the oscillator are now computed by self-consistently coupling back the *polarization bubble* $\check{\Pi}$ via the Boson–Dyson equation

$$\check{D} = \check{D}_0 + \check{D}_0 \circ \check{\Pi} \circ \check{D} \tag{3.74}$$

The oscillator self-energy, $\check{\Pi}(\omega)$, defined below will encode how electron–hole excitation modifies the state of the oscillator. We will see that the superconducting state on the level introduces a new mode for the oscillator that will be dependent on the phase difference between the two superconducting leads.

The final structure of the equations in electron–hole space suggests a convenient factorization of the electronic Green's functions (and the self-energies) as $\hat{G}^X = G^X_+\hat{\eta}_+ + G^X_-\hat{\eta}_-$ where $X = R, A, K$, and the basis matrices are $\hat{\eta}_\pm = \frac{1}{2}(\hat{\tau}_3 \pm i\hat{\tau}_2)$. Here $\hat{1}$ is the unit matrix, and $\hat{\tau}_j$ ($j = 1, 2,$ and 3) are Pauli matrices in electron–hole space. The retarded scalar component of Green's function G^R_s (the index $s = \pm 1$ will be referred to as the branch index) is

$$G^R_s(\varepsilon) = \frac{\bar{\Omega}^R(\varepsilon)}{z^R_s(\varepsilon)\varepsilon^R + s\Delta\cos(\varphi/2)} \tag{3.75}$$

with the energy renormalization

$$z^R_s(\varepsilon)\varepsilon^R = \tilde{\varepsilon}^R + \bar{\Omega}^R(\varepsilon)\left(\varepsilon^R - \Sigma^R_{-s}\right) \tag{3.76}$$

where $\bar{\Omega}^R(\varepsilon) = \sqrt{|\Delta|^2 - (\tilde{\varepsilon}^R)^2}/2\Gamma$ and $\tilde{\epsilon}^R = \epsilon^R + i\gamma$. The corresponding oscillator component is

$$D^R(\omega) = \frac{2\omega_o}{(\tilde{\omega}^R)^2 - \omega_o^2 - 2\omega_o \Pi^R(\omega)} \tag{3.77}$$

with $\tilde{\omega}^R = \omega^R + i\kappa$. Above γ and κ are phenomenological parameters describing inelastic scattering in the electronic leads ($\gamma \ll \Delta$) and a finite (but high) quality factor Q of the oscillator with $\kappa = \omega_0/Q \ll \omega_0$. Advanced components are given by complex conjugation of the retarded. The Keldysh components of the Green's functions (and the self-energies) can be divided into a spectral part, $G_s^{sp} = -2\,\mathrm{Im}G_s^R$ and an anomalous part, G_s^a, as $G_s^K = -iG_s^{sp}\, n_e + G_s^a$, where $n_e(\varepsilon) = \tanh(\varepsilon/2T_e)$. Analogously for the oscillator Green's function, $D^K = -iD^{sp}n_b + D^a$, where $n_b(\omega) = \coth(\omega/2T_b)$. The electron and the oscillator reservoirs may be at two different temperatures T_e and T_b. If the two reservoirs are in thermal equilibrium, then $T_e = T_b \equiv T$. The equations for the anomalous parts read

$$G_s^a = G_s^R \Sigma_{-s}^a G_s^A \tag{3.78}$$

$$D^a = D^R \Pi^a D^A \tag{3.79}$$

These will not contribute in this calculation as we will keep the electronic and the oscillator system in equilibrium and at the same temperature.

The equations for the self-energies entering in the expressions above read

$$\begin{aligned}
\Sigma_s^R(\varepsilon) &= i\frac{\lambda^2}{2}\left[D^K(\omega)\circ G_s^R(\varepsilon-\omega) + D^R(\omega)\circ G_s^K(\varepsilon-\omega)\right] \\
\Sigma_s^K(\varepsilon) &= i\frac{\lambda^2}{2}\left[D^K(\omega)\circ G_s^K(\varepsilon-\omega) - D^{sp}(\omega)\circ G_s^{sp}(\varepsilon-\omega)\right]
\end{aligned} \tag{3.80}$$

$$\begin{aligned}
\Pi^R(\omega) &= -i\lambda^2\sum_{s=\pm1}\left[G_{-s}^R(\varepsilon)\circ G_s^K(\varepsilon-\omega) + G_{-s}^K(\varepsilon)\circ G_s^A(\varepsilon-\omega)\right] \\
\Pi^K(\omega) &= -i\lambda^2\sum_{s=\pm1}\left[G_{-s}^K(\varepsilon)\circ G_s^K(\varepsilon-\omega) + G_{-s}^{sp}(\varepsilon)\circ G_s^{sp}(\varepsilon-\omega)\right]
\end{aligned} \tag{3.81}$$

where the convolution is defined as $a(x) \circ b(x-y) = \int_{-\infty}^{\infty} \frac{dx}{2\pi} a(x)b(x-y)$. We note that the Hartree term, which should be included in Σ^R when studying molecular contacts [50], vanishes by symmetry for $\varepsilon_d = 0$. Equations 3.80 and 3.81 are solved self-consistently by numerical iteration. Upon reaching self-consistency, the Josephson current is calculated by summing up the contribution from each branch ($s = \pm 1$)

$$j = \frac{e}{2\hbar}\Delta \sin\frac{\varphi}{2} \sum_{s=\pm 1} \int_{-\infty}^{\infty} \frac{d\varepsilon}{2\pi} \left[j_s^{sp}(\varepsilon) n_e(\varepsilon) + j_s^a(\varepsilon) \right] \tag{3.82}$$

where the kernels are

$$j_s^{sp}(\varepsilon) = 2s \,\mathrm{Im}\left[\frac{G_s^R(\varepsilon)}{\bar{\Omega}^R(\varepsilon)}\right] \tag{3.83}$$

$$j_s^a(\varepsilon) = -s \,\mathrm{Re}\left[\frac{iG_s^a(\varepsilon)}{\bar{\Omega}^R(\varepsilon)}\right] \tag{3.84}$$

The last expression is added for completeness but will, as stated above, not contribute to the current in the limit we are studying.

3.5.2 *One-Iteration Approximation*

To get an analytic handle on our calculation, we make a one-iteration approximation of Eqs. 3.80 and 3.81 keeping subgap energies only. We write $G_s^{R(0)}(\varepsilon) = \mathcal{A}_E/(\varepsilon^R + sE_A)$ for vanishing electron–oscillator coupling and consider thermal equilibrium. The ABS energy E_A is a function of phase difference and of the tunneling rate and is given by the poles of $G_s^R(\varepsilon)$ in Eq. 3.80, which in general must be determined numerically. The residue

$$\mathcal{A}_E = \frac{\Omega^2(E_A)}{\Omega^2(E_A) - E_A^2 + 2\Gamma\Omega(E_A)}, \tag{3.85}$$

where $\Omega = 2\Gamma\bar{\Omega}$, gives a measure of the fraction of the normal state dot level that resides in the superconducting ABS. In the limit $\Gamma \ll \Delta$, $\mathcal{A}_E \approx 1$ and also phase independent. In this limit, the one-iteration approximation describes the self-consistent solution well because the continuum states give only small contributions to the integrals (Eqs. 3.80 and 3.81). In the opposite limit $\Gamma \gg \Delta$, $\mathcal{A}_E \propto \Delta/\Gamma$ and strongly phase dependent. In this case, the continuum states give large corrections. For the parameters used in the figures $\mathcal{A}_E \approx 0.34$, i.e., in between the two limits, but to 10% accuracy phase independent. The result of one iteration is

$$\Sigma_s^R(\varepsilon) = \frac{\lambda^2}{2}\left[\frac{\mathcal{A}_E \Phi_{-s}}{\varepsilon^R - \omega_o + sE_A} + \frac{\mathcal{A}_E \Phi_s}{\varepsilon^R + \omega_o + sE_A}\right] \tag{3.86}$$

where $\Phi_s = n_b(\omega_o) + s n_e(E_A)$ is a thermal factor. The corresponding expression for the oscillator self-energy reads

$$\Pi^R(\omega) = 2\lambda^2 \frac{\mathcal{A}_E^2}{\omega^{R2} - 4{E_A}^2} n_e(E_A) \tag{3.87}$$

3.5.3 *Results*

Without coupling to the oscillator, the Josephson effect in this system is well known [41]. The Green's function-amplitudes, G_s, describe two different quasi-particle branches that form on the dot, one having an ABS below ($s = +1$) and one having an ABS above ($s = -1$) the Fermi level (see Fig. 3.14) The two branches contribute to the Josephson current (Eq. 3.82) in opposite manner. For sub-gap energies, $G_s(|\varepsilon| < \Delta)$ contributes to the current in the positive direction for $s = 1$ (left to right over the junction) and vice versa for the $s = -1$ branch. The continuum part of $G_s(|\varepsilon| > \Delta)$ contributes to the current in the opposite direction compared with its corresponding sub-gap part.

The effect of the electron–oscillator interaction on the electron system is to couple the two quasi-particle branches. This is seen explicitly in the energy-renormalization factor z_s of one branch in Eq. 3.76, which is modified by the self-energy Σ_{-s} of the other branch. In Fig. 3.15, we show a fully self-consistent calculation of the Andreev spectrum as a function of superconducting phase-difference. There are two different regions: the off-resonant region $2E_A(\varphi) \neq \omega_o$ and the resonant region $2E_A(\varphi) \approx \omega_o$, with the resonance appearing at $\varphi \approx 0.9\pi$ for the chosen parameters. Away from resonance, the ABS is shifted by

$$E_A \to \bar{E}_A = E_A + \frac{\lambda^2 \mathcal{A}_E^2}{2} \left(\frac{\Phi_+}{2E_A + \omega_o} + \frac{\Phi_-}{2E_A - \omega_o} \right) \tag{3.88}$$

as compared with the case without coupling to the oscillator, but with retained phase-dependent spectral weight $\mathcal{A}_E$ of the state at $\varepsilon = \pm E_A(\varphi)$. The effective electron–oscillator coupling in the sub-gap region is given by the product $\lambda \mathcal{A}_E$. In Eq. 3.88, thermal occupation factors enter the combinations $\Phi_\pm = \coth(\omega_o/2T) \pm \tanh(E_A/2T)$. Apart from the shifted ABS, we find satellite resonances at $\varepsilon = s\bar{E}_A \pm \omega_o$ with spectral weights $\frac{1}{2}\lambda^2 \mathcal{A}_E^3 \Phi_{\pm s}/$

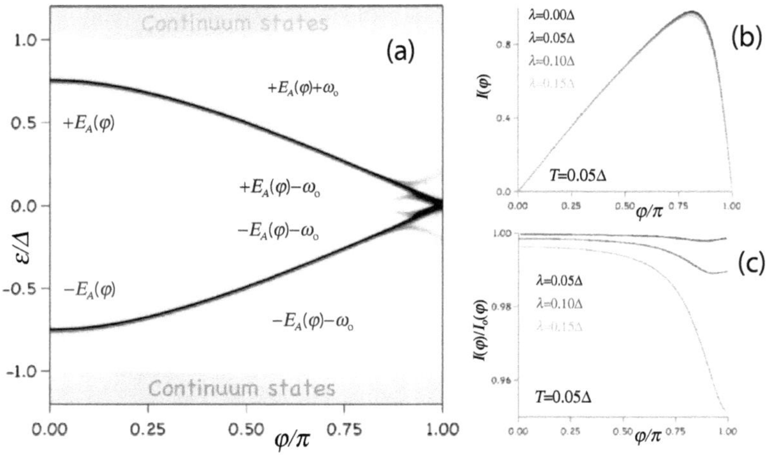

Figure 3.15 The electronic subsystem. (a) Self-consistently determined spectrum of Andreev bound states as a function of superconducting phase-difference (φ) for electron–oscillator coupling $\lambda = 0.1\Delta$ and oscillator frequency $\omega_o = 0.2\Delta$. The tunneling rate is $\Gamma = \Delta$, and the temperature is $T = 0.05\Delta$. The side bands to each ABS are due to the dressing of the ABS by the electron–oscillator coupling. The color coding of the labels denotes quasi-particle branch (red for $s = +1$ and blue for $s = -1$) of the ABS and the side bands. (b) The current-phase relation for different electron-oscillator couplings for $T = 0.05\Delta$. (c) The reduction of current due to coupling to the oscillator as a function of phase for the same temperature and electron–oscillator couplings as in panel (b).

$(2sE_A \pm \omega_o)^2$. Around resonance $2E_A \approx \omega_o$, the satellite on either branch with spectral weight $\propto \Phi_-$ interferes with the main ABS of the same branch index.

The Josephson current-phase relation is presented in (b) and (c) of Fig. 3.15 for the case that the electron and the oscillator systems are in thermal equilibrium. Because the interaction with the oscillator involves a change of branch index, the nearest satellites to each ABS contribute to the current in the opposite direction to the ABS itself (see Fig. 3.14). Due to the different magnitudes of the thermal factors $\Phi_\pm$, the current contribution of the satellites is dominated by the satellite with weight $\propto \Phi_+$. This satellite reduces the Josephson current by $\sim \lambda^2 \mathcal{A}_E^2/(2E_A + \omega_o)^2$, which is of the order of a few percent of the full current for our parameter values. There is no dramatic signature in the current-phase relation of an emerging

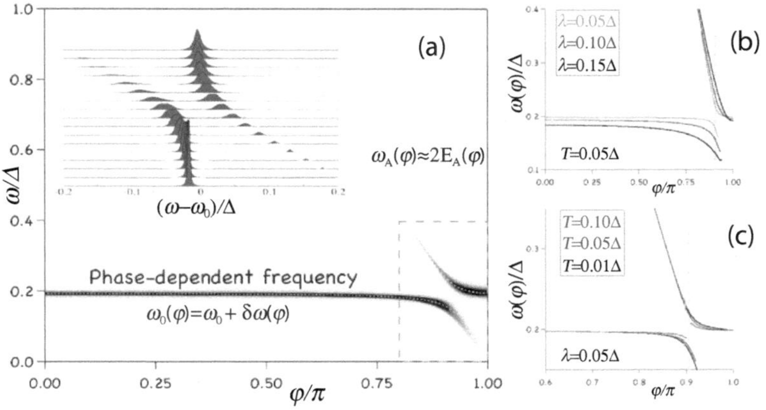

Figure 3.16 Oscillator subsystem. (a) The self-consistently determined spectral function $D^{sp}(\omega)$ of the oscillator plotted as a function of superconducting phase-difference with the same parameters as in Fig. 3.15. A collective mode emerges at $\omega_A(\varphi) = 2E_A(\varphi)$, and this mode sharpens up to a polariton when in resonance with the oscillator (inset). The dashed box indicates where the spectra in the inset are taken. Away from resonance, there is a Stark shift $\delta\omega(\varphi)$ of the base frequency. (b) The Stark shift is shown at different electron–oscillator coupling strengths for $T = 0.05\Delta$ and (c) at different temperatures for an electron–oscillator coupling strength $\lambda = 0.05\Delta$. In both cases, $\omega_o = 0.2\Delta$. The dashed lines in panels (b) and (c) indicate the resonant phase.

anticrossing at resonance because interference occurs between states carrying current in the same direction, and moreover, these states have spectral weights shared between them drawn from the original ABS, and the populations of the states are largely phase-independent.

The Josephson current through the level modifies the oscillator spectrum giving it a phase-dependence shown in Fig. 3.16a for the same parameters as for the ABS spectrum in Fig. 3.15. The polarization $\Pi(\omega)$ gives the possible collective excitations that are supported by the electronic system. Iterating the self-consistency equations once, we get

$$\Pi(\omega) = 2\lambda^2 \frac{\mathcal{A}_E^2}{\omega^2 - 4{E_A}^2} n_e(E_A) \tag{3.89}$$

for the retarded phonon self-energy. As may be expected, there is a mode with the phase dispersion $\omega_A = 2E_A(\varphi)$ originating from the transitions between the two ABS and the subsequent emission or absorption of the energy ω_A. A spectral weight $\sim 2\lambda^2\mathcal{A}_E^2[4\omega_o^2/(4E_A^2-\omega^2)^2]n_e(E_A)$ for this collective excitation is vanishingly small away from resonance. Near resonance, the oscillator mode and the excitation interact strongly and an avoided crossing appears with a frequency split at $2E_A = \omega_o$

$$\omega_o(\varphi) = \omega_o + \frac{\delta}{2} \pm \sqrt{\left(\frac{\delta}{2}\right)^2 + 2\lambda^2\mathcal{A}_E^2 n_e(E_A)} \tag{3.90}$$

where $\delta = 2E_A - \omega_o$ is the detuning. The analytic estimate of the split, $2\lambda\mathcal{A}_E\sqrt{2n_e(E_A)} \approx 0.1$, for a nearly phase-independent $\mathcal{A}_E \sim 0.34$ when $\Gamma = \Delta$, is in good agreement with the self-consistently determined split, $\approx 0.13\Delta$, extracted from the inset in Fig. 3.16a at $\varphi = 0.9\pi$. It is important to note that this split is significantly larger than the intrinsic broadening of the spectral features ($\sim\kappa$) and signals a strong coupling regime in the sense of cavity-QED, i.e., $\lambda\mathcal{A}_E \gg \kappa$.

Computing the self-energies self-consistently is important as self-consistency gives additional modes in the oscillator spectrum around resonance. These modes are due to the allowed transitions between the ABS peaks and satellites or between satellites of the dressed ABS. There are selection rules for the transitions as only electronic Green's functions of opposite branch index s couple. Possible transitions in the electron spectrum and the corresponding modes may be extracted from calculations and are shown in Fig. 3.17. The intensities of the transitions are related to the imaginary part of the self-energy $\Pi(\omega)$ and shown in panel (b) of Fig. 3.17. The first two phase-dependent higher-order modes $\omega_B(\varphi)$ and $\omega_C(\varphi)$ become visible at phase-differences corresponding to the resonance condition. Here the sidebands to the ABS have sizable spectral weight.

Away from resonance, the electron–oscillator coupling gives a negative phase-dependent Stark shift of the oscillator base frequency. We plot this shift in Fig. 3.16b,c both as a function of electron–oscillator coupling strength λ for $T = 0.05\Delta$ and as a function of temperature for $\lambda = 0.05\Delta$. Our analytic estimate in

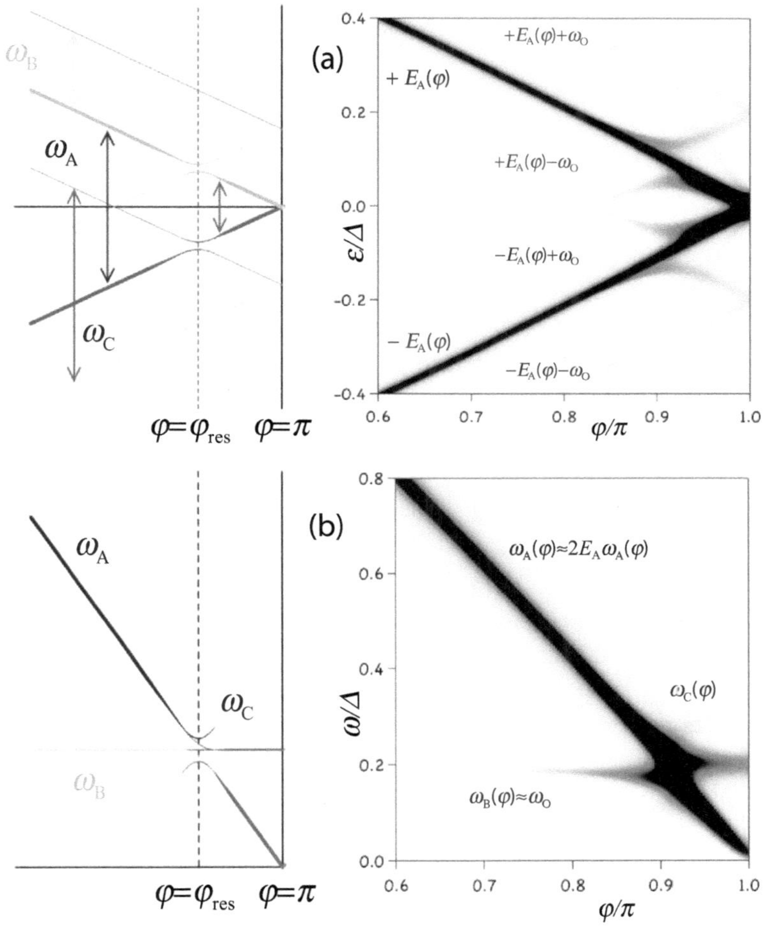

Figure 3.17 Allowed transitions between ABS and ABS-supported boson modes. (a) The allowed transitions between ABS, indicated by arrows, are shown schematically (left) and compared to the computed spectra (right) at phase-differences in vicinity of resonance. Transitions occur only between states of different branch index s. (b) A schematic view of the boson modes $\omega_\alpha(\varphi)$ supported by the ABS (left) compared with the imaginary part of the computed phonon self-energy(right). The color coding connects the phase-dependence of an ABS-transition, an arrow in panel (a), with its corresponding boson mode, a full line in panel (b).

Eq. 3.90 gives in the limit of large detuning

$$\omega_o(\varphi) = \omega_o + \delta\omega(\varphi) = \omega_o - 2\frac{\lambda^2 \mathcal{A}_E^2 n_e(E_A)}{2E_A - \omega_o} \tag{3.91}$$

This analytical expression for $\delta\omega(\varphi)$ is a good approximation when the continuum contributions can be neglected, i.e., for $\Gamma/\Delta \ll 1$, while in general one must use numerics to extract $\delta\omega(\varphi)$. Dispersive measurement of the acquired phase-dependent resonance frequency of the oscillator, $\omega_o(\varphi)$, gives a possibility to detect the position of the polariton resonance and hence define the energy of the ABS. Furthermore, by sweeping the base frequency of the oscillator, the phase dispersion of the ABS energy can be detected.

3.6 Summary

In this chapter, we have reviewed the basic theory of electron transport through molecular scale systems with an aim to introduce the subject to advanced students. We focused on quantum coherent transport within a Hamiltonian description of the device, including aspects of the tight-binding theory. The theory was formulated in terms of Green's functions, which allowed us to include a variety of phenomena central within molecular electronics, such as level shifts, level broadening, quantum interference, coupling to internal degrees of freedom in the molecule (phonons), and the influence of the contacts. The latter can be in the form of surface states (for instance graphene zigzag edge states hybridizing with the molecular orbitals) or superconducting leads that turn the molecular transistor into a Josephson junction with special properties. Below are a few problems of increasing difficulty that can be used to work with the material that has been presented here.

Acknowledgments

We gratefully acknowledge financial support from the Swedish Research Council, the European Union through FP7 STREPs SINGLE and ConceptGraphene, and DFG and SFB 767.

Problems

1. Starting from Eq. 3.35, show that the conductance $d\mathcal{J}/dV$ at zero temperature and low voltage takes the Landauer–Büttiker form for conductance
$$G = \frac{2e^2}{h} T(E_{\mathrm{F}}) \tag{3.92}$$
where E_{F} is the Fermi energy of the leads. What is the form of $T(E_{\mathrm{F}})$?
2. Starting from Eq. 3.35, sketch the current–voltage characteristics $\mathcal{J}(V)$ and the conductance $d\mathcal{J}(V)/dV$, assuming that the level is located at E_r and that the lead coupling functions are equal, $\Gamma_L^R = \Gamma_R^R = \Gamma$, and independent of energy (i.e., Γ is a complex number). For this symmetric junction, the voltage can be assumed to drop symmetrically (see Fig. 3.18). Consider and discuss in particular the high temperature limit $T \gg |\mathrm{Im}(\Gamma)|$ and the low temperature limit $T \ll |\mathrm{Im}(\Gamma)|$.

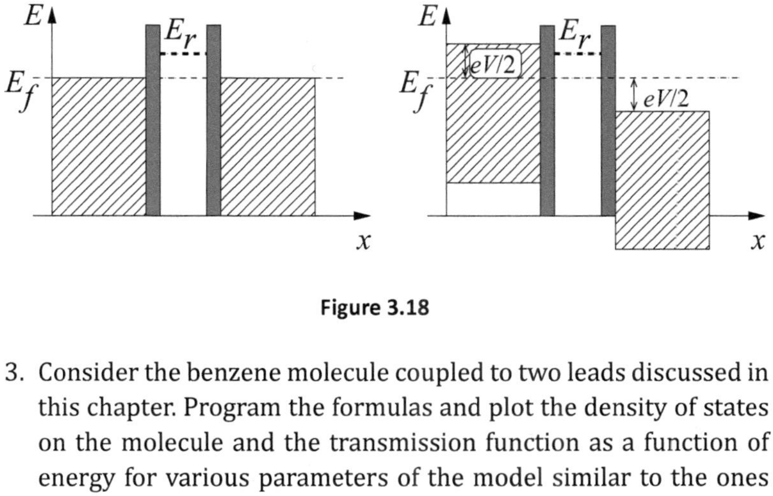

Figure 3.18

3. Consider the benzene molecule coupled to two leads discussed in this chapter. Program the formulas and plot the density of states on the molecule and the transmission function as a function of energy for various parameters of the model similar to the ones chosen in the text. Thereby you should be able to reproduce Figs. 3.8–3.13. For graphene leads, you may use Eq. 3.66.
4. Using the superconducting Green's function given by Eq. 3.73 in the text, derive a formula for the Josephson current over a single molecular level as a function of a symmetric coupling strength, i.e., $v_L = v_R = v$. Explore the current-phase relation

and the Andreev-level spectrum as a function of coupling v and temperature. Assume that the superconducting order-parameter has a temperature dependence given by $\Delta(T) = \Delta_0 \tanh[(1 - T/T_c)/0.3]$, where $\Delta_0 = 1.76 k_B T_c$ is the BCS-value of the order-parameter amplitude at zero temperature.

References

1. Bruus, H., and Flensberg, K. (2004) *Many Body Quantum Theory in Condensed Matter Physics: An Introduction*, Oxford University Press, Oxford.
2. Thijssen, J. M., and van der Zant, H. S. J. (2005) Charge transport and single-electron effects in nanoscale systems. *Phys. Stat. Sol. (b)* **245**, 8, 1455.
3. Bergvall, A., Berland, K., Hyldgaard, P., Kubatkin, S., and Löfwander, T. (2011) Graphene nanogap for gate-tunable quantum-coherent single-molecule electronics. *Phys. Rev. B* **84**, 155451
4. Datta, S. (1995) *Electron Transport in Mesoscopic Systems*, Cambridge University Press, Cambridge.
5. Zahid, F., Paulsson, P., and Datta, S. (2003) Electrical conduction through molecules. In *Advanced Semiconductors and Organic Nano-Techniques*, Morkoc, H. (ed), Academic Press, (available for download at nanoHUB.org).
6. Cuniberti, G., Fagas. G, and Richter, K. (eds). (2010) *Introducing Molecular Electronics (Lecture Notes in Physics)*, Springer.
7. Datta, S. (2005) *Quantum Transport: Atom To Transistor*, Cambridge University Press, Cambridge.
8. Haug, H., and Jauho, J.-P. (2007) *Quantum Kinetics in Transport and Optics of Semiconductors*, Springer Series in Solid-State Sciences.
9. Cuevas, J. C., and Scheer, E. (2010) *Molecular Electronics*, World Scientific Publishing
10. Caroli, C., Combescot, R., Nozieres, P, and Saint-James, D. (1972) *J. Phys. C: Solid States Phys.* **5**, 21.
11. Meir, Y., and Wingreen, N. S. (1992) *Phys. Rev. Lett.* **68**, 2512.
12. Jauho, A. P., Wingreen, N. S., and Meir, Y. (1994) Time-dependent transport in interacting and noninteracting resonant-tunneling systems. *Phys. Rev. B* **50**, 5528.

13. Cuevas, J. C., Martín-Rodero, A., and Levy Yeyati, A. (1996) Hamiltonian approach to the transport properties of superconducting quantum point contacts. *Phys. Rev. B* **54**, 7366.
14. Cuevas, J. C., and Fogelström, M. (2001) Quasiclassical description of transport through superconducting contacts. *Phys. Rev. B* **64**, 104502.
15. Datta, S. (2004) Electrical resistance: An atomistic view *Nanotechnology* **15**, 7, S433.
16. Thouless, D. J., and Kirkpatrick, S. (1981) Conductivity of the disordered linear chain *J. Phys. C.: Solid State Phys.* **14**, 235.
17. Ferry, D. K., and Goodnick, S. M. (1997) *Transport in Nanostructures*, Cambridge University Press, Cambridge.
18. Löfwander, T., San Jose, P., and Prada, E. (2013) Quantum Hall effect in graphene with twisted bilayer stripe defects. *Phys. Rev. B* **87**, 205429.
19. Kazymyrenko, K., and Waintal, X. (2008) Knitting algorithm for calculating Green functions in quantum systems. *Phys. Rev. B* **77**, 115119.
20. Castro Neto, A. H., Guinea, F., Peres, N. M. R., Novoselov, K. S., and Geim A. K. (2009) The electronic properties of graphene. *Rev. Mod. Phys.* **81**, 109.
21. Kubatkin, S., Danilov, A., Hjort, M., Cornil, J., Bredas, J.-L., Stuhr-Hansen, N., Hedegard, P., and Bjornholm, T. (2003) Single-electron transistor of a single organic molecule with access to several redox states. *Nature* **425**, 698.
22. Cardamone, D. M., Stafford, C. A., and Mazumdar, S. (2006) Controlling quantum transport through a single molecule. *Nano Lett.* **6**, 2422.
23. Motta, C., Trioni, M. I., Brivio, G. P., and Sebastian, K. L. (2011) Conductance of a photochromic molecular switch with graphene leads. *Phys. Rev. B* **84**, 113408.
24. Carrascal, D., Carcia-Suarex, V. M., and Ferrer, J. (2012) Impact of edge shape on the functionalities of graphene-based single-molecule electronic devices. *Phys. Rev. B* **85**, 195434.
25. Ryndyk, D. A., Bundesmann, J., Liu, M.-H., and Richter, K. (2012) Edge state effects in junctions with graphene electrodes. *Phys. Rev. B* **86**, 195425.
26. Pshenichnyuk, I. A., Coto, P. B., Leitherer, S., and Thoss, M. (2013) Charge transport in pentacene-graphene nanojunctions. *J. Phys. Chem. Lett.* **4**, 809.
27. Kazumov, A. Yu., Deblock, R., Kociak, M., Reulet, B., Bouchiat, H., Khodos, I. I., Gorbatov, Yu. B., Volkov, V. T., Journet, C., and Burghard, M.

(1999) Supercurrents through single-walled carbon nanotubes. *Science* **284**, 1508.

28. Buitelaar, M. R., Bachtold, A., Nussbaumer, T., Iqbal, M., and Schönenberger, C. (2002) Multiwall carbon nanotubes as quantum dots. *Phys. Rev. Lett.* **89**, 256801.
29. Buitelaar, M. R., Belzig, W., Nussbaumer, T., Babic, B., Bruder, C, and Schönenberger, C. (2003) Multiple Andreev reflections in a carbon nanotube quantum dot. *Phys. Rev. Lett.* **91**, 057005.
30. Jarillo-Herrero, P., van Dam, J. A., and Kouwenhoven, L. P. (2006) Quantum supercurrent transistors in carbon nanotubes. *Nature* **439**, 953.
31. Jørgensen, H. I., Grove-Rasmussen, K., Novotny, T., Flensberg, K., and Lindelof, P. E. (2006) Electron transport in single-wall carbon nanotube weak links in the Fabry–Pérot regime. *Phys. Rev. Lett.* **96**, 207003.
32. Cleuziou, J. P., Wernsdorfer, W., Bouchiat, V., Ondarçuhu, T., and Monthioux, M. (2006) Carbon nanotube superconducting quantum interference device. *Nature Nanotechnology* **1**, 53.
33. Pillet, J.-D., Quay, C. H. L., Morfin, P., Bena, C., Levy Yeyati, A., and Joyez, P. (2010) Andreev bound states in supercurrent-carrying carbon nanotubes revealed. *Nat. Phys.* **6**, 965.
34. Dirks, T., Hughes, T. L., Lal, S., Uchoa, B., Chen, Y.-F., Chialvo, C., Goldbart, P. M., and Mason, N. (2011) Transport through Andreev bound states in a graphene quantum dot. *Nat. Phys.* **7**, 386.
35. van Dam, J. A., Nazarov, Y. V., Bakkers, E. P. A. M., De Franceschi, S., and Kouwenhoven, L. P. (2006) Supercurrent reversal in quantum dots. *Nature* **442**, 667.
36. Xiang, J., Vidan, A., Tinkham, M., Westervelt, R. M., and Lieber, C. M. (2006) Ge/Si nanowire mesoscopic Josephson junctions. *Nat. Nanotechnol.* **1**, 208.
37. Marchenkov, A., Dai, Z., Donehoo, B., Barnett, R. N., and Landman, U. (2007) Alternating current Josephson effect and resonant superconducting transport through vibrating Nb nanowires. *Nat. Nanotechnol.* **2**, 481.
38. Andreev, A. F. (1966) Electron spectrum of the intermediate state of superconductors. *Zh. Eksp. Teor. Fiz.* **49**, 665 [*Sov. Phys. JETP* **22**, 455 (1966)].
39. Kulik, I. O. (1969) Macroscopic quantization and the proximity effect in S-N-S junctions. *Zh. Eksp. Teor. Fiz.* **57**, 1745 [*Sov. Phys. JETP* **30**, 944 (1970)].

40. Ishii, C. (1970) Josephson currents through junctions with normal metal barriers. *Prog. Theor. Phys.* **44**, 1525.

41. Beenakker, C. W. J., and van Houten, H. (1992) Resonant Josephson current through a quantum dot. In *Single-Electron Tunneling and Mesoscopic Devices.*, Koch, H. and Lübbig, H. (eds), pp. 175–179, Springer, Berlin.

42. Zazunov, A., Shumeiko, V. S., Bratus', E. N., Lantz, J., and Wendin, G. (2003) Andreev level qubit. *Phys. Rev. Lett.* **90**, 087003.

43. Zazunov, A., Shumeiko, V. S., Wendin, G, and Bratus', E. N. (2005) Dynamics and phonon-induced decoherence of Andreev level qubit. *Phys. Rev. B.* **71**, 214505.

44. Sköldberg, J., Löfwander, T., Shumeiko, V. S., and Fogelström, M. (2008) Spectrum of Andreev bound states in a molecule embedded inside a microwave-excited superconducting junction. *Phys. Rev. Lett.* **101**, 087002.

45. Sandberg, M. Wilson, C. M., Persson, F., Johansson, G., Shumeiko, V., Duty, T., and Delsing, P. (2008) In-situ frequency tuning of photons stored in a high-Q microwave cavity. *Appl. Phys. Lett.* **92**, 203501.

46. Wallquist, M., Shumeiko, V.S., and Wendin, G (2006) Selective coupling of superconducting qubits via tunable stripline cavity. *Phys. Rev. B* **74**, 224506.

47. Gorelik, L. Y., Shumeiko, V. S., Shekhter, R. I., Wendin, G., and Jonson, M. (1995) Microwave-induced "Somersault effect" in flow of Josephson current through a quantum constriction. *Phys. Rev. Lett.* **75**, 1162.

48. Della Rocca, M. L., Chauvin, M., Huard, B., Pothier, H., Esteve, D., and Urbina, C. (2007) Measurement of the current-phase relation of superconducting atomic contacts. *Phys. Rev. Lett.* **99**, 127005.

49. Wallraff, A., Schuster, D. I., Blais, A., Frunzio, L., Huang, R.-S., Majer, J., Kumar, S., Girvin, S.M., and Schoelkopf, R. J. (2004) Strong coupling of a single photon to a superconducting qubit using circuit quantum electrodynamics. *Nature* **431**, 165.

50. Hyldgaard, P., Hershfield, S., Davies, J. D., and Wilkins, J. W. (1994) Resonant tunneling with an electron-phonon interaction. *Ann. Phys.* **236**, 1.

51. Mitra, A., Aleiner, I., and Millis, A. J. (2004) Phonon effects in molecular transistors: Quantal and classical treatment. *Phys. Rev. B* **69**, 245302.

52. Viljas, J., Cuevas, J. C., Pauly, F., and Häfner, M. (2005) Electron-vibration interaction in transport through atomic gold wires. *Phys. Rev. B* **72**, 245415.

53. Novotný, T., Rossini, A., and Flensberg, K. (2005) Josephson current through a molecular transistor in a dissipative environment. *Phys. Rev. B* **72**, 224502.

54. Zazunov, A., Egger, R., Mora, C., and Martin, T. (2006) Superconducting transport through a vibrating molecule. *Phys. Rev. B* **73**, 214501.

Chapter 4

First-Principles Simulations of Electron Transport in Atomic-Scale Systems

Thomas Frederiksen[a,b]

[a] *Donostia International Physics Center (DIPC), Paseo Manuel de Lardizabal 4, E-20018 Donostia-San Sebastian, Spain*
[b] *IKERBASQUE, Basque Foundation for Science, E-48011 Bilbao, Spain*
thomas_frederiksen@ehu.es

An important theoretical challenge in the field of single-molecule electronics is to develop general methods for quantitative simulations of real devices taking into account the atomistic details. This chapter describes an approach toward this goal based on density-functional theory (DFT) for the electronic structure in combination with nonequilibrium Green's functions (NEGF) for the transport. We also address recent developments based on the DFT+NEGF approach to describe electron-vibration interactions in molecular junctions, local heating effects, and inelastic signatures in device current–voltage and shot noise characteristics.

4.1 Introduction

The widespread use of the density-functional theory (DFT) in solid state physics is due to a good overall description of chemical bonding

Handbook of Single-Molecule Electronics
Edited by Kasper Moth-Poulsen

ISBN 978-981-4463-38-6 (Hardcover), 978-981-4463-39-3 (eBook)
www.panstanford.com

and atomic structure; at the same time, it is realistic to apply the methods to systems involving hundreds to thousands of atoms [1, 2]. Typical nanoscale junctions are, due to the involvement of two or more electrodes connecting the actual device, often at least of this size. Therefore, DFT can be considered an attractive starting point to describe the electronic structure at an atomistic level.

On the other hand, it is clear that DFT is, by construction, targeting the total energy of the system (in equilibrium) and not electron transport (where the system is out of equilibrium). Nevertheless, during the last decades, DFT-based methods for transport have been developed, and it is an ongoing effort to establish what systems and transport properties the methods can describe adequately. The most common approach is based on a combination of standard DFT methods for the electronic structure supplemented with the nonequilibrium Green's function (NEGF) theory for the transport. Here we will refer to this approach as the DFT+NEGF method. More sophisticated approaches, such as those based on time-dependent DFT [3] or the so-called GW approximation [4–7], have also been developed. However, these more advanced treatments are often significantly more limited in terms of the system sizes that can be handled as compared with DFT+NEGF. In the context of molecular electronics and actual devices, it is, therefore, interesting to investigate to what extent the DFT+NEGF method yields accurate results as compared with available experiments.

In this chapter, we describe the DFT+NEGF transport method and discuss some recent applications and extensions to describe both elastic transport as well inelastic effects due to interactions with vibrations in the molecular device. Specifically, we will show that the DFT+NEGF method gives a satisfactory description of electron transport in single-C_{60} junctions. We also discuss how simulations of vibrational effects on the current–voltage and shot noise characteristics can provide insight into the structural composition of molecular devices. These developments are exemplified by considering the electron transport in prototype alkanethiol junctions as well as atomic wires and point contacts of gold.

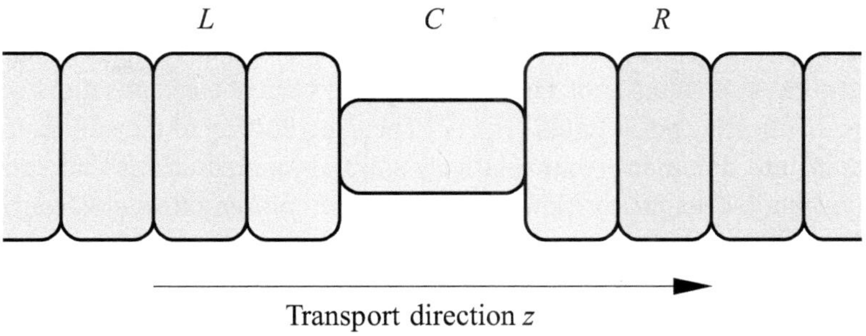

Figure 4.1 Generic setup in the DFT+NEGF approach to quantum transport. A central region C (blue) is coupled to left L and right R electrodes (yellow), which are composed of principal layers repeated semi-infinitely in the transport direction z. Periodic boundary conditions are generally assumed in the xy-plane perpendicular to z.

4.2 The DFT+NEGF Approach

Let us consider the generic two-terminal transport setup illustrated in Fig. 4.1: a left semi-infinite electrode (L), a contact region (C), and a right semi-infinite electrode (R). The problem of a large number of interacting electrons and nuclei in such a setup is highly nontrivial at the level of first principles (i.e., one where only atom types and positions are inputs). Probably the most widely used method in condensed matter physics to attack this problem is based on DFT [1, 2]. The basis is the Hohenberg–Kohn theorem [8], which states that there is a one-to-one correspondence between the external potential due to the nuclei and the ground-state electron density n_0 of the system. As the external potential uniquely defines the system, it follows that any equilibrium property can be considered a functional of n_0. Specifically, the total energy of the system is a functional of the density. Thus, instead of solving the Schrödinger equation for the electronic many-particle wave function $\Psi(r_1, r_2, \ldots)$ (which depends on $3N$ variables for an N-electron system), one can determine the ground state by the variational principle applied to the total energy with respect to the electron density $n = n(r)$ (which only depends on three variables) [1].

The Hohenberg–Kohn theorem is mostly a formal result, and to proceed further, one widely applies the Kohn–Sham (KS) scheme [9] involving two parts: (1) The ansatz that the ground-state electron density of the interacting system of electrons and nuclei is also the ground-state density of a noninteracting system of electrons moving in an effective potential $V_{\text{eff}}(r)$. (2) The contribution to the energy is split into dominant—but relatively simple—contributions that can be handled explicitly (kinetic energy of noninteracting electrons, Hartree electron–electron interactions, ion–ion interactions) and the subtle "rest" (exchange and correlation), which are dealt with approximately. The widespread use, nowadays, of KS-DFT is due to the existence of pretty successful approaches such as the local density approximation or the generalized gradient approximation.

In most implementations of KS-DFT, the electronic structure problem is solved only for the valence electrons (representing the core electrons by pseudopotentials) and by choosing some basis set $\{\phi_i\}$ to expand the wave functions:

$$\psi_n(r) = \sum_i c_{ni}\phi_i(r) \tag{4.1}$$

A convenient choice for the generic LCR-setup is an atomic-centered localized basis set since this allows to divide space (Hilbert space) into separate pieces and calculate the electronic current flowing between these pieces.

In the SIESTA DFT-code [10], one makes use of strictly confined atom-like basis orbitals, which are products of a numerical radial function and a spherical harmonic. For each atom I, located at R_I, one has a set of basis orbitals:

$$\phi_{Ilmn}(r) = \phi_{Iln}(|r - R_I|)Y_{lm}(r - R_I) \tag{4.2}$$

where ϕ_{Iln} is a radial function (which is strictly zero beyond some characteristic cutoff radius r_c) and Y_{lm} is the real spherical harmonics with angular momentum labeled by l, m. In this basis, the Hamiltonian H and overlap matrix S become:

$$H_{ij} = \langle\phi_i|\hat{H}|\phi_j\rangle = \int dr\phi_i^*(r)\left(-\frac{1}{2}\nabla^2 + V_{\text{eff}}[n_0]\right)\phi_j(r) \tag{4.3}$$

$$S_{ij} = \langle\phi_i|\phi_j\rangle = \int dr\phi_i^*(r)\phi_j(r) \tag{4.4}$$

In matrix notation, one can write:

$$H = \begin{pmatrix} H_L & V_L & 0 \\ V_L^\dagger & H_C & V_R \\ 0 & V_R^\dagger & H_R \end{pmatrix} \tag{4.5}$$

$$S = \begin{pmatrix} S_L & S_{LC} & 0 \\ S_{CL} & S_C & S_{CR} \\ 0 & S_{RC} & S_R \end{pmatrix} \tag{4.6}$$

where the basis orbitals and size of the central region are chosen such that there is no direct coupling between L and R. This is important as we do not want currents to bypass the device region C. In practice, with the TranSIESTA[11] DFT+NEGF code, this is easily satisfied if C contains a molecule and a few atomic layers from each electrode. The semi-infinite matrices $H_{L/R}$ have a tri-diagonal structure stemming from the layer-assumption of the electrodes, e.g.,

$$H_R = \begin{pmatrix} h_R & v_R & 0 & 0 \\ v_R^\dagger & h_R & v_R & 0 \\ 0 & v_R^\dagger & h_R & \ddots \\ 0 & 0 & \ddots & \ddots \end{pmatrix} \tag{4.7}$$

The usual periodic boundary conditions used in SIESTAneed to be relaxed in the transport direction. To be able to handle the infinite dimensionality of the problem, a convenient approach is to use NEGF [11–13]. This theory is particularly suitable as it also allows to (1) take into account finite voltages applied to the electrodes, (2) calculate electron currents flowing through the central region, and (3) include particle interactions (beyond the KS-DFT mean-field description) such as the perturbations from electron-vibration couplings (see Section 4.4).

The retarded Green's function matrix g defined for the whole LCR system [13]:

$$g(E) = [(E + i0^+)S - H]^{-1} \tag{4.8}$$

is impractical to invert because of the infinite dimension of the system. However, due to the tri-diagonal and periodic structure of the electrode Hamiltonians $H_{L/R}$, one can calculate the surface

Green's functions $g_{L,R}^{\mathrm{surf}}(E)$ (i.e., the Green's function of the isolated electrode at the surface block layer) using recursive methods [14]. This allows one to write the Green's function matrix g_C for the central part as:

$$g_C(E) = [(E + i0^+)S_C - H_C - \Sigma_L(E) - \Sigma_R(E)]^{-1} \tag{4.9}$$

where

$$\Sigma_{L/R}(E) = (V_{L,R})^\dagger g_{L,R}^{\mathrm{surf}}(E) V_{L,R} \tag{4.10}$$

is the (retarded) self-energy due to each electrode. Roughly speaking, the real part of $\Sigma_{L/R}(E)$ describes the change in energy levels in region C due to the bond formation with the L/R electrodes, whereas the imaginary part is a current operator describing the decay (inverse life time or level broadening) of electronic states located inside C. The latter part is directly related to the electron transport and conventionally denoted:

$$\Gamma_{L/R}(E) = i[\Sigma_{L/R}(E) - \Sigma_{L/R}^\dagger(E)] \tag{4.11}$$

Based on the quantities above, the elastic transmission probability of electrons through C can be expressed as $\mathrm{Tr}[T_0]$, where the transmission matrix T_0 is given by [13]:

$$\begin{aligned} T_0(E) &= g(E)\Gamma_L(E)g^\dagger(E)\Gamma_R(E) = A_L(E)\Gamma_R(E) \\ &= g(E)\Gamma_R(E)g^\dagger(E)\Gamma_L(E) = A_R(E)\Gamma_L(E) \end{aligned} \tag{4.12}$$

where

$$A_{L/R}(E) = g(E)\Gamma_{L/R}(E)g^\dagger(E) \tag{4.13}$$

represents the partial spectral density of states in C due to scattering states originating in the L/R electrode. The product $A_L(E)\Gamma_R(E)$ can thus be interpreted as a probability that states filled with electrons from L are transferred to electrode R. The elastic steady-state current is then given by integrating the transmission function $\mathrm{Tr}[T_0(E)]$ over the bias window, as given by the Landauer–Büttiker expression [15]:

$$I(V) = \frac{G_0}{e}\int_{-\infty}^{\infty} \mathrm{Tr}[T_0(E)]\{n_F(E-\mu_L) - n_F(E-\mu_R)\}dE \tag{4.14}$$

where $G_0 = 2e^2/\hbar$ is the conductance quantum:

$$n_F(E) = 1/(\exp(E/k_B T) + 1) \tag{4.15}$$

is the Fermi–Dirac distribution function with inverse temperature β, and $\mu_{L/R}$ the chemical potentials of the leads.

The zero-bias conductance is simply given by the total transmission at the Fermi level times the conductance quantum:

$$G(V) = G_0 \mathrm{Tr}[T_0(E_F)] \tag{4.16}$$

We conclude this section by mentioning that there exist several different implementations for electronic transport, which are based on a combination of DFT and NEGF. Some examples are included in Ref. [16–22].

4.3 Application: Conductance of a Single C_{60} Molecule Junction—Atom-by-Atom Engineering of the Electrode Interface

As an illustration of the DFT+NEGF method, we will consider electron transport in single-C_{60} junctions. More specifically, we review recent work on the role of the atomic-scale interface of the electrode using a method to probe the conductance of the junction while changing, one by one, the number of atoms in the electrode that are in contact with the molecule [23]. The basic motivation is to investigate if the usual concepts of "good" and "bad" contacts, commonly used in macro- and mesoscopic physics, also can be applied at the single-molecule level. To do so, one needs atomic-scale control of the junction geometry.

In contrast to the mechanically controllable break junction (MCBJ) technique [24], where the geometry of a molecular junction can only be inferred indirectly, the scanning tunneling microscope (STM) offers the ability to image structures before and after using the mechanical control of the tip position to contact single molecules adsorbed directly on metal surfaces [24–29].

The idea of the experiments discussed in the following, performed using a homemade ultrahigh vacuum STM operated at 5 K, was to combine (1) atom manipulation techniques to assemble clusters of a well-defined number of atoms on a pristine surface, (2) STM tip functionalization by attaching rigidly a C_{60} molecule to the tip apex, and (3) finally use the functionalized tip to make contact to each of the clusters (as well as the pristine surface) and at the same time record the current flowing through the junction [23]. In this way, the detailed atomic structure of the interface to the C_{60} molecule is controlled and its role on the junction conductance can be explored.

Typical experimental conductance traces as a function of C_{60} tip approach are shown in Fig. 4.2. The tip was first used to contact the pristine Cu(111) surface and then to each atomic cluster Cu_N with $N = 1, \ldots, 5$. The right part of the traces corresponds to the tunneling range. The inflection of the conductance curves is characteristic of contact formation and defines the conductance at contact G_c, and the contact distance z_c [26]. The origin of the abscissa ($z = 0$) is set to z_c of the bare surface. Therefore, for $N = 1, \ldots, 5$, z_c is the z-position of the C_{60} tip in contact with the atomic cluster relative to the underlying surface (see figure inset).

The measured conductance values G_c at contact for the different cluster sizes are shown in Fig. 4.3. Panel (a) shows the data (black symbols) for different experimental series (i.e., different C_{60} tips) on an absolute scale. For all data sets, $G_c(N)$ increases with N. The highest contact conductance was always observed at the bare copper surface and varied from 1.0 G_0 to 1.6 G_0 with different C_{60} tips. This variation probably also reflects differences in the inaccessible tip-molecule geometry. Panel (b), therefore, shows $G_c(N)$ normalized to the conductance measured at the clean surface with the same tip. This enables a more direct comparison of the different series as it removes effects related to the tip-molecule interface. As shown, all data collapses to a single curve. Finally, panel (c) shows the normalized conductance *per atom* in the contact. This quantity reveals that $G_c(N)/N$ versus N has a subtle maximum observed around $N \approx 5$.

To interpret these experimental observations, electronic structure calculations were performed with SIESTA/TranSIESTA [10, 11] for supercells such as the one shown in Fig. 4.4 Relaxed structures representing the experimental junctions were considered for two different electrode separations ($L_1 = 17.2$ Å and $L_2 = 18.0$ Å measured between the second-topmost electrode layers), corresponding to different molecule–cluster interaction regimes around the point of contact. These structures were obtained

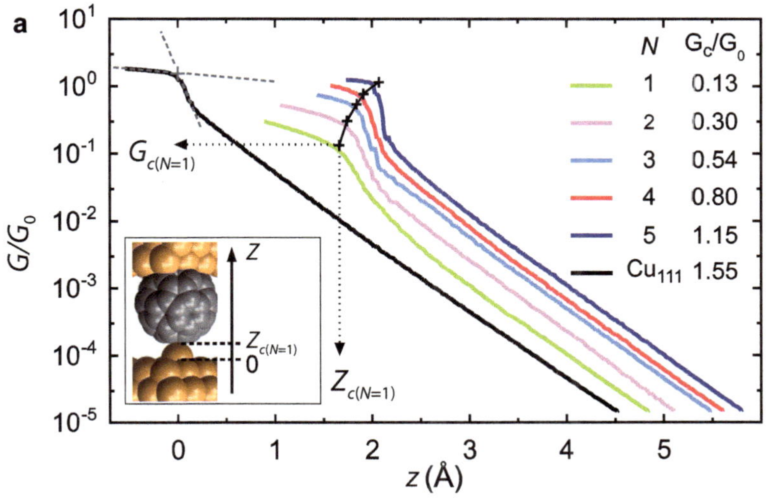

Figure 4.2 Transport measurements of a single C_{60} molecule in contact with an increasing number of atoms. Conductance in units of G_0 (on a logarithmic scale) versus distance for a C_{60} tip approaching a bare Cu(111) surface (black line) and a surface covered with Cu_N clusters (colored lines). One of the hexagons of the C_{60} cage on the STM tip is facing the surface. The distance scale was established from the apparent height of each cluster in STM images recorded at $I = 100$ pA and $V_s = -0.1$ V. Hereafter, with the same settings, the STM feedback was opened for collecting conductance data. Crosses mark the experimental contact points defined as the intersection of the contact and transition regimes (dashed grey lines indicate the bare surface data). Insets: geometry (bottom left) and list (right) of the experimental conductances at contact (G_c) derived from the conductance traces. z_c is the contact distance. Reprinted with permission from Macmillan Publishers Ltd: Nature Nanotechnology, copyright (2010) Ref. [23].

in the following way: An initial structure consisting of a C_{60} molecule adsorbed with a hexagon on an HCP hollow site on a 4×4 representation of a slab containing seven Cu(111) layers was fully relaxed, until all forces on the molecule and the surface layer were smaller than 0.02 eV/Å. Based on this configuration, larger supercells were built with 13 layers in the slab and N atoms positioned on the reverse side of the film. These adatoms and the underlying surface layer were then relaxed to 0.02 eV/Å for the two different electrode separations. The zero-bias conductance G

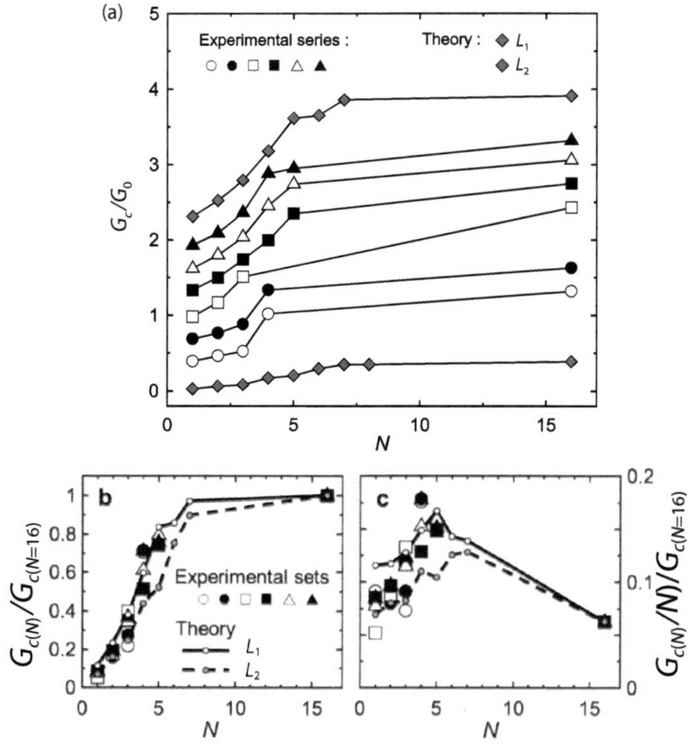

Figure 4.3 Contact conductance $G_c(N)$ vs cluster size N corresponding to experimental and theoretical data sets. The experimental series comprises data from six different molecular tips, and the theoretical series shows results for two electrode separations $L_1 = 17.2$ Å and $L_2 = 18.0$ Å (measured between the second-topmost layers) around the point of contact. (a) Absolute conductances; for clarity the curves have been offset in units of 0.3 G_0 with respect to the L_2 data. (b) Conductance at contact normalized to the value for a single C_{60} molecule in contact with a 16-atom cluster (flat surface). (c) Normalized conductance at contact per atom in the cluster. Reprinted with permission from Macmillan Publishers Ltd: Nature Nanotechnology, copyright (2010) Ref. [23].

was derived from the transmission at the Fermi energy E_F, sampled over a k-point mesh of 6×6 for the two-dimensional Brillouin zone transverse to the transport axis. A direct comparison between the calculations and experiments is given in Fig. 4.3 In all aspects, the DFT+NEGF calculations nicely reproduce the experimental data.

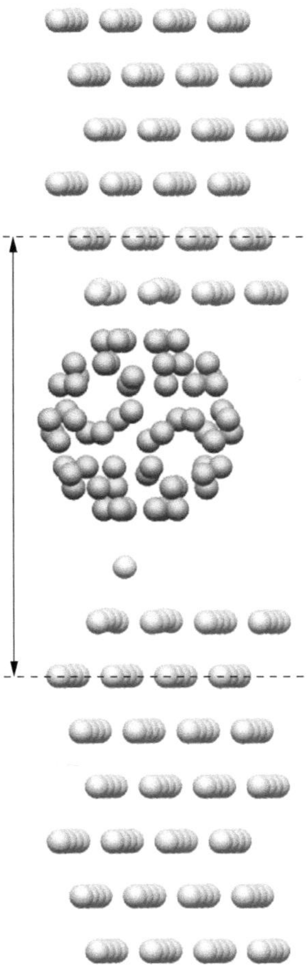

Figure 4.4 Supercell setup used in the SIESTA/TranSIESTA electronic structure calculations. One C_{60} molecule is adsorbed with a hexagon on an hexagonal close-packed (HCP) hollow site on a 4×4 representation of a slab containing 13 Cu(111) layers. Cu_N clusters are positioned on the reverse side of the film. The electrode separation L is measured between the second-topmost layers of the electrodes. Reprinted with permission from Macmillan Publishers Ltd: Nature Nanotechnology, copyright (2010) Ref. [23].

Some general insight into the transport problem can be inferred from having a look at the transmission function $\mathrm{Tr}[T_0(E)]$ as shown in Fig. 4.5 This figure reveals clear resonances in the transmission, which can be traced back to the molecular orbitals of C_{60}, specifically

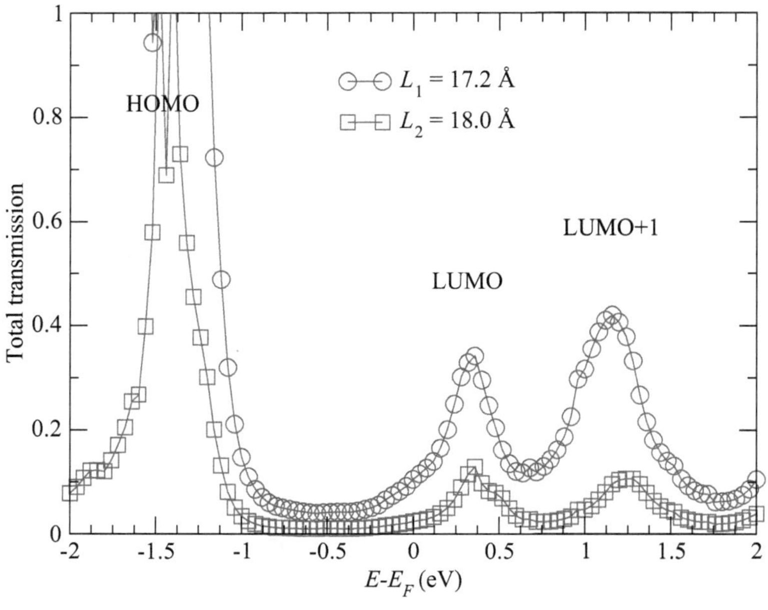

Figure 4.5 Sample transmission functions computed with TranSIESTA for two different electrode separations (L_1 and L_2). The C_{60} tip is here approached directly above a Cu_1 adatom on the Cu(111) surface as shown in Fig. 4.4. The C_{60} molecular resonances are clearly visible in the transmission spectrum. At small voltages, the conductance is dominated by states derived by the lowest unoccupied molecular orbital (LUMO).

the highest occupied molecular orbital, LUMO, and LUMO+1. At small applied voltages, the electron conduction is only involving the LUMO-derived states.

The eigenchannel scattering states, i.e., "nonmixing" electron waves propagating through the nanoscale contact, can also be computed and visualized within the DFT+NEGF method [30]. In essence, the transmitted part of the electron wave is given by the transmission probability for the corresponding eigenchannel. Hence, looking at these scattering states, one can gain insight into where the electrons are flowing through the junction.

Figure 4.6 visualizes isosurfaces of the current-carrying eigenchannels in the considered C_{60} junctions. Here electrons are incident

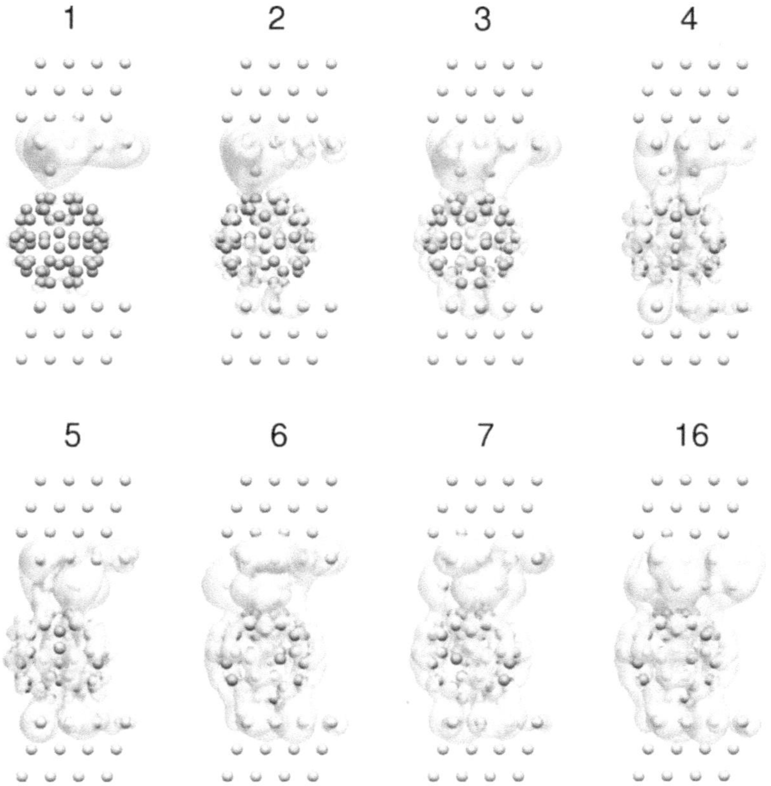

Figure 4.6 Scattering at the contact between a single C_{60} molecule and clusters of copper atoms. Visualization of the scattering states at the Fermi energy for different numbers of contacting atoms. The calculated isosurfaces (shown in grey) represent the electron density in the junction at the center of the Brillouin zone for a sum over the first three (most transmitting) eigenchannels (electron waves coming from below). Data were calculated for clusters with $N = 1, \ldots, 7$ and 16 adatoms at an electrode separation $L_1 = 17.2$ Ang (measured between the second-topmost layers). Not all copper atoms in the cluster are visible. The scattering states are only calculated in the region of space defined by the topmost copper surface layers. Reprinted with permission from Macmillan Publishers Ltd: Nature Nanotechnology, copyright (2010) Ref. [23].

from the top electrode. As the isosurface value is taken to be the same for each of the structures, the magnitude of the transmitted part represents the relative transmissions. For $N = 1$, the incoming waves are strongly reduced at the single atom contact to the molecule, which acts as a bottleneck for electron transport. As N increases, the density at the bottom metal-molecule interface gradually increases (corresponding to a larger part of the waves being transmitted). At $N \approx 5$, the electron density plots do not single out a specific part of the junction as a limiting bottleneck. Instead, the molecular orbitals near the Fermi level control the number of transmission eigenchannels, and conductance is limited by scattering at the molecule. A further increase in the number of atomic connections to the molecule does not significantly affect the conductance, which explains why G_c/N decreases with N in this range. This also confirms that the tip-molecule interface, where a large number of connections is expected, plays little role in the variation of G_c with N. Finally, the maximum in the conductance per atom at $N \approx 5$ marks the crossover between cluster-size-limited ('bad contact' to the molecule) and molecule-limited ('good contact') transport regimes.

In summary, the conductance of a C_{60} molecule varies by a factor of up to 20, depending on the cross section of the electrodes, illustrating that the conductance is not an intrinsic property of a molecule. Furthermore, the presented results demonstrate the capability of the DFT+NEGF method to reproduce quantitatively the measured conductances for the Cu–C_{60}–Cu single-molecule junctions.

4.4 Electron-Vibration Interactions

Up to now, we have considered the transport of electrons as coherent waves that are being transmitted or reflected as they propagate through the nanoscale junction. However, the traversing electrons may scatter inelastically against atoms in the nanostructure and thereby interchange energy with the atomic vibrations.

Such inelastic scattering events are interesting as they provide a way to characterize the structure of molecular junctions via measurements of current–voltage ($I - V$) characteristics as follows: At low temperatures, typically below 10 K, the vibrational motion of molecules is almost completely frozen. However, when electrons pass through the junction, the molecular vibrations can be excited by the electrons. Since vibrational motion is quantized, this only happens when an electron has sufficient energy to deposit a quantum of vibrational energy $\hbar\Omega$. Due to the Pauli principle, this does not occur in equilibrium and at low temperatures since all electronic states are occupied up to the Fermi level. In other words, there are no empty final states for the electrons to scatter into. This situation changes with an applied voltage across the junction (Fig. 4.7). Then electrons are injected from one side with energies above the chemical potential of the other electrode. Within this bias window, an electron may, therefore, find an empty final state when the excess energy eV provided by the voltage V exceeds the vibrational quantum, i.e., when $|eV| > \hbar\Omega$. Inelastic scattering is, therefore, expected to show up as nonlinearities in the $I - V$ characteristics of molecular devices.

In fact, the detection of vibronic signatures in bias spectroscopy is commonly referred to as inelastic electron tunneling spectroscopy (IETS) [31] in the tunneling regime or as point contact spectroscopy (PCS) [32] for metallic point contacts. Both techniques date way back before it was conceived feasible to probe inelastic signals in the conductance through single molecules [24, 33–37].

As illustrated in Fig. 4.7, when the bias voltage exceeds a characteristic vibrational frequency of the junction, a change in the current may be detected. Such a change in the current will appear as a step in the differential conductance dI/dV, or as a peak or dip in the second derivative d^2I/dV^2. As the inelastic scattering cross section is rather low, say of the order of a few percent of the passing electrons, it is usually more accurate to measure experimentally these inelastic signals in the derivatives using the so-called lock-in technique.

A recent example of high-resolution IETS recorded with a low-temperature STM on an alkanethiol self-assembled monolayer on Au(111) is shown in Fig. 4.8 [38]. Note that the vertical axis shows d^2I/dV^2 normalized by dI/dV. This is a common strategy as it cancels effects of the tip-substrate distance etc. [39]. The IETS

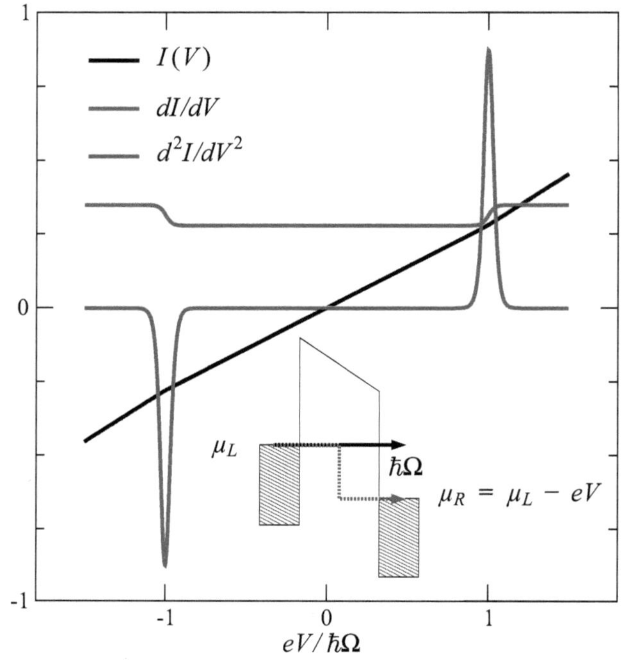

Figure 4.7 Sketch of inelastic electron tunneling spectroscopy. The $I - V$ characteristics (black curve) display a kink at the emission threshold $|eV| = \hbar\Omega$. The corresponding signals in dI/dV (red curve) and d^2I/dV^2 (blue curve) are steps and peaks/dips, respectively. The inset shows an elastic tunneling event (black path) as well as an inelastic tunneling event (red dotted path). The latter can only occur when the bias voltage exceeds the vibrational energy quantum $\hbar\Omega$.

reveals distinct signals in the range 50–200 meV associated with C-S stretch, CH_2 wag, and C-C stretch vibrations as well as the high-energy C-H stretch modes around 360–370 meV.

Thus, this kind of vibrational spectroscopy is important as it can probe characteristic frequencies of the atomic configuration of the junction. Obviously, different molecules have different characteristic vibrations, and such differences may show up directly in their measured IETS spectra. But the technique, in combination with atomistic modeling, may also go beyond that and reveal features that can be traced back to the molecular bonding motifs to the electrodes and/or to the amount of strain that is applied to the molecule.

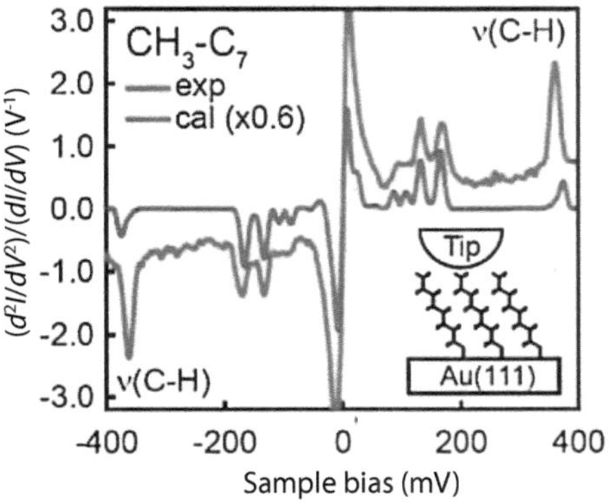

Figure 4.8 Comparison of experimental (red line) and calculated (blue line) inelastic electron tunneling spectroscopy for $CH_3(CH_2)_7S$/Au(111) SAM (CH_3-C_7). The broadening in both experiment and theory is determined by the temperature (4.4 K) and the lock-in modulation voltage ($V_{\mathrm{mod}} = 7.2$ mV). From Ref. [38]. Copyright (2010) by the American Physical Society.

The study of inelastic effects in electron transport is not only interesting as a spectroscopic tool. It is interesting as it is also intrinsically related to local heating of the junction. Depending on the detailed balance of energy dissipation in the junction and vibrational relaxation to the heat sinks provided by the electrodes, this can lead to rich phenomena such as conformational changes, induced chemical reactions, adatom movement, and electromigration. Obviously, it also affects device stability and, ultimately, device malfunctioning.

4.5 Inelastic Transport with DFT+NEGF

As illustrated in the previous section, inelastic effects in electron transport through nanoscale junctions can reveal interesting information about the microscopic details. From the theoretical side, various simulation strategies have been developed [29, 40–48]. In this section, we describe a way to include such inelastic effects in transport simulations based on DFT+NEGF. First we note that the probability for an electron to emit or absorb a quantum of vibration during the passage of the junction is rather low. Typically, the change in conductance around the onset of vibrational emission is only of the order of a few percent. Under such conditions, it is a natural starting point to use perturbation theory to include the electron-vibration interaction.

To extend the DFT+NEGF picture for elastic transport presented in Section 4.2, we first need to reconsider the assumption that the atomic coordinates are frozen. To relax this picture, we allow atoms in the junction to oscillate with a small amplitude Q around their equilibrium positions R_0 (the harmonic approximation):

$$Q(t) = R(t) - R_0 = Qe^{i\omega t} \tag{4.17}$$

From the Born–Oppenheimer total energy surface $E(R)$, i.e., here the total energy E calculated by DFT for some atomic arrangement R, one defines the matrix of interatomic force constants (usually called the Hessian or spring constant matrix) as:

$$C_{I\nu,J\mu} = \left.\frac{\partial^2 E(R)}{\partial R_{I\mu} \partial R_{J\nu}}\right|_{R=R_0} \tag{4.18}$$

As the atomic forces $F_{I\nu} = -\partial E(R)/\partial R_{I\nu}$ are directly obtained by SIESTA along with the total energy calculation, the dynamic matrix can readily be constructed by finite differences by performing a series of calculations where one atom at a time is slightly displaced [49]:

$$C_{I\nu,J\mu} \approx -\frac{F_{I\nu}(Q_{J\mu}) - F_{I\nu}(-Q_{J\mu})}{2Q_{J\mu}} \tag{4.19}$$

With the force constants matrix and Newton's second law of motion, an eigenvalue problem is defined:

$$(\omega^2 - W)v = 0 \tag{4.20}$$

where $W_{I\nu,J\mu} = C_{I\nu,J\mu}/\sqrt{M_I M_J}$ is the mass-scaled matrix of interatomic force constants and where $v_I = \sqrt{M_I} Q_I$ is the mass-scaled displacement. Importantly, the eigenvalues ω_λ^2 and eigenvectors v^λ characterize each mode λ by the mode frequency ω_λ and normal coordinate v^λ.

The corresponding electron-vibration coupling to each mode λ is then given as the linear term in an expansion of the electronic Hamiltonian operator:

$$\hat{H}(Q_\lambda) \approx \hat{H}(R_0) + \frac{\partial \hat{H}_C}{\partial Q_\lambda} \cdot Q_\lambda \tag{4.21}$$

or simply in the chosen DFT basis:

$$H(Q_\lambda) \approx H(R_0) + M^\lambda(b_\lambda + b_\lambda^\dagger) \tag{4.22}$$

where the displacements have been quantized and written in terms of the standard vibration annihilation (creation) operator b_λ ($b_\lambda^\dagger$), and the electron-vibration coupling matrix M^λ is defined with matrix elements:

$$M_{ij}^\lambda = \sum_{I\nu} \langle \phi_i | \frac{\partial \hat{H}_C}{\partial Q_{I\nu}} | \phi_i \rangle v_{I\nu}^\lambda \sqrt{\frac{\hbar}{2M_I \omega_\lambda}} \tag{4.23}$$

The electron-vibration coupling matrices $\{M^\lambda\}$ can also be quantified by finite displacements. However, it should be noted that when one uses atom-centered orbitals for the basis, such as in the SIESTA code, the finite difference scheme involves some additional technical steps, which are omitted here. The interested reader is referred to Ref. [49, 50].

Let us, therefore, now consider that we have determined the electronic Hamiltonian matrix $H(R_0)$ as well as a set of vibrational frequencies ω_λ and coupling matrices M_λ for vibrations in the nanoscale contact. Then comes the question how to incorporate the electron-vibration interaction in the description of the electronic transport. One starting point is the Meir–Wingreen formula [51], which states that the electronic current flowing from the left lead L into the central region C in Fig. 4.1 (where the interactions are assumed to be localized) is given by:

$$I_L(V) = \frac{G_0}{e} \int_{-\infty}^{\infty} \mathrm{Tr}[\Sigma_L^<(E) g_C^>(E) - \Sigma_L^>(E) g_C^<(E)] \tag{4.24}$$

where $\Sigma_L^{\lessgtr}(E)$ are the lesser/greater self-energies from the left lead and $g_C^{\lessgtr}(E)$ are the lesser/greater Green's function in the central region. Loosely speaking, $\Sigma_L^<(E)$ describes the in-scattering rate of electrons at energy E and $g_C^>(E)$ the empty states in C. Conversely, $\Sigma_L^>(E)$ describes the out-scattering rate of electrons and $g_C^<(E)$ the occupied states in C. In this way, the Meir–Wingreen formula expresses the current as the difference between electrons flowing in and out between the electrode and the central region.

The quantities:

$$\Sigma_L^<(E) = i n_F(E - \mu_L)\Gamma_L(E) \tag{4.25}$$

$$\Sigma_L^>(E) = i(1 - n_F(E - \mu_L))\Gamma_L(E) \tag{4.26}$$

are easy as they do not depend on the electron-vibration interactions in C, cf. Eq. (4.11). The tricky part is to calculate $g_C^{\lessgtr}(E)$ in the presence of the electron-vibration interactions. Using diagrammatic techniques of the NEGF theory, one can derive the following set of (Dyson and Keldysh) equations [12]:

$$g_C(E) = g_C^0(E) + g_C^0(E)\Sigma_{\text{ph}}(E)g_C(E) \tag{4.27}$$

$$g_C^{\lessgtr}(E) = g_C(E)[\Sigma_L^{\lessgtr}(E) + \Sigma_R^{\lessgtr}(E) + \Sigma_{\text{ph}}^{\lessgtr}(E)]g_C^{\dagger}(E) \tag{4.28}$$

and

$$\Sigma_{\text{ph}}(E) = \frac{1}{2}[\Sigma_{\text{ph}}^{>}(E) - \Sigma_{\text{ph}}^{<}(E)] - \frac{i}{2}\mathcal{H}[\Sigma_{\text{ph}}^{>}(E) - \Sigma_{\text{ph}}^{<}(E)] \tag{4.29}$$

$$\Sigma_{\text{ph}}^{\lessgtr}(E) = \sum_{\lambda} M^{\lambda}[(n_{\text{B}}^{\lambda} + 1)g_C^{\lessgtr}(E \pm \hbar\omega_{\lambda}) + n_{\text{B}}^{\lambda} g_C^{\lessgtr}(E \mp \hbar\omega_{\lambda})]M^{\lambda} \tag{4.30}$$

where g^0 is the unperturbed retarded Green's function [i.e., without the electron-vibration interaction, cf. Eq. (4.9)], $n_{\text{B}}^{\lambda} = (e^{\beta\hbar\omega_{\lambda}} - 1)^{-1}$ is the Bose–Einstein factor, and $\mathcal{H}$ is the Hilbert transform.

Solving these equations self-consistently is called the self-consistent Born approximation (SCBA). In practice, this is an extremely tough numerical problem at the DFT+NEGF level as all Green's functions and self-energies are matrices in orbital space and further need to be resolved on a fine energy grid. A brute-force approach is, therefore, of limited value.

4.6 Lowest-Order Expansion Approach

A way to dramatically reduce numerical challenge of SCBA consists of invoking the following two approximations [49, 50, 52, 53]: The first one, sometimes called the extended wideband limit (EWBL), consists of approximating the noninteracting retarded Green's function $g_C^0(E)$ as well as the level broadenings $\Gamma_{L/R}$ with their values at the Fermi energy E_F, i.e.,

$$g_C(E) \approx g_C(E_F) \equiv g_C \tag{4.31}$$

$$\Gamma_{L/R}(E) \approx \Gamma_{L/R}(E_F) \equiv \Gamma_{L/R} \tag{4.32}$$

Physically, this is motivated by the fact that in many real systems, the electronic spectral properties typically vary slowly on the scale of a few phonon energies and applied voltages. The second approximation is to expand the Meir–Wingreen expression for the current to second order in the electron-vibration coupling. This is justified for relatively weak interactions. Together these two approximations are sometimes denoted the "lowest-order expansion" (LOE). As a result, the current can now be written as:

$$I(V) \approx I_0(V) + \sum_\lambda \delta I_\lambda(V), \tag{4.33}$$

where

$$I_0(V) = G_0 \mathrm{Tr}[T_0(E_F)]\, V \tag{4.34}$$

is the usual Landauer–Büttiker part with transmission T_0 given by Eq. (4.12) and:

$$\delta I_\lambda(V) = \delta I_{\mathrm{el},\lambda}(V) + \delta I_{\mathrm{inel},\lambda}(V) \tag{4.35}$$

$$\delta I_{\mathrm{el},\lambda}(V) = G_0\{(1 + 2n_B^\lambda)\mathrm{Tr}[T_\lambda^{(0)}] + 2n_B^\lambda \mathrm{Tr}[T_\lambda^{(1)}]\}V \tag{4.36}$$

$$\delta I_{\mathrm{inel},\lambda}(V) = G_0 \mathrm{Tr}[T_\lambda^{(1)}] g_\lambda(V) \tag{4.37}$$

represent correction terms from each mode λ with microscopic factors given by traces over the following quantities:

$$T_\lambda^{(0)} = \Gamma_L\big(g_C M^\lambda g_C^{\mathrm{Re}} M^\lambda A_R + \mathrm{H.c.}\big) \tag{4.38}$$

$$T_\lambda^{(1)} = \Gamma_L g_C \big\{M^\lambda A_R M^\lambda - \frac{i}{2}\big(M^\lambda A M^\lambda g_C \Gamma_R - \mathrm{H.c.}\big)\big\} g_C^\dagger \tag{4.39}$$

and the voltage-dependent universal functions given by:

$$g_\lambda(V) = V + \frac{1}{2}\{U(eV - \hbar\omega_\lambda) - U(eV + \hbar\omega_\lambda)\} \tag{4.40}$$

$$U(eV) = V \coth(\beta eV/2) \tag{4.41}$$

In the above equations, g_C^{Re} denotes the real part of g_C and:

$$A_{L/R} = g_C \Gamma_{L/R} g_C^\dagger \tag{4.42}$$

is the partial spectral function corresponding to lead L/R. The total spectral function is simply $A = A_L + A_R$.

Upon differentiation of Eq. (4.35) with respect to voltage, one obtains the correction of the conductance $\delta G(V) = \partial_V(\delta I(V))$. In the zero-temperature limit, $\delta G(V)$ is discontinuous at the inelastic threshold. The corresponding jump in the inelastic correction to the

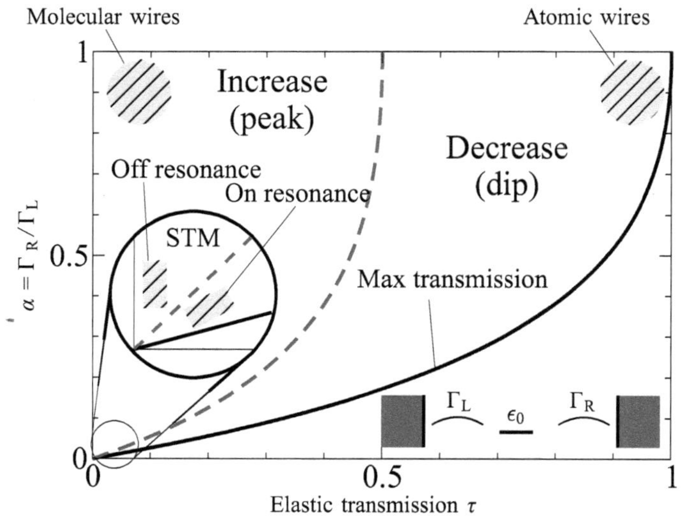

Figure 4.9 Phase diagram for a one-level model (inset) illustrating the sign of the conductance change at the onset of phonon emission [i.e., sign of $T_\lambda^{(1)}$ in Eq. (4.39)]. At a given asymmetry factor, the elastic transmission has an upper bound max (black line), and the inelastic conductance change undergoes a sign change at crossover $\tau_{\max} = \tau/2$ (dashed line). From Ref. [54]. Copyright (2008) by the American Physical Society.

conductance $\Delta G_\lambda \equiv \lim_{\eta \to 0^+}\{\delta G(\hbar\omega_\lambda + \eta) - \delta G(\hbar\omega_\lambda - \eta)\}$ at the threshold voltage corresponding to mode λ is thus simply:

$$\Delta G_\lambda = G_0 \mathrm{Tr}[T_\lambda^{(1)}]. \tag{4.43}$$

Each mode thus contributes with a step in the differential conductance at the inelastic threshold, where the sign and magnitude are controlled by the microscopic factor $T_\lambda^{(1)}$. From the numerical point of view, Eq. (4.35) is a vast simplification over SCBA as there are no energy integrations to be carried out. All system-dependent quantities are simple matrix products that can be performed numerically very rapidly.

As mentioned, the conductance step at the inelastic threshold can be either positive or negative. In the simplest possible case of a one-level model for the device region, a phase diagram for the sign can be rationalized as shown in Fig. 4.9 [54]. This shows that conductance decreases (increases) are generally expected for on- (off-) resonant transport.

For the specialist reader, we finally note that the expressions given in Eq. (4.35)–(4.39) only cover special cases of a more general formulation [52]. In particular, (1) asymmetric contributions to the conductance (with respect to voltage) have been ignored [52, 55, 56], (2) phonon populations n_B^λ are kept in equilibrium (regime of equilibrated phonons), and (3) the so-called Hartree self-energy is disregarded.

4.7 Application: Inelastic Conductance Signals in Atomic Gold Chains

To illustrate the methodology of the preceding section, we consider here electron transport through monatomic gold chains such as shown in Fig. 4.10. Such thin wires can be achieved utilizing the tip of an STM to first make contact with a gold surface and then slowly withdraw the tip such that the gold bridge thins out. This may lead to the formation of a chain of single atoms of up to around seven atoms in length. Alternatively, these wires can also be fabricated using the MCBJ technique, consisting of a macroscopic gold wire mounted on a flexible substrate, which is bent until the wire breaks and exposes clean fracture surfaces. By controlling the bending, it is possible to repeatedly form contacts and sometimes to pull chains several atoms long [57].

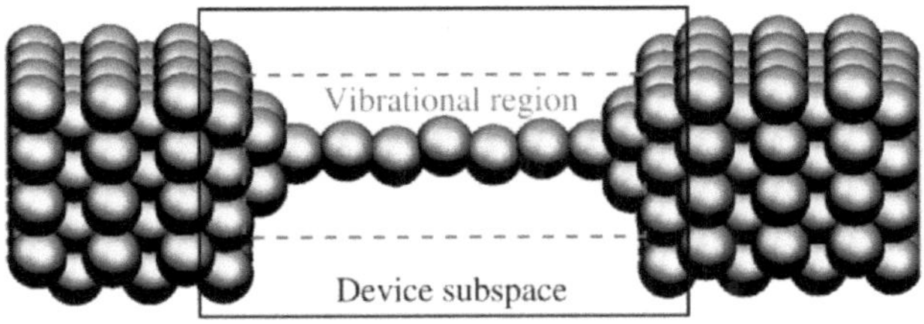

Figure 4.10 Generic transport setup in which a relaxed wire geometry, here a seven-atom chain with $L = 29.20$ Ang, is coupled to semi-infinite electrodes. As indicated on the figure, the vibrational region is taken to include the atoms in the pyramidal bases and the chain itself, whereas the device region (describing the e-ph couplings) includes also the outermost surface layers. From Ref. [49]. Copyright (2007) by the American Physical Society.

These thin metallic wires are interesting for several reasons. They are nearly ideal realizations of the perfectly transmitting one-dimensional conductor and have a conductance close to the quantum $G_0 = 2e^2/h$ due to a single completely open transmission channel. Also their mechanical and chemical properties are very different from that of bulk gold due the low coordination of chain atoms. Despite the probability of forming a long wire is low, the chains are remarkably stable once they are formed; experimentally they can be held stable for hours and sustain enormous current densities and voltages up to 2 V.

The first report on energy dissipation and phonon scattering in atomic gold wires was reported more than a decade ago [33]. The researchers used a cryogenic STM to first create an atomic gold wire between the tip and the substrate surface, and then to measure the conductance against the displacement of the tip. From the length of the observed conductance plateau around G_0, one can determine the approximate size as well as the level of strain of the created wire. Under these conditions, they then performed PCS and found that the conductance decreases a few percent around a particular tip-substrate voltage (symmetric around zero bias) coinciding with the natural frequency of a certain vibrational mode of the wire ($\hbar\omega \sim 15$ meV). The experimental data is shown in Fig. 4.11 as thin noisy curves.

Let us next consider computer simulations based on DFT+NEGF within the LOE for the electron-vibration interaction. The results for the seven-atom Au chain (geometry in Fig. 4.10) are included in Fig. 4.11 (colored lines). In very good agreement with the experiment, a conductance reduction of about 1% is obtained when the applied voltage exceeds the threshold $|V| = 13$ mV, associated with the coupling to the longitudinal alternating-bond length (ABL) vibrational mode of the chain [58].

To explain why the drops in the observed symmetric conductance relate to phonon scattering, it is useful to consider the allowed transitions in the electronic bands for an *infinite* linear atomic wire. Figure 4.12 shows a representative band structure from DFT calculated with SIESTA. The filled d states are positioned just below the Fermi energy, leaving effectively a single half-filled s band crossing the Fermi level [59]. If an electric field is now applied along

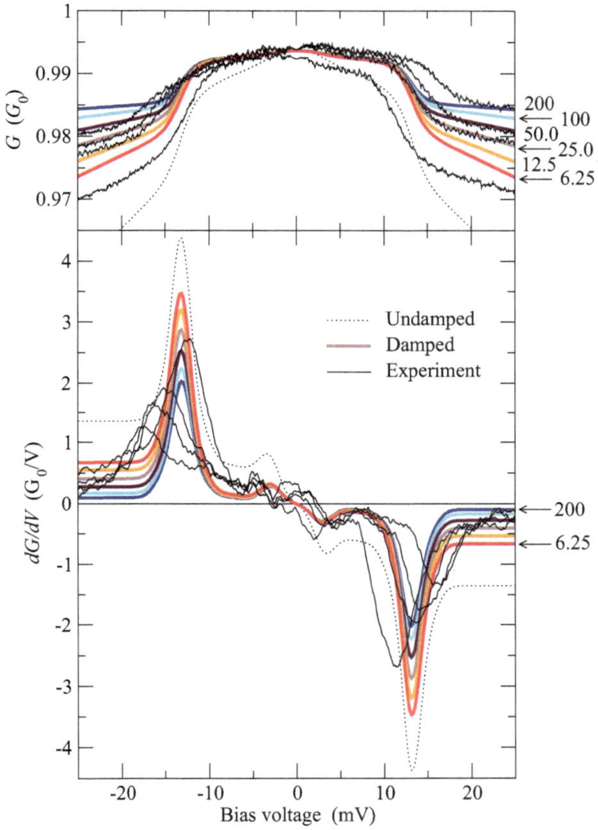

Figure 4.11 Comparison between theory and experiment [33] for the inelastic conductance of an atomic gold wire. The measured characteristics (noisy black curves) correspond to different states of strain of wire (around seven atoms long). The calculated results (smooth colored lines) are for the seven-atom wire using different values for the external damping γ_d as indicated in the right side of the plot (in units of μeV/$\hbar$). The dashed curve is the calculated result in the externally undamped limit ($\gamma_d = 0$). The lower plot is the numerical derivative of the conductance. The temperature is $T = 4.2$ K, and the lock-in modulation voltage is $V_{\mathrm{rms}} = 1$ meV. From Ref. [49]. Copyright (2007) by the American Physical Society.

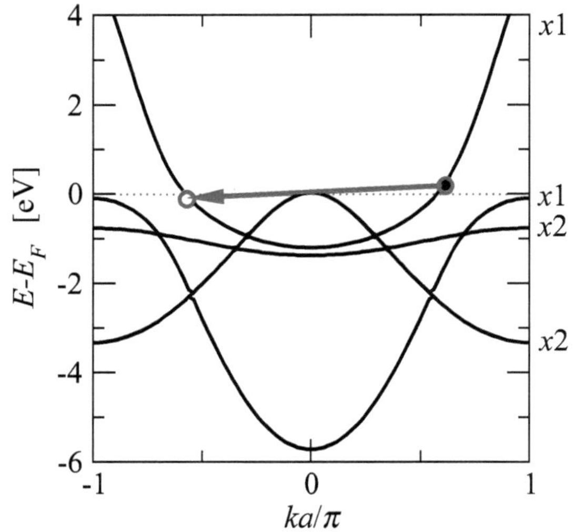

Figure 4.12 Band structure calculation from DFT for an infinite linear gold chain (interatomic distance $a = 2.5$ Å). In the ground state, the electron states are occupied up to the Fermi level. If an electric field is applied along the wire, the electrons will be accelerated and hence populate forward ($k > 0$) and backward ($k < 0$) moving states differently. When an electron has gained sufficient energy, it may scatter inelastically in the emission process sketched in the figure (red arrow). In this process, the electron changes its crystal momentum and loses the small amount of energy corresponding to the phonon quantum (hence, the process is only approximately horizontal on the energy scale of the band structure). The band degeneracy is indicated on the right side of the graph.

the wire, the electrons will be accelerated and start to populate forward ($k > 0$) and backward ($k < 0$) moving states differently. At some point, an electron has gained enough energy to emit a phonon and scatter into a state with lower energy. Due to the Pauli principle, the only available electron states are those of the opposite momentum. On the energy scale of variations in the electronic band structure, the phonon energies (up to around 20 meV) are so small that the electron scattering process will appear as a horizontal transition at the Fermi energy (shown with a red arrow in Fig. 4.12). Momentum conservation further implies that the wavenumber q of the involved vibration matches the change in electronic momentum, i.e., that $q = 2k_{\mathrm{F}} \approx \pi/a$. It is thus concluded that the inelastic scattering in infinite gold chains only involves the zone-boundary phonon of a two-atom Brillouin zone (BZ), corresponding to a wavelength of $2a$ [33]. It turns out that this simple picture actually applies well to describe *finite* ballistic gold chains, i.e., to explain why only a single vibration mode is active in the bias spectroscopy.

Figure 4.11 also illustrates an interesting point about energy dissipation and local heating of the wire. If one considers that the local vibrational mode not only interchanges energy with the current-carrying electrons, but is also coupled to a bath to which it can dissipate its energy (characterized by a parameter $\tau_d = 1/\gamma_d$), one can write a simple rate equation for the nonequilibrium mode population n_λ and solve for the steady state [49]:

$$n_\lambda(V) = n_{\rm B}(\hbar\omega_\lambda) + \frac{1}{\gamma_d}\frac{p_\lambda(V)}{\hbar\omega_\lambda} \tag{4.44}$$

where $p_\lambda(V)$ is the power dissipated by the electrons at a given voltage V. Within the LOE approach, $p_\lambda(V) \propto |eV|\theta(|eV|/\hbar\omega - 1)$, hence the phonon population $n_\lambda(V)$ grows linearly beyond the threshold voltage with a slope inversely proportional to the damping γ_d. As seen in Fig. 4.11 in the limit of relatively strong damping (large γ_d), the conductance beyond the emission threshold voltage is essentially constant (no local heating). Contrary, if the external damping is weaker (small γ_d), a certain slope in the conductance develops with increasing voltage, which is an indication of vibrational heating: Not only are electrons back-scattered by the vibrational emission process, they can also scatter by absorption if the oscillator is pumped out of its ground state by the current. In this limit, the vibrational mode is heated up.

4.8 Inelastic Effects in Shot Noise

As we have seen in the previous sections, vibrational spectroscopy in the current–voltage characteristics is a powerful technique to gain insight into the atomic structure of a nanoscale junction. It is so because distinct vibrational modes may leave signatures in the electronic current.

Shot noise is another important quantity that characterizes electron transport in nanoscale junctions. It describes the time-dependent current fluctuations due to the discreteness of the electron charge; these fluctuations occur due to the statistical nature of quantum mechanics in which electrons are either transmitted or reflected at the junction with certain probabilities [57, 60]. Therefore, shot noise measurements can reveal insight into the transmission coefficients of the individual conductance eigenchannels. This is in contrast to the (mean) conductance that only depends on the total transmission (i.e., the sum of transmission coefficients).

The correlation function of the current operator evaluated at zero frequency, e.g., the (bare) shot noise characteristics $S_0(V)$, is given by the simplified expression [60]

$$S_0(V) = G_0 \frac{2}{\beta} \mathrm{Tr}[T_0^2] + G_0 \mathrm{Tr}[T_0(1 - T_0)] U(eV) \qquad (4.45)$$

where the function $U(eV)$ is given in Eq. (4.41). At low temperatures ($k_B T \ll eV$), the shot noise is, therefore, expected to be proportional to the (mean) current:

$$S_0(V) \approx 2eF\, I(V) \qquad (4.46)$$

where $F = \mathrm{Tr}[T_0(1 - T_0)]/\mathrm{Tr}[T_0]$ is the so-called Fano factor.

However, several groups predicted recently that nontrivial corrections from this linear dependence should occur in the shot noise when the bias voltage exceeds the threshold for the excitation of a vibration mode [61–63]. Such signatures of interactions with vibrations were, shortly after, also observed experimentally in atomic gold wires [64].

At second-order perturbation theory in the e-ph coupling strength, the corrections at finite temperatures and within the EWBL were derived in Ref. [52]. Since the expressions are rather lengthy and complicated, we just state here the simpler result for the zero-temperature limit and equilibrated vibrations [65]:

$$S(V) = S_0(V) + \delta S(V) \qquad (4.47)$$

$$\delta S(V) = \delta S_{\mathrm{el}}(V) + \delta S_{\mathrm{inel}}(V) \qquad (4.48)$$

$$\delta S_{\mathrm{el}}(V) = G_0 \sum_\lambda \mathrm{Tr}[(1 - 2T_0) T_\lambda^{(0)}] |eV| \qquad (4.49)$$

$$\delta S_{\mathrm{inel}}(V) = G_0 \sum_\lambda \mathrm{Tr}[(1 - 2T_0) T_\lambda^{(1)} + Q_\lambda] \times (|eV| - \hbar\omega_\lambda)\theta(|eV| - \hbar\omega_\lambda) \qquad (4.50)$$

with

$$Q_\lambda = -g_C^\dagger \Gamma_L g_C \{ M^\lambda A_R \Gamma_L A_R M^\lambda \\ + M^\lambda A_R \Gamma_L g_C M^\lambda g_C \Gamma_R + \text{H.c.} \} \tag{4.51}$$

Analogous to the inelastic corrections to the current (presented in Section 4.6), the inelastic noise corrections can thus be written as products of universal voltage-dependent functions with the microscopic factors (traces over matrices involving T_0, $T_\lambda^{(0)}$, $T_\lambda^{(1)}$, and Q_λ in electronic space).

The term $\delta S_{\text{el}}(V)$ [Eq. (4.49)] represents an elastic correction to the noise, whereas $\delta S_{\text{inel}}(V)$ [Eq. (4.50)] is related to inelastic signatures of phonon activation in the shot noise.

4.9 Application: Inelastic Shot Noise Signals in a Gold Point Contact

In this section, we discuss an application of the inelastic shot noise corrections to DFT+NEGF simulations for the gold point contact shown in Fig. 4.13a.

The electronic structure, vibrational modes, and electron-vibration coupling were calculated using the SIESTA/TranSIESTA codes as outlined in Section 4.2 for different electrode separations L [65]. For simplicity, we only consider that only the two apex atoms can vibrate, leaving us with six characteristic vibrational modes in the device. This assumption is made to facilitate a fundamental understanding of the inelastic signals; including more atoms (pyramidal bases and surface Au atoms) in the numerics is straightforward.

Figure 4.13b,c shows the dependence of vibrational frequencies $\hbar\omega_\lambda$ and the total electron transmission probability τ as a function of L. The conductance through the point contact is essentially made up by a single channel originating in the 6*s* valence electrons of Au.

The simulated corrections in the conductance $\delta G/\tau(V)$ and in the noise change $\delta \dot{S}/\tau(V)$ are shown in Fig. 4.14 for a range of distances spanning both tunneling and contact regimes. Figure 4.14a illustrates that for each of the considered geometries, the correction to the reduced conductance exhibits a threshold-like character

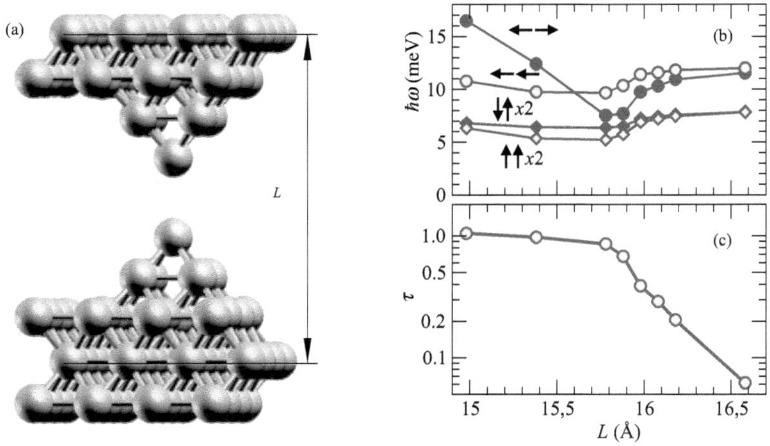

Figure 4.13 (a) Setup for the calculation of structural properties of the atomic gold point contact. The periodic supercell consists of a 4×4 representation of two Au(100) surfaces sandwiching two pyramids pointing toward each other. The characteristic electrode separation L is measured between the second-topmost surface layers (since the surface layer itself is relaxed and hence deviates on the decimals from the bulk values). (b) Vibrational modes and frequencies as a function of L considering only the two apex atoms free to vibrate. (c) Elastic transmission probability versus L, displaying an exponential (linear) dependence on distance in the tunnel (contact) regime. From Ref. [65]. Copyright (2012) by the American Physical Society.

around $eV \sim \hbar\omega \approx 10$ meV corresponding to the out-of-phase longitudinal vibrational mode [66]. The signals from the other five modes are so small that they are hardly visible. For the tunneling setups ($\tau < 1/2$), the activation of phonon emission processes above the inelastic threshold opens a new channel for conduction, thus increasing the conductance compared to its elastic background value. In contrast, in the contact regime ($\tau > 1/2$), the activation of inelastic scattering processes reduces the conductance (back-scattering), i.e., it results in a negative jump.

Figure 4.14b shows the corresponding noise change $\delta\dot{S}(V) \equiv \partial_V(S(V))$ characteristics, which also exhibit a threshold response for voltages close to the phonon frequency. But in contrast to the conductance curves, the finite-temperature effect is not only to smooth the jump but also to produce some small downturn in the

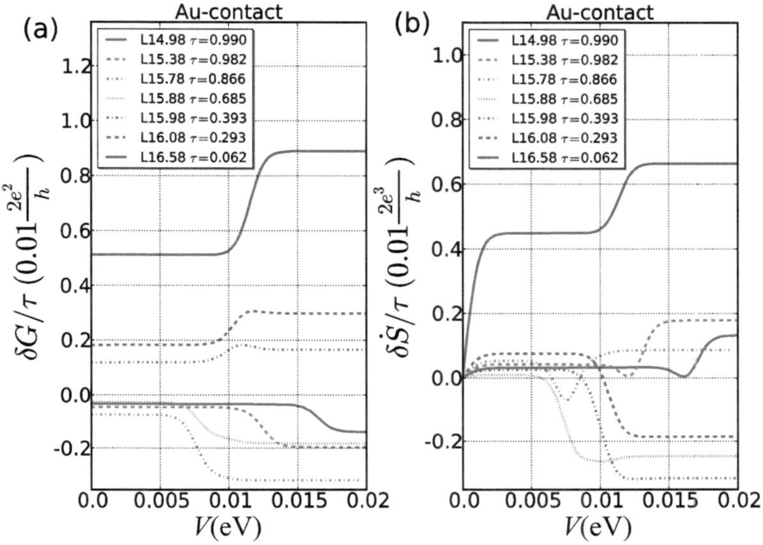

Figure 4.14 Inelastic conductance and noise corrections for Au atomic point contacts with different electrode separations L in the regime of equilibrated phonons. (a) Conductance corrections $\delta G/\tau(V)$ induced by e-ph interactions as a function of voltage V. (b) Derivative of the shot noise with respect to voltage $\delta\dot{S}/\tau(V)$ induced by e-ph interactions. For each geometry, six eigenmodes are considered as only the two apex atoms are vibrating. The calculations are performed at $T = 4.2$ K. From Ref. [65]. Copyright (2012) by the American Physical Society.

vicinity of the inelastic threshold. The sign of the jumps is consistent with the predictions based on "single-level, single-mode" models [61–63], i.e., it is negative only in the region $0.15 < \tau < 0.85$ and positive elsewhere.

The detection of vibrational signals in the shot noise can, therefore, serve as an additional spectroscopic tool to probe vibrations in molecular junctions. It is also worth to point out that, depending on the elastic transmission, the vibrations might leave more visible signals in the noise than in the conductance. In Fig. 4.14, this is for instance the case for the geometry $L = 15.98$ Å ($\tau \approx 0.4$): While the change at the threshold in the conductance is weak, the same mode leaves a strong signal in the shot noise.

4.10 Summary

In this chapter, we have described the DFT+NEGF transport method and discussed some recent extensions to describe inelastic effects due to interactions with vibrations in the molecular device as observed in conductance or shot noise spectroscopy.

Through presentation of some applications to real molecular devices and nanoscale contacts, we have seen that DFT+NEGF-based simulations are capable of quantitative predictions for the electronic transport starting from realistic models for the atomic structure of the junction. This highlights that the combination of computer simulations and detailed experiments can be used to reach detailed and consistent pictures for electron transport processes at the molecular scale.

Problems

1. Based on the EWBL/LOE approach, show that the IETS crossover at $T = 0$ for a single-level model symmetrically coupled to the electrodes $\Gamma_L = \Gamma_R$ occurs at a transmission of $\tau = 1/2$ (no vibrational heating).
2. Show that the corresponding crossovers in the inelastic corrections to the shot noise occur at transmission values $\tau \approx 0.15$ and $\tau \approx 0.85$.
3. What is assumed about the magnitude of the vibrational frequencies in the EWBL/LOE approach?
4. Derive the Landauer–Buttiker result for the current Eq. (4.14) starting from the more general Meir–Wingreen formula Eq. (4.24).
5. Derive Eq. (4.20) from Newton's second law of motion.
6. The DFT+NEGF approach is based on the assumption that the Kohn–Sham mean-field theory provides an adequate description of the actual (physical) electronic structure. Give examples relevant for molecular electronics (e.g., actually fabricated molecular devices) where this situation is expected to be reasonable as well as examples where it is expected to fail. For instance, consider

off-resonant tunneling situations, Coulomb blockade, the Kondo effect, nanoscale metallic conductors, etc.

References

1. W Kohn. Nobel lecture: Electronic structure of matter-wave functions and density functionals. *Rev. Mod. Phys.*, 71: 1253–1266, 1999.
2. R M Martin. *Electronic Structure: Basic Theory and Practical methods.* Cambridge University Press, 2004.
3. S Kurth, G Stefanucci, C-O Almbladh, A Rubio, and E K U Gross. Time-dependent quantum transport: A practical scheme using density functional theory. *Phys. Rev. B*, 72: 035308, 2005.
4. P Darancet, A Ferretti, D Mayou, and V Olevano. Ab initio *gw* electron-electron interaction effects in quantum transport. *Phys. Rev. B*, 75: 075102, 2007.
5. J B Neaton, M S Hybertsen, and S G Louie. Renormalization of molecular electronic levels at metal-molecule interfaces. *Phys. Rev. Lett.*, 97: 216405, 2006.
6. M Strange, C Rostgaard, H Hakkinen, and K S Thygesen. Self-consistent gw calculations of electronic transport in thiol- and amine-linked molecular junctions. *Phys. Rev. B*, 83: 115108, 2011.
7. K S Thygesen and A Rubio. Conserving gw scheme for nonequilibrium quantum transport in molecular contacts. *Phys. Rev. B*, 77: 115333, 2008.
8. P Hohenberg and W Kohn. Inhomogeneous electron gas. *Phys. Rev.*, 136: B864–B871, 1964.
9. W Kohn and L J Sham. Self-consistent equations including exchange and correlation effects. *Phys. Rev.*, 140: A1133–A1138, 1965.
10. J M Soler, E Artacho, J D Gale, A García, J Junquera, P Ordejón, and D Sánchez-Portal. The siesta method for ab initio order-n materials simulation. *J. Phys.: Condens. Matter*, 14: 2745–2779, 2002.
11. M Brandbyge, J L Mozos, P Ordejón, J Taylor, and K Stokbro. Density-functional method for nonequilibrium electron transport. *Phys. Rev. B*, 65: 165401, 2002.
12. H Haug and A-P Jauho. *Quantum Kinetics in Transport and Optics of Semiconductors.* Springer-Verlag, 1996.

13. S Datta. *Electronic Transport in Mesoscopic Systems*. Cambridge University Press, 1995.
14. M P L Sancho, J M L Sancho, and J Rubio. Highly convergent schemes for the calculation of bulk and surface green-functions. *J. Phys. F: Met. Phys.*, 15: 851–858, 1985.
15. M Büttiker, Y Imry, R Landauer, and S Pinhas. Generalized many-channel conductance formula with application to small rings. *Phys. Rev. B*, 31: 6207–6215, 1985.
16. J J Palacios, A J Perez-Jimenez, E Louis, E SanFabian, and J A Verges. First-principles approach to electrical transport in atomic-scale nanostructures. *Phys. Rev. B*, 66: 035322, 2002.
17. F Pauly, J K Viljas, U Huniar, M Häfner, S Wohlthat, M Bürkle, J C Cuevas, and G Schön. Cluster-based density-functional approach to quantum transport through molecular and atomic contacts. *New J. Phys.*, 10: 125019, 2008.
18. A R Rocha, V M García-Suarez, S Bailey, C Lambert, J Ferrer, and S Sanvito. Spin and molecular electronics in atomically generated orbital landscapes. *Phys. Rev. B*, 73: 085414, 2006.
19. J Taylor, H Guo, and J Wang. Ab initio modeling of quantum transport properties of molecular electronic devices. *Phys. Rev. B*, 63: 245407, 2001.
20. K S Thygesen and K W Jacobsen. Molecular transport calculations with Wannier functions. *Chem. Phys.*, 319: 111–125, 2005.
21. J Tomfohr and O F Sankey. Theoretical analysis of electron transport through organic molecules. *J. Chem. Phys.*, 120: 1542–1554, 2004.
22. Y Q Xue and M A Ratner. Microscopic study of electrical transport through individual molecules with metallic contacts. I. Band lineup, voltage drop, and high-field transport. *Phys. Rev. B*, 68: 115406, 2003.
23. G Schull, T Frederiksen, A Arnau, D Sanchez-Portal, and R Berndt. Atomic-scale engineering of electrodes for single-molecule contacts. *Nat. Nanotechnol.*, 6: 23–27, 2011.
24. J C Cuevas and E Scheer. *Molecular Electronics: An Introduction to Theory and Experiment*. World Scientific, 2010.
25. C Joachim, J K Gimzewski, R R Schlittler, and C Chavy. Electronic transparency of a single C_{60} molecule. *Phys. Rev. Lett.*, 74: 2102–2105, 1995.
26. N Néel, J Kröger, L Limot, T Frederiksen, M Brandbyge, and R Berndt. Controlled contact to a C_{60} molecule. *Phys. Rev. Lett.*, 98: 065502, 2007.

27. G Schull, Y J Dappe, C González, H Bulou, and R Berndt. Charge injection through single and double carbon bonds. *Nano Lett.*, 11(8): 3142–3146, 2011.
28. G Schull, T Frederiksen, M Brandbyge, and R Berndt. Passing current through touching molecules. *Phys. Rev. Lett.*, 103: 206803, 2009.
29. G Schulze, K J Franke, A Gagliardi, G Romano, C S Lin, A L Rosa, T A Niehaus, Th Frauenheim, A Di Carlo, A Pecchia, and J I Pascual. Resonant electron heating and molecular phonon cooling in single C_{60} junctions. *Phys. Rev. Lett.*, 100: 136801, 2008.
30. M Paulsson and M Brandbyge. Transmission eigenchannels from nonequilibrium Green's functions. *Phys. Rev. B*, 76: 115117, 2007.
31. R C Jaklevic and J Lambe. Molecular vibration spectra by electron tunneling. *Phys. Rev. Lett.*, 17: 1139–1140, 1966.
32. A G M Jansen, A P Vangelder, and P Wyder. Point-contact spectroscopy in metals. *J. Phys. C: Solid State Phys.*, 13: 6073–6118, 1980.
33. N Agraït, C Untiedt, G Rubio-Bollinger, and S Vieira. Onset of energy dissipation in ballistic atomic wires. *Phys. Rev. Lett.*, 88: 216803, 2002.
34. J G Kushmerick, J Lazorcik, C H Patterson, R Shashidhar, D S Seferos, and G C Bazan. Vibronic contributions to charge transport across molecular junctions. *Nano Lett.*, 4: 639–642, 2004.
35. R H M Smit, Y Noat, C Untiedt, N D Lang, M C van Hemert, and J M van Ruitenbeek. Measurement of the conductance of a hydrogen molecule. *Nature (London)*, 419: 906–909, 2002.
36. B C Stipe, M A Rezaei, and W Ho. Single-molecule vibrational spectroscopy and microscopy. *Science*, 280: 1732–1735, 1998.
37. W Y Wang, T Lee, I Kretzschmar, and M A Reed. Inelastic electron tunneling spectroscopy of an alkanedithiol self-assembled monolayer. *Nano Lett.*, 4: 643–646, 2004.
38. N Okabayashi, M Paulsson, H Ueba, Y Konda, and T Komeda. Inelastic tunneling spectroscopy of alkanethiol molecules: High-resolution spectroscopy and theoretical simulations. *Phys. Rev. Lett.*, 104: 077801, 2010.
39. L J Lauhon and W Ho. Effects of temperature and other experimental variables on single molecule vibrational spectroscopy with the scanning tunneling microscope. *Rev. Sci. Instrum.*, 72: 216–223, 2001.
40. M Alducin, D Sánchez-Portal, A Arnau, and N Lorente. Mixed-valency signature in vibrational inelastic electron tunneling spectroscopy. *Phys. Rev. Lett.*, 104: 136101, 2010.

41. Y C Chen, M Zwolak, and M Di Ventra. Inelastic current–voltage characteristics of atomic and molecular junctions. *Nano Lett.*, 4: 1709–1712, 2004. Erratum, Nano. Lett. **5**, 813 (2005).
42. M Galperin, M A Ratner, and A Nitzan. Molecular transport junctions: Vibrational effects. *J. Phys.: Condens. Matter*, 19: 103201, 2007.
43. J Jiang, M Kula, W Lu, and Y Luo. First-principles simulations of inelastic electron tunneling spectroscopy of molecular electronic devices. *Nano Lett.*, 5: 1551–1555, 2005.
44. N Lorente and M Persson. Theory of single molecule vibrational spectroscopy and microscopy. *Phys. Rev. Lett.*, 85: 2997–3000, 2000.
45. N Mingo and K Makoshi. Calculation of the inelastic scanning tunneling image of acetylene on cu(100). *Phys. Rev. Lett.*, 84: 3694–3697, 2000.
46. H Nakamura, K Yamashita, A R Rocha, and S Sanvito. Efficient ab initio method for inelastic transport in nanoscale devices: Analysis of inelastic electron tunneling spectroscopy. *Phys. Rev. B*, 78: 235420, 2008.
47. G C Solomon, A Gagliardi, A Pecchia, T Frauenheim, A Di Carlo, J R Reimers, and N S Hush. Understanding the inelastic electron-tunneling spectra of alkanedithiols on gold. *J. Chem. Phys.*, 124: 094704, 2006.
48. A Troisi and M A Ratner. Modeling the inelastic electron tunneling spectra of molecular wire junctions. *Phys. Rev. B*, 72: 033408, 2005.
49. T Frederiksen, M Paulsson, M Brandbyge, and A P Jauho. Inelastic transport theory from first principles: Methodology and application to nanoscale devices. *Phys. Rev. B*, 75: 205413, 2007.
50. J K Viljas, J C Cuevas, F Pauly, and M Hafner. Electron-vibration interaction in transport through atomic gold wires. *Phys. Rev. B*, 72: 245415, 2005.
51. Y Meir and N S Wingreen. Landauer formula for the current through an interacting electron region. *Phys. Rev. Lett.*, 68: 2512–2515, 1992.
52. F Haupt, T Novotný, and W Belzig. Current noise in molecular junctions: Effects of the electron-phonon interaction. *Phys. Rev. B*, 82: 165441–, 2010.
53. M Paulsson, T Frederiksen, and M Brandbyge. Modeling inelastic phonon scattering in atomic- and molecular-wire junctions. *Phys. Rev. B*, 72: 201101, 2005.
54. M Paulsson, T Frederiksen, H Ueba, N Lorente, and M Brandbyge. Unified description of inelastic propensity rules for electron transport through nanoscale junctions. *Phys. Rev. Lett.*, 100: 226604, 2008.
55. R Egger and A O Gogolin. Vibration-induced correction to the current through a single molecule. *Phys. Rev. B*, 77: 113405, 2008.

56. O Entin-Wohlman, Y Imry, and A Aharony. Voltage-induced singularities in transport through molecular junctions. *Phys. Rev. B*, 80: 035417, 2009.

57. N Agraït, A L Yeyati, and J M van Ruitenbeek. Quantum properties of atomic-sized conductors. *Phys. Rep.*, 377: 81–279, 2003.

58. T Frederiksen, M Brandbyge, N Lorente, and A-P Jauho. Inelastic scattering and local heating in atomic gold wires. *Phys. Rev. Lett.*, 93: 256601, 2004.

59. D Sánchez-Portal, E Artacho, J Junquera, P Ordejón, A García, and J M Soler. Stiff monatomic gold wires with a spinning zigzag geometry. *Phys. Rev. Lett.*, 83: 3884–3887, 1999.

60. Ya M Blanter and M Büttiker. Shot noise in mesoscopic conductors. *Phys. Rep.*, 336: 1–166, 2000.

61. R Avriller and A Levy Yeyati. Electron-phonon interaction and full counting statistics in molecular junctions. *Phys. Rev. B*, 80: 041309, 2009. Erratum Phys. Rev. B **81**, 089901(E) (2010).

62. F Haupt, T Novotný, and W Belzig. Phonon-assisted current noise in molecular junctions. *Phys. Rev. Lett.*, 103: 136601, 2009.

63. T L Schmidt and A Komnik. Charge transfer statistics of a molecular quantum dot with a vibrational degree of freedom. *Phys. Rev. B*, 80: 041307, 2009.

64. M Kumar, R Avriller, A Levy-Yeyati, and J M van Ruitenbeek. Detection of vibration-mode scattering in electronic shot noise. *Phys. Rev. Lett.*, 108: 146602, 2012.

65. R Avriller and T Frederiksen. Inelastic shot noise characteristics of nanoscale junctions from first principles. *Phys. Rev. B*, 86: 155411, 2012.

66. T Frederiksen, N Lorente, M Paulsson, and M Brandbyge. From tunneling to contact: Inelastic signals in an atomic gold junction from first principles. *Phys. Rev. B*, 75: 235441, 2007.

Chapter 5

Controlling the Molecule–Electrode Contact in Single-Molecule Devices

Joshua Hihath
Department of Electrical and Computer Engineering, University of California, Davis, One Shields Ave Davis, CA 95616, USA
jhihath@ucdavis.edu

Descriptions of transport in single-molecule devices often focus on the properties of the molecular backbone, the energy separation between the highest occupied molecular orbital (HOMO) and the lowest unoccupied molecular orbital (LUMO), molecular conjugation, device size, or the addition of ions to the molecule. Although these features are undeniably important, the conductance, device lifetime, mechanical stability, and energy-level alignment all depend critically upon the molecule–electrode contact. The ability to create stable, long-lifetime, and reproducible contacts between a single molecule and two electrodes is necessary to enable technologically practical devices, but these capabilities also allow the exploration of unique device paradigms where mechanically perturbing a single molecule controls the electronic system. This chapter will discuss the molecule–electrode contact by developing a simple transport model to demonstrate how contact affects the transport

Handbook of Single-Molecule Electronics
Edited by Kasper Moth-Poulsen

ISBN 978-981-4463-38-6 (Hardcover), 978-981-4463-39-3 (eBook)
www.panstanford.com

properties, explore the use of various molecule–electrode linker chemistries, and also discuss controlling the molecule–electrode contact mechanically to control the device's charge transport properties.

5.1 Introduction and Background

Molecules represent a unique class of electronic materials that are inherently quantum mechanical in nature, can be designed and constructed with atomic-level precision, are natural one-dimensional transport devices, and possess unique opportunities for device fabrication, function, and implementation [1]. However, despite the plethora of exciting features of these systems, many of the investigations of the electronic properties of molecular systems have focused on mimicking the functions of conventional semiconductor devices such as transistors, diodes, and wires rather than exploring the distinctive properties of molecules that may allow the development of novel device functions and operational paradigms [2–4]. Development of these novel systems and devices requires the capability of creating stable, long-lifetime devices and also requires a thorough understanding of how the electrical and mechanical contact between the molecule and the outside world can be controlled. These features are tightly coupled in a single-molecule device, and often the strength of binding relates directly to the contact resistance in the system. However, this is not always the case, and the creation of devices with long lifetimes opens up new avenues to control the transport properties in a single-molecule device.

As an example, consider the electromechanical properties of a single-molecule system. If the binding between the molecule and the electrode is strong enough, one can apply a tensile or compressive force to the molecular system. If a strain or stress is applied to conventional three-dimensional materials, two possible effects can result in what is typically a small change in conductance. First, strain can cause a geometry change, which will result in a change in either the device cross section or length, resulting in a change in the total conductance; this is typically the case in metallic conductors. Additionally, strain can cause a small perturbation of the crystal structure, resulting in a change in the band structure, which in turn causes a change in mobility and, therefore, the conductivity. This process is commonly seen in semiconductors and has been used in complementary metal–oxide–semiconductor technologies to enhance device performance. In both of these cases, the conductance change is typically a small percentage of the original conductance. Alternatively, the electronic properties of a one-dimensional conjugated system are exquisitely sensitive to any structural perturbation that affects the lattice spacing and equilibrium atomic distances [5]. This disturbance can significantly change the energy gap between the highest occupied molecular orbital (HOMO) and the lowest unoccupied molecular orbital (LUMO) and can, therefore, greatly affect the device conductance.

The reason that the electronic properties of a one-dimensional system are so sensitive to a structural perturbation is straightforward to understand. An infinite conjugated chain of atoms equally spaced by a distance a will produce a system with a half-filled energy band and become a metallic conductor. However, in 1955 Peierls demonstrated that this structure was unstable and that the energy of the system would be reduced if every other atom in the chain was moved by a small distance Δr [6]. The displacement of half of the atoms changes the lattice constant of the system from a to $2a$. This change in the unit cell results in a band structure with one completely filled band and one empty band and thus creates an insulator or semiconductor. This conclusion was published in Peierls' 1955 book, "Quantum Theory of Solids" [6], but was not widely considered because there were no natural one-dimensional electronic systems to test. Eventually, organic crystals with quasi-one-dimensional transport properties were synthesized and studied [7, 8]. The conductivity of these systems tended to decrease with decreasing temperature, and it was realized that this effect was due to the Peierls distortion [5]. The Peierls distortion clearly demonstrates how strongly coupled the electronic and structural/mechanical properties of a molecular system are. In this case, an extremely small change in lattice spacing changes the system from a metal-like conductor to an insulator or semiconductor.

Expanding on this example, it is apparent that other structural or mechanical changes in a single-molecule device will have a profound impact on the electronic properties of the system. For instance, changes in the molecule–electrode contact will play a strong role in the transport properties [9–12], as this will affect not only the mechanical coupling, but also the electronic coupling between the molecule and the electrodes. Also, if the molecular orbitals are not perpendicular to the electrode surface, then overlap between the molecular orbitals and the electronic states of the electrodes can drastically affect the electronic properties of the system, and this effect has been predicted and observed in several molecular systems [13–16]. Furthermore, polarization of the molecule as the electrodes are separated can also affect the alignment between the molecular orbitals' energy level and the Fermi energy of the electrodes [17, 18]. All these effects are unique to molecular systems but depend on developing the ability to directly control the molecule–electrode contact, either chemically, electrically, or mechanically. As such, this chapter is devoted to exploring this important feature of molecular-electronic systems.

The remainder of this chapter is divided into four sections; we will begin by developing a simple model for tunneling (super-exchange) transport in molecular devices with an emphasis on determining the contact resistance. This model will then be used to aid in our analysis of the transport properties of the alkane molecular family, which will be the topic of Section 5.3. This family of molecules has been studied in detail by a number of research groups, and the transport properties are well understood and can be qualitatively compared to the simple model of Section 5.2. Moreover, a variety of molecule–electrode linkers have been studied for this family of molecules, and thus this system provides unique insights into the role of contact on one-dimensional transport problems. The final two sections will focus on some specific examples of controlling the molecule–electrode contact mechanically and discussing how these mechanical perturbations impact transport. Section 5.4 discusses systems in which the overlap between the molecular orbitals and the electrodes can be controlled, and Section 5.5 discusses force-induced changes in the electronic structure of the molecule to create changes in the transport properties.

5.2 Contact Resistance of Molecular Wires

To begin, we need a simple, analytical model to describe transport in molecular systems to be able to compare results across molecular families, research groups, and molecule–electrode contact chemistries or geometries. Since the HOMO–LUMO gap is typically large in small molecules, it is not surprising that tunneling is often the dominant transport mechanism. However, to capture the underlying physics of through-bond transport and molecule–electrode coupling, a transport model beyond the square-barrier tunneling process presented in most introductory quantum mechanics textbooks is required (for a refresher on square-barrier tunneling, see Ref. [19]). Thus, in the following, we will develop a model that accounts for many of the unique features of a molecular tunneling barrier, which is typically referred to as superexchange where a series of discrete energy states exist between the two electrodes (see Fig. 5.1a). In this model, we will pay special attention

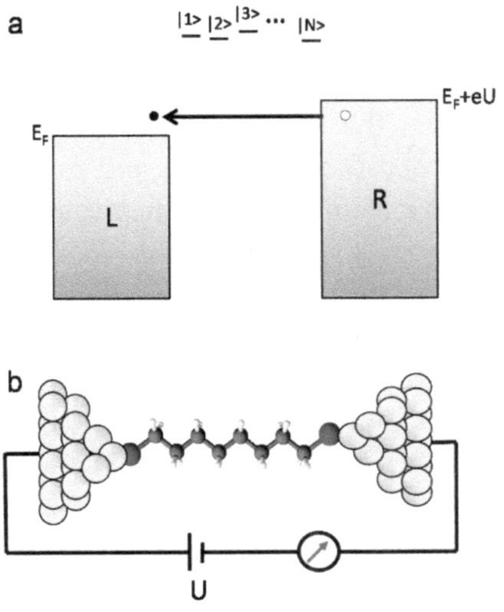

Figure 5.1 Schematic of a single-molecule device. (a) The energy band diagram we will use to describe superexchange transport in a single-molecule junction. Two electron reservoirs with a bias U are connected by a series of bridge sites N. (b) Schematic of single octanedithiol molecular junction.

to understanding the molecule–electrode contact. The chemistry used to bind the molecule to two electrodes, the binding site, and the contact geometry define many important features of the molecular junction. These features dictate the electrical transport properties through the contact conductance G_C, the lifetime and stability of the molecular junction (due to the binding energy), and the alignment between the molecular orbital energy levels and the Fermi energy of the electrodes. Furthermore, it is particularly important to have a robust contact between the molecule and the electrodes if one is to explore electromechanical effects. Thus, this section will focus on developing a framework for describing the role of the molecule–electrode contact on the charge transport properties of the system. To achieve this goal, we will follow closely the works of Mujica et al. [20, 21] and Nitzan [22], who have employed Green's function techniques to describe the transport properties of a single molecule bound to two electrodes.

To begin we will start from Fermi's Golden Rule expression for the charge transfer rate, w, between two electrodes:

$$w = \frac{2\pi}{\hbar} \sum_{i,f} f\left(E_i\right)\left[1 - f\left(E_f\right)\right] \left|T_{if}\right|^2 \delta\left(E_i - E_f\right) \tag{5.1}$$

where $f(E)$ represents the Fermi distribution of electrons in each electrode, i and f represent the initial and final states, $|T_{if}|^2$ is the transition operator from scattering theory, and $\delta(E)$ is the delta function.

The transfer rate, w, in Eq. 5.1 is defined for the transfer between two states and requires that the energy of the initial state be equal to the final state before transfer can occur. This point is ensured by the delta function and guarantees conservation of energy. Equation 5.1 also requires that the initial state be full and the final state empty, which is ensured by the Fermi statistics. However, the transfer rate we have discussed is only defined for a single-electron event; if we wish to expand this expression to describe current where the initial states are "instantaneously" refilled by an external bias, we need to define the current, and thus the conductance of the molecular device. To this end, we recognize that a current can be defined as $I = ew$, where e is the electron charge, this expression gives us units of C/s, or amps. Furthermore, the bias (U) must be included in the

Fermi statistics, and the resulting expression is:

$$I = \frac{2\pi e}{\hbar} \sum_{i,f} f(E_i + eU)\left[1 - f(E_f)\right] \left|T_{if}\right|^2 \delta(E_i - E_f) \quad (5.2)$$

Assuming a low bias voltage so that $|T_{if}|^2$ does not change over the bias range and a low temperature so that the Fermi distributions can be treated as step functions allows us to simplify Eq. 5.2 to:

$$I = \frac{2\pi e^2}{\hbar} U \sum_{i,f} \left|T_{if}\right|^2 \delta(E_i - E_F)\delta(E_f - E_F) \quad (5.3)$$

where E_F is the Fermi energy of the electrodes with no bias applied. The current in a one-dimensional system is uniquely tuned to the transition probability from the initial state to the final state. So the question that must be answered to accurately determine the conductance of a molecular junction is to define $|T_{if}|^2$ for any particular case. T is defined from scattering theory as:

$$T = V + VGV \quad (5.4)$$

where V is the overlap between states and G is the Green's function matrix defined as:

$$G = (E - H \pm i\eta)^{-1} \quad (5.5)$$

Here H is the molecular Hamiltonian, E is a continuous variable of energy, and η is an infinitesimally small value that prevents singularity in the case when the energy is precisely equal to the Hamiltonian energy.

To demonstrate a simple example, we will use the system shown in Fig. 5.1a to model our molecular system. Here, we have two electrodes, L and R, with final states $|f\rangle$ and initial states $|i\rangle$, respectively. The electrodes are separated by several intervening bridge sites denoted as $|1\rangle$ to $|N\rangle$. The left electrode will only couple to the first molecular bridge site $|1\rangle$, and the right electrode will only couple to the final site $|N\rangle$. With this, our transition operator can be written as:

$$T_{if} = V_{if} + V_{iN}G_{N1}V_{1f} \approx V_{iN}G_{N1}V_{1f} \quad (5.6)$$

The approximation on the far right can be made when the molecular bridge is sufficiently large that the direct coupling between the initial and final states, V_{if}, is negligible. This is typically a good approximation, but this term can be included if the molecule is short, and direct through-space transfer becomes important [20].

Equation 5.6 can be substituted into Eq. 5.3. By dividing the bias to obtain conductance (g) and including a factor of 2 for spin degeneracy, one obtains:

$$g = \frac{8\pi^2 e^2}{h} |G_{N1}|^2 \sum_i V_{iN}^2 \delta (E_i - E_F) \sum_f V_{1f}^2 \delta (E_f - E_F) \quad (5.7)$$

If the molecular state $|1\rangle$ interacts with all possible final states, $|f\rangle$, in the L electrode the same way, then V_{1f} can be replaced with V_{1L} and pulled out of the summation. The same is true for the $|N\rangle$ states on the R side to obtain V_{RN}. With this we are left with the coupling term and the density of states for each electrode. It is now possible to define the broadening of the molecular states due to the interaction with the electrodes. This is defined as:

$$\Gamma_L = 2\pi V_{1L}^2 \sum_i \delta (E_i - E_F), \quad \Gamma_R = 2\pi V_{RN}^2 \sum_f \delta (E_f - E_F) \quad (5.8)$$

Substituting these terms in Eq. 5.7 yields:

$$g = \frac{2e^2}{h} |G_{N1}|^2 \Gamma_L \Gamma_R \quad (5.9)$$

The prefactor term is often referred to as the conductance quantum ($G_0 = 2e^2/h$), and those readers familiar with Landauer theory will quickly recognize this as the conductance value for a single perfectly transmitting channel. However, despite a few notable exceptions [23, 24], very few molecular systems are near this conductance value, so it is necessary to accurately describe the Green's function term to describe the transport accurately.

Although we have discussed the broad features of the system depicted in Fig. 5.1a, other than assuming that the electrodes only couple to the end states of the molecule, we have not invoked this structure to make any significant assumptions in the derivation. Now, to create a more specific transport model, we will assume that the system is a molecular wire (meaning that it is a linear chain), that the states in the wire exhibit only nearest neighbor coupling (tight-binding), and that the coupling between neighbors is weak. From this it is possible to define a Hamiltonian for the molecular system, H_M. For this, we will start from the familiar expression:

$$H_M = H_M^0 + V_M \tag{5.10}$$

Using the tight-binding approximation, each term is defined as:

$$H_M^0 = \sum_{\alpha=1}^{N} |\alpha\rangle\, E_\alpha\, \langle\alpha|; \quad V_M = \sum_{\alpha}^{N} |\alpha\rangle\, V_{\alpha,\alpha\pm1}\, \langle\alpha \pm 1| \tag{5.11}$$

The α terms represent summations over the bridge states, E_α is the energy of each bridge site, and $V_\alpha, \alpha_{\pm1}$ is the coupling between sites. From these we can define a Green's function matrix from Eq. 5.5 for the extended molecular Hamiltonian as:

$$G = \begin{bmatrix} E - E_1 - \Sigma_1 & -V_{1,2} & 0 & \cdots & & 0 \\ -V_{2,1} & E - E_2 & & & & \vdots \\ 0 & & \ddots & & & 0 \\ \vdots & & & E - E_{N-1} & & -V_{N-1,N} \\ 0 & \cdots & 0 & -V_{N,N-1} & & E - E_N - \Sigma_N \end{bmatrix}^{-1} \tag{5.12}$$

Note that for completeness, we have also included the self-energy terms (Σ_a), which represent both the broadening and shifting of the molecular energy levels due to the interaction with the continuum of states in the electrodes. These terms are only present in the first and last terms since we have made the assumption that only these states interact with the reservoirs. For a thorough discussion of the self-energy terms due to electrode interactions, see Chapter 9 of Datta's book [25]; here we will only note the final result:

$$\Sigma_{\alpha\alpha'} = \delta_{\alpha,\alpha'} \left(\delta_{\alpha,1} + \delta_{\alpha,N}\right) \left[\Lambda_\alpha - \frac{i}{2}\Gamma_\alpha\right] \tag{5.13}$$

Here, $\Gamma_\alpha = \Gamma_L + \Gamma_R$ (as defined in Eq. 5.8) is the complex component of the self-energy and represents the broadening of the energy level, and as such represents the finite lifetime of a charge on the molecular bridge. Λ, the real part, represents any shifts in energy and is given by:

$$\Lambda_\alpha = \frac{P}{2\pi} \int_{-\infty}^{\infty} \frac{\Gamma_\alpha(E')}{(E - E')} dE' \tag{5.14}$$

where P is known as the principle Cauchy value. With these terms in hand, the transport problem has been reduced to inverting the matrix in Eq. 5.12 to obtain the relevant matrix element, which reduces to:

$$G_{N1} = \frac{V_{N,N-1}}{(E - E_1 - \Sigma_1)(E - E_N - \Sigma_N)} \prod_{\alpha=2}^{N-1} \frac{V_{\alpha,\alpha-1}}{E - E_\alpha} \tag{5.15}$$

This result can then be inserted into Eq. 5.9. Then, if we further assume that all coupling terms, energy levels, and self-energies are identical ($V_{\mathrm{M}} = V_a$, a_{-1}, $E_a = E_M$, and $\Sigma_1 = \Sigma_{\mathrm{N}} = \Sigma_{\mathrm{M}}$) (this assumption simply means that each bridge site is the same), and additionally set $E_M - E = \Delta E$, we arrive at:

$$g = \frac{2e^2}{h}\left[\frac{V_M}{(-\Delta E - \Sigma_M)^2}\left(\frac{V_M}{-\Delta E}\right)^{N-2}\right]^2 \Gamma_L \Gamma_R \tag{5.16}$$

Given reasonable parameters for the coupling terms and energies, it is straightforward to model molecular transport using this form. However, in the interest at arriving at a simple analytical form that can be qualitatively compared to experimental results, we will make the additional assumption that $\Delta E >> \Sigma_M$ yielding:

$$g = \frac{2e^2}{h}\frac{\Gamma_L \Gamma_R}{V_M^2}\left(\frac{V_M}{\Delta E}\right)^{2N} \tag{5.17}$$

It should be noted that this is not typically a valid assumption as the self-energy plays an important role in the transport characteristics. Nevertheless, it allows us to qualitatively compare results across molecular families and experiments performed from different groups, allowing analysis of important trends for various linker groups and molecules as is done for the alkanes in the next section.

Next, by recognizing that if the spacing between molecular energy levels is a constant, d, then the length of the molecular system is simply $L = Nd$. We can then recast Eq. 5.17 as:

$$g = \frac{2e^2}{h}\frac{\Gamma_L \Gamma_R}{V_M^2}\exp\left(-\frac{2}{d}\ln\left(\frac{\Delta E}{V_M}\right)L\right) \tag{5.18}$$

By assigning two additional terms:

$$\beta = \frac{2}{d}\ln\left(\frac{\Delta E}{V_M}\right); \quad G_C = \frac{2e^2}{h}\frac{\Gamma_L \Gamma_R}{V_M^2} \tag{5.19}$$

we arrive at the traditional form of the expression for the conductance of a tunneling barrier:

$$g = G_C \exp(-\beta L) \tag{5.20}$$

Here, β is the exponential decay factor for the tunneling system, which depends primarily on the height of the energy barrier (ΔE) and on the coupling between levels within the molecule. It should also be noted that the derivation given here, where the transport is through molecular bonds that each have their own states, is typically referred to as superexchange to emphasize the difference between this process and tunneling through a square barrier as is often seen in traditional solid-state systems.

The other important term in Eq. 5.20 is G_C, which is the contact conductance (contact resistance is $R_C = 1/G_C$) for the molecular system. It includes the quantum conductance term G_0, as mentioned above, but this term is also weighted by broadening terms Γ_L and Γ_R, which in turn depend on the coupling of the molecular states to the reservoirs, and the density of states in the electrodes. Importantly, V_M, the intramolecular coupling term, is also present in the contact resistance. The appearance of this term in the contact resistance is interesting because it implies that if two different molecular species have the same linker groups to attach to the electrodes, and the same electrode material is used, the contact resistance can still be different. That is to say, the contact resistance depends not only on the contact itself but on the energy levels within the molecule. Equation 5.20 will be referred to repeatedly in the following section to describe how molecule–electrode coupling affects the conductance properties of a single-molecule device.

5.3 Molecular Linkers and Contact Geometry

To produce reliable electronic devices from a single-molecule system, it is necessary to have a robust, reproducible, and long-lifetime contact between the molecule and the electrodes. This problem becomes even more important if one desires to mechanically control the electronic properties of these systems. As such, creating stable, low contact resistance bonds between a molecule and two electrodes has been a major focus in the field for the past several years [26–30]. To explore different chemical linkers, binding geometries, and binding strengths, much of the work has focused on the alkane chains [9–11, 14, 31–39]. The alkanes are saturated, linear carbon chains that have a large HOMO–LUMO energy gap and are, therefore, expected to transport charge via tunneling (see Fig. 5.1b). Also, a variety of chemical end groups can be integrated to vary the contact resistance and orbital alignment in these systems. The chemical and transport simplicity of this system enables the use of the qualitative model described above to compare linker groups and understand what parameters control the transport properties, the mechanical stability, and the electromechanical coupling.

An additional important point is that if one wants to consider electromechanical effects, then it is necessary to have not only electrical contact to the molecule of interest, but it must also be possible to apply a mechanical perturbation. For this reason, the remainder of this chapter will focus on the use of break-junction techniques—the mechanically controllable break junction (MCBJ) and the scanning tunneling microscope (STM) break junction—to measure the transport properties of single-molecule junctions. Thus, Section 5.3.1 provides a brief discussion of the break-junction techniques, Section 5.3.2 discusses the alkanedithiol molecular family, and Section 5.3.3 ends the section with a discussion of the alkane chains with different linker groups, such as amines, carboxylic acid, and methyl sulfides.

5.3.1 *Break-Junction Techniques for Single-Molecule Junctions*

Since the MCBJ and STM break junction are described in detail in Chapter 2, the following provides only a brief overview of the way these systems are used to measure the conductance of a molecular junction.

The break-junction techniques generally rely on having a linker group on each end of the molecule under test to create a bond with each of the electrodes. The measurement proceeds by bringing the two electrodes into close proximity while in the presence of these

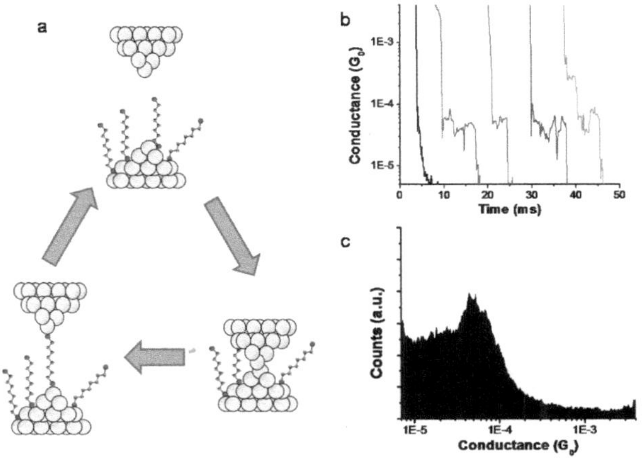

Figure 5.2 Break-junction experiments. (a) Experimental sequence followed during break-junction experiments. Two electrodes are brought into contact and separated repeatedly in the presence of molecules capable of binding to both. When this occurs, steps appear in the current vs. distance trace as shown in (b) [traces in (b) offset horizontally for clarity]. By repeating this process thousands of times, a conductance histogram can be constructed as shown in (c). This histogram has a clear peak indicating the most likely conductance of a molecular junction.

molecules. The basic criteria is that the two electrodes must be close enough together for the linkers to bind to both electrodes and create a "molecular bridge" (see Fig. 5.2). After this approach process is completed, the separation between the electrodes is increased and the current recorded until it reaches zero. The separation between the electrodes at this point depends on the resolution of the current amplifier. Three different outcomes are often observed during the electrode separation process. In the first scenario, no molecules bind to both electrodes, and through-space tunneling occurs during the entire withdraw process resulting in an exponential decay in the current vs. distance trace (Fig. 5.2b, black trace). In the second scenario, one or more molecules bind to both electrodes. In this case, the current vs. distance trace reveals a series of steps as molecule–electrode contacts break, or the junction reconfigures. Eventually, a final plateau level is reached, indicating the formation of a single-molecule junction, and upon further separation as this molecular junction is destroyed, the current drops to zero. Several current vs. distance traces exemplifying this type of behavior are shown in Fig. 5.2b. The third possibility is the occurrence of extremely noisy curves. These can occur due to mechanical instabilities in the break-junction system, sudden bindings of molecules, etc. This approach/withdraw cycle can be repeated hundreds or even thousands of times to create a large dataset for statistical analysis.

Since most molecular junctions are inherently quantum mechanical transport devices, small changes in binding site, angle, configuration, strain, etc., can drastically affect the transport properties. As such, it is common to see a large dispersion in the conductance of a single-molecule device. For this reason, large numbers of break-junction measurements are made, and the results are presented in what is referred to as a conductance histogram. These histograms are constructed by counting the amount of time spent in each conductance interval. By adding these counts up from each of the thousands of curves acquired under a specific bias condition, one can obtain a histogram as shown in Fig. 5.2c.

Beyond obtaining simple conductance histograms, this technique also enables studies of I–V characteristics [40] and electrochemical gating [41]. Additionally, the break-junction methods are directly applicable to electromechanical studies because once the molecular junction is formed, the separation between the two electrodes can be changed in real time to control either the strain or the angle of the molecular orbitals with the electrode surfaces. As such, unless otherwise noted, the measurements and results discussed in the remainder of this chapter will be obtained with some variety of the break-junction method.

5.3.2 *Alkanedithiols*

Alkanedithiols are an important class of molecules within the field. The transport properties of this molecular family were studied both theoretically [42, 43] and experimentally [44–47] before it became possible to perform single-molecule measurements. By the advent of single-molecule measurements, it was well known that these molecules transported charge via tunneling, that the transport was temperature independent, and that the HOMO–

LUMO gap was sufficiently large that the transport barrier height did not change significantly with molecular length [48, 49]. In addition, since the system is well characterized, the alkanes have provided a unique and important test bed for studying different linker chemistries, and different characterization techniques such as inelastic electron tunneling spectroscopy (IETS) [50, 51], transition voltage spectroscopy (TVS) [52–54], AC-mechanical perturbation techniques [55, 56], and many more. Despite the apparent simplicity of this molecular species and its widespread use in single-molecule studies, it still required many years before the conductance properties were understood, and even now there are still open questions remaining about the transport properties of these devices.

Xu and Tao performed the first break-junction measurement on alkanedithiols in 2003 [40]. Their work studied hexanedithiol (C6), octanedithiol (C8), and decanedithiol (C10) and obtained conductance values of 1.23×10^{-3} G_0, 2.5×10^{-4} G_0, and 2.0×10^{-5} G_0, respectively (see Fig. 5.3) [40]. These values were significantly different from single-molecule studies performed using Au nanoparticles to connect to the molecules, and a conducting AFM tip to connect to the nanoparticle [57]. The difference was ascribed simply to differences in the contact resistance until Haiss et al. also performed a study of C6, C8, and C10 using a different implementation of the STM to perform single-molecule conductance measurements and obtained values of $\sim 3.2 \times 10^{-4}$ G_0, 1.27×10^{-5} G_0, and 6.6×10^{-6} G_0, respectively [36]. Because of the similarity of the techniques in the Haiss et al. [36] and Xu et al. [40] studies, one might expect that the conductance values would be the same, but they were in fact different by about one order of magnitude in each case. Even more importantly, in applying the model from Eq. 5.20, the exponential decay factors, β-values, were significantly different in each study.

In 2006, several experimental groups published break-junction-based single-molecule studies of alkanedithiols in attempts to understand this discrepancy. Li et al. published a study showing two different conductance values for each of the alkanedithiols and suggested that the difference was due to different molecule–electrode geometries [38]. In this case, a molecular junction with the thiol bound to a hollow site on one gold electrode and a top site on

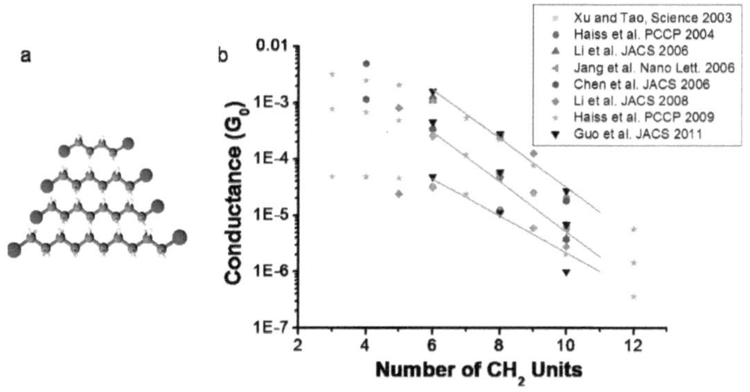

Figure 5.3 Conductance values for the alkanedithiol molecular junctions. (a) Chemical structure of several alkane chains from butanedithiol to decanedithiol in the all-*trans* configuration. Grey atoms are carbon, red are sulfur, and white are hydrogen. (b) Log of conductance vs. number of methylene units for this molecular family. Three sets of conductance values have been reproduced by several groups. Grey lines are a guide to the eye to show the different β-values for the different conductance values in the C6–C10 range. Data compiled from several Refs. [10, 11, 35, 36, 38–40, 54].

the other would result in a higher conductance value than the case where the molecule was bound to top sites on both ends. Fujihira et al. published a work indicating three different conductance values for C6. They suggested that gauche defects had an important impact on the conductance, and indicated that there may be even more possible conductance values [58]. Ulrich et al. were unable to obtain clear conductance peaks in the histograms and concluded that there were too many combinations of molecular conformations and binding sites available to be able to clearly resolve single-molecule conductance values for alkanedithiols [31]. Two other papers by Jang et al. and González et al. introduced algorithms to identify conductance values for single-molecule devices, and the values in these works agreed with the lower conductance values of Li et al. [39, 59].

In 2007, Li et al. published a combined experimental and theoretical study of alkanedithiols [11]. This work included wide-range current sensitivity for the break-junction system, and it was concluded that three sets of peaks could be discerned in the conductance histograms: *H* (high), *M* (medium), and *L* (low). The first two sets of peaks corresponded with those obtained in Ref. [38], and that the lowest value corresponds to the values of Haiss et al. [36]. Furthermore, the β-values for the first two sets of peaks were high: 0.96/CH_2 and 0.94/CH_2, respectively, while the lowest set had a β-value of 0.45, again matching the previous results from the other groups. The differences in the first set of peaks were attributed to differences in gold–thiol bonding geometries, but the third set was attributed to the presence of gauche defects in the alkane backbone. This work provided a framework for rationalizing the various different results while demonstrating that all the conductance values could be obtained within a single experimental setup.

However, this does not end the story of alkanedithiols. In 2009, Haiss et al. published two additional experimental studies which investigated a wider range of alkane chains ranging from pentanedithiol (C3) to dodecanedithiol (C12) [35, 60]. Interestingly, they found that if the molecular length was increased, all three sets of conductance values had β-values on the order of $\sim 1/CH_2$. Furthermore, the conductance behavior deviated from the predicted exponential behavior of Eq. 5.20 as the length decreased to $N \leq 5$ (N equals the number of CH_2 units in the backbone). This effect was most pronounced for the lowest set of conductance values, suggesting that this deviation from the expected exponential behavior is the reason for the large differences in measured β-values for the low set of peaks.

The reason for this deviation in the short molecules is not known, though it has been hypothesized that the short chains are more susceptible to image charge effects from the electrodes, and as such the effective tunneling barrier is lower than in the longer chains [35].

Unfortunately, it is not possible to determine the barrier height directly from conductance measurements, and additional information is needed to determine the difference between the molecular orbital energy and the electrodes' Fermi energies. To this end, a modified break-junction system was introduced by Guo et al. in 2011 [54]. This approach allowed fast I–V sweeps to be made of the single-molecule junction during the tip withdraw cycle. This additional capability allows one to directly extract bias effects on the conductance from a single measurement. Furthermore, this method can also be used to perform TVS to directly measure the difference in energy between the dominant transport level (HOMO or LUMO) and the electrode's Fermi energy [53]. This experiment demonstrated that the transition voltages for the high and medium conductance values were similar for C6, C8, and C10, thus indicating that the β-value in Eq. 5.20 is constant for these conductance peaks. Because the β-value is similar in these cases, these experiments also indicate that the change in conductance between the medium and high peaks comes from the contact conductance G_C. This indicates that the conductance change comes from a change in molecule–electrode binding, a supposition that has been suggested by several authors [11, 27, 28, 38].

The origin of the low-conductance peaks is less obvious. Even though the conductance value is significantly lower, the measured transition voltage and, therefore, the β-value is lower. At first glance, this observation may seem counterintuitive, but it is consistent with the length-dependent studies of alkanedithiols, which show that the low-conductance peaks yield a lower β-value [11, 36]. As such, it is difficult to attribute this lower conductance set directly to a change in contact configuration, and it may in fact be due to gauche-defects as has been suggested previously [11]. Continued studies exploring the transition voltage of shorter and longer chains, mechanical binding effects, breakdown forces, or inelastic tunneling spectra may yield more direct evidence on the binding geometry and molecular configuration to determine the structural properties that determine these low-conductance values.

The results from several studies on alkanedithiols have been summarized in Fig. 5.3 [10, 11, 35, 36, 38–40, 54]. This figure includes many of the systematic length studies on the alkanedithiol family, and several important trends are obvious. First, it is clearly possible to delineate three sets of conductance peaks from this variety of results. Second, the first two sets of peaks behave similarly with length, while the lowest set of conductance values deviates from the expected exponential dependence more drastically. Third, although it is possible to extrapolate the contact resistance for the longer length molecules ($N > 6$), it is not possible to use the resulting β and G_C values to predict the conductance of the shorter chains. Finally, even though the systematic study of short alkanedithiols has not yet been reproduced by other groups, the change in length dependence for these shorter chains is striking, and understanding the physical mechanism behind this change is still an open question.

Despite the remaining outstanding issues with alkanedithiols, the abundance of experimental studies on this family has led to a new understanding of single-molecule transport phenomena and has enabled the development or application of a variety of new techniques and methods for studying single-molecule devices. However, it is also apparent that if one wishes to develop technologically relevant devices from single molecules, it is necessary to have systems that will provide single contact points to the electrodes, and ideally the system will have strong binding and long lifetime at room temperature. To achieve these goals, a number of groups have begun studying different molecule–electrode contacts to improve these parameters.

5.3.3 *Varying the Molecule–Electrode Contact*

Thiols provide a strong bond to the gold electrodes and enable reproducible single-molecule conductance studies. However, from a technology standpoint, the appearance of multiple conductance values for a molecular junction is disconcerting. Clearly, to be able produce reliable and reproducible molecular devices, it is necessary to link the molecule to the gold in the same way every time. To this end, soon after the variability in conductance values was observed, groups began exploring other linker groups to bind the molecules to gold electrodes to study single-molecule transport [9, 10, 61]. Many examples of different linkers were studied either in films or in single-molecule devices, including amines, carboxylic acids, isocynides, selenium, etc. [49, 52, 62, 63]. In the single-molecule devices, the alkanes still provided an important test bed for exploring these different linker systems because it allows simple comparison of two values, G_C and β, across different linker groups. This allows us to explore difference in binding potential, and how the contact affects the alignment between the molecular orbitals and the electrodes' Fermi energies.

One of the first single-molecule systems to be studied beyond alkanedithiols using the break-junction approach was the alkane-

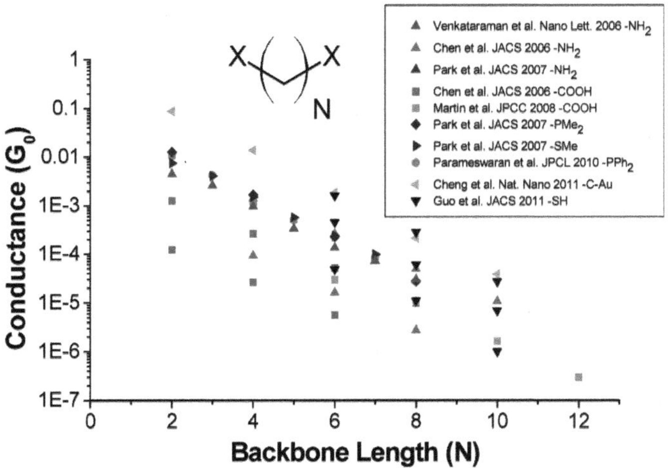

Figure 5.4 Conductance values of alkanes with different linker chemistries. The alkane chains have been terminated with different linker groups, and different conductance values have been observed. The exponential trend is still evident in all these cases though the contact resistance can vary by several orders of magnitude. Data compiled from several references [9, 10, 32, 33, 54, 61, 64].

diamines (see Fig. 5.4). Venkataraman et al. suggested that the amines provide more reproducible contact to the electrodes than the thiols, with a lower overall dispersion [9]. The cost of this decrease in dispersion was weaker binding between the gold and the molecule resulting in a larger contact resistance and lower breakdown force. Interestingly, the β-value for these chains was also different ($\beta = 0.78/CH_2$), indicating the importance of the binding in controlling the orbital/Fermi energy alignment. Shortly thereafter, alkanes with carboxylic acid linkers were also measured; these demonstrated an even higher contact resistance [10, 61]. In 2007, Chen et al. demonstrated that multiple conductance peaks were also possible for each of these linker groups [10].

Work on different linkers for the alkanes continued with diphenylphosphines (PPh_2) [33], dimethylphosphines (PMe_2), and methylsulfides (SMe) [32]. These groups bind more weakly to gold than sulfur does, show similar trends for β-values, and show contact resistances that are higher than in the thiol case. In 2011, Cheng et al.

demonstated that $-SnMe_3$ groups at the ends of alkane chains could be cleaved off in the presence of gold to provide a direct Au–C bond and allow creation of single-molecule junctions with the carbon backbone directly bound to the electrodes [64]. This process yielded strong binding between the molecule and the electrode (low contact resistance), and a β-value of $0.97/CH_2$, similar to the H-conductance values for alkanedithiols.

To summarize this variety of results, Fig. 5.4 plots the log of conductance versus the alkane-chain length for several of the alkane-linker combinations discussed above [9, 10, 32, 33, 61, 64], along with one set of alkanedithiol data for comparison [54]. A couple of points should be noted about this figure. First, in an attempt to compare similar lengths of the backbones regardless of linker types, any atom involved in the binding is not included in the total length represented here (N). This normalization simply means that octanedithiol (8 carbons), octanediamine (8 carbons), decanedioic acid (10 carbons), and 1,10-bis(trimethylstannyl)decane (10 carbons) are all plotted at the $N = 8$ position because there are 8 atoms in the methyl chain not connected to the gold electrodes. Second, it should be pointed out that some points were extracted from graphs in the original references, and not more than one significant digit can be estimated.

From this variety of experimental results, some important conclusions can be drawn as a general guideline, weaker binding results in a larger contact resistance. The β-values can change somewhat with binding, but in all cases the transport still appears to be dominated by off-resonant tunneling behavior, and Eq. 5.20 is applicable to all the cases presented here. Furthermore, although there may be some small changes in the variety of conductance values obtained with different linker groups, in general the conductance histograms still yield large dispersions. In fact, this trend still continues down to low temperatures [50, 65] indicating that the contact and junction configurations can still sample a large energy surface for conductance.

A variety of experiments performed with different linker groups clearly indicate that the linkers play a vital role in the transport properties of the single-molecule junction, but they are also important for device stability. Conductive AFM break-junction studies have been performed on many of linker systems [66–69]. These measurements have demonstrated the short lifetime and low breakdown force of most of the molecule–electrode contact combinations. However, if the molecule–electrode binding is stronger than the Au–Au bond, then the upper limit for device lifetime is determined by the lifetime of the bond between gold-apex atom and the rest of the gold electrode. This bond has repeatedly been shown to break down at $\sim$1.5 nN [66, 70] and has short lifetimes at room temperature [69, 71, 72]. As such, single-molecule junctions relying on gold electrodes will not have sufficient stability to create long-lifetime technological devices. These systems will remain useful for understanding physical and chemical processes, performing proof-of-principle experiments, or testing various device paradigms, but creating a device that will have a lifetime of years will require stronger binding and longer lifetimes than a single Au–Au bond can provide.

To begin creating stable, long-lifetime devices, several groups have begun looking at carbon electrodes to bind to the molecular systems [73–76]. Although still in their infancy and currently limited by device yield, these systems provide significant promise to create stable devices with reproducible contacts, smaller dispersions in the conductance measurements, and the tantalizing possibility of demonstrating integrated molecular circuits.

5.4 Mechanical Control of Molecule–Electrode Coupling

In our discussion so far, the molecule–electrode coupling has been determined by the linker group that provides the mechanical contact between the molecule and the electrodes. This is not always the case, and to create functional molecular devices, it is important to control not just the mechanical stability of the device, but also the electronic properties of the systems, and as discussed in Section 5.2, the contact resistance plays a substantial role in this process. One of the ways in which electrical contact can be formed without a robust adsorption on the electrode surface is through lateral coupling of the molecular orbitals to the electrode surfaces. This effect was first discussed theoretically by Kornilovitch

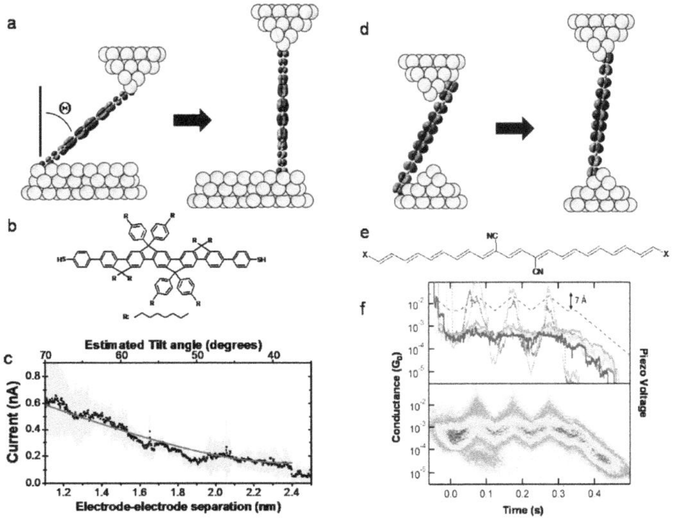

Figure 5.5 Controlling the coupling of molecular orbitals to the electrodes. (a) Controlling the tilt angle between the molecule and the electrode surface will change the coupling between the molecular energy states and those in the electrodes. (b) Molecule studied in Ref. [15] to explore this effect. (c) The conductance of the molecular junction (black curve) changes as $(\sin\Theta)^4$ as predicted theoretically (blue curve). (d) Alternative mechanism for obtaining overlap between molecular orbitals and the electrodes. The rough surface provides the coupling and yields an exponential change in the conductance. (e) Molecule studied in Ref. [80] to study this mechanism. (f) Experimental results showing clear change in the conductance as electrode separation is controlled. Part (f) from Ref. [80]; copyright (2011) American Chemical Society. Reprinted with permission.

and Bratkovsky in 2001 [77]; a model for molecular π-orbitals interacting with the electrode surfaces was proposed as seen in Fig. 5.5a. By explicitly including the angle between the molecular π-orbitals and the surface normal, Θ, the authors predicted that the device conductance would change as $G \propto (\sin\Theta)^4$ in the nonresonant tunneling case. Subsequent theoretical studies on transport in molecular devices have supported this initial model [78, 79]; however, it is not the only mechanism whereby the molecular orbitals can overlap the electrode surface states.

An alternative model for coupling between molecular orbitals and the electrode surface is shown in Fig. 5.5d. Here the molecular orbitals overlap with the rough electrode surfaces, and as the electrodes are withdrawn, the linker groups slide along the rough surface decreasing the overlap between the electrodes and the molecular orbitals. In this model, the conductance would decrease as $G = G_C \cdot \exp(-\beta L)$. L increases as the electrodes are separated; β is determined by the energy difference between the molecular orbitals and the electrode Fermi energy; and G_C is determined by the coupling between the molecular orbitals and the electrode surface rather than the molecule–electrode linker. Regardless of the model, in both these cases, the linker group provides mechanical coupling between the molecule and the electrodes, but the electrical circuit is completed by the direct electronic contact between the electrode and the molecular orbitals.

In 2006, Haiss et al. provided the first study of the tilt-angle dependence on the transport properties of a single-molecule device [16]. In this study, a variation of the STM break-junction system was used in which a constant separation between the STM tip and the substrate is maintained. In these experiments, the substrate is prepared with a low coverage of the molecules of interest, and then occasional "blinking" events occur where the current suddenly increases when a molecule binds, and then returns to the previous value when the molecule–electrode contact is broken. This procedure allows one to systematically set the distance between the electrodes and measure a statistically significant number of blinking events at each separation value. Clearly, this technique enables variation of the tilt angle, and it was demonstrated that for the molecule studied (1,4-bis[4-(acetylsulphanyl)phenylethynyl]-2,6-dimethoxybenzene), the conductance increased as the separation decreased (Θ increased) as expected [16].

The results described above provided evidence that molecular orbitals could provide sufficient coupling to the electrodes to complete the electronic circuit in a single-molecule device. But these measurements were statistical in nature, and it was not apparent whether a single molecule could be swept through a range of separations without experiencing breakdown, after all a change in angle is necessarily accompanied by some movement of the linker group across one of the electrode surfaces. However, in 2011 two separate groups demonstrated that the conductance of a single-molecule device could be continuously controlled by controlling the separation between the electrodes [15, 80].

In the first study, Díez-Pérez et al. [15] used an approach similar to that of Haiss et al. described above [16], but once the blinking event was observed, the authors controllably separated the two electrodes over a set distance. Using the molecule shown in Fig. 5.5b, the authors were able to demonstrate that the conductance of a single molecule could be decreased and increased as the separation between the electrodes was continuously controlled. This process was repeated on a number of junctions, and the change in the conductance with tilt angle matched the predicted $(\sin\Theta)^4$ behavior.

Alternatively, Meisner et al. used a long conjugated polyolefin chain (see Fig. 5.5e) with the STM break-junction method to observe coupling between the molecular orbitals and the electrode surface [80]. In this case, the break-junction cycle was modified so that instead of continuously separating the two electrodes, a triangle wave with an amplitude of ~7 Å was applied to the piezoelectric transducer during the withdraw cycle. Then when the molecule was bound, this triangle wave resulted in a continuous change in the single-molecule conductance. By repeating this process thousands of times, the authors demonstrated that the conductance change as a function of distance did not match the $(\sin\Theta)^4$ behavior, but instead decayed exponentially with distance with a β-value similar to that found when systematically changing the length of the oligoene chain. This result suggests that the second model (Fig. 5.5d) is the more likely mechanism for controlling the molecule–electrode coupling and the conductance in this case.

These two experimental methods demonstrated that electronic coupling between the molecule and the electrode can be controlled via the coupling of the orbitals of the molecular backbone to the electronic states of the metal. Although the two groups observed different dependencies of junction separation on the conductance and, therefore, invoked different models to describe the results, both results are consistent with their respective methods. In the case of the rigid pentaphenylene molecule, the junction is formed over an atomically flat terrace on the substrate electrode, implying that the molecule will have some angle with the surface [15]. In the case of the polyolefin molecule, the linkers used are expected to bind preferentially with under-coordinated gold atoms, and the electrodes are forced into hard contact before withdraw. This process will result in considerable surface roughness on both electrodes, and as such the proposed sliding model appears reasonable [80].

Regardless of the specific mechanism, the ability to mechanically control the conductance of a single-molecule device in real time represents an important capability for creating new types of molecular-electronic systems and devices, and will also continue to provide important information about the physical and chemical processes that control charge transport in molecular-scale systems.

5.5 Mechanical Control of Molecular Energy Levels

Thus far we have discussed molecule–electrode contacts as primarily affecting the contact resistance of the junction. However, in many cases, the contact between the molecule and the electrode directly affects the energy alignment between the dominant molecular orbital and the Fermi energy of the electrodes. The reason that the contact binding changes the energy level alignments is straightforward to understand. The binding of the linker group to the electrode necessarily involves some charge transfer to create a strong binding between the two. This potential shift also causes a change in the energy levels of the molecule in question. Because of this shift in energy levels upon binding, it is common practice to consider an "extended molecule" in molecular transport models, which includes not only the molecule, but several gold atoms on either side. This method provides a more accurate picture of the energies of the molecular system. Also, if different linkers are used with the same molecular backbone, there will be different amounts of charge transfer and a different shift in the energy levels. Thus, the molecule–electrode linker can play a pivotal role in the transport properties of the molecular system not simply because of stability and contact resistance, but also in determining the energy alignment.

Although the discussion above focuses on how different molecular linkers can affect the alignment in molecular devices, it is also important to consider what happens in a single-molecule device if the contact is changed due to either a compressive or a tensile force. The effects of a tensile strain are particularly important because many of the single-molecule measurements are performed in break-junction devices, which implicitly apply a strain to the molecular junction. In these cases, as the junction is stretched, the bond between the molecule and the gold electrodes will also be strained; this will be associated with some charge transfer and again result in a change in the potential profile and energy alignment. The archetypical example of this process occurring in a single-molecule device is in benzendithiol (BDT).

BDT has a storied history in molecular electronics with many theoretical works devoted to it, and a myriad of conductance studies [17, 26, 29, 30, 81–87]. In 2003, Xue and Ratner studied the effects of molecule–electrode separation on the conductance of BDT [30]. Their theoretical study compared the change in coupling between the molecule and the electrode as the electrodes were separated versus the change in orbital/Fermi energy alignment during this process. Since the spin-singlet state of BDT is near the Fermi energy of gold, the difference between the molecular orbital energy and the Fermi energy will decrease, resulting in a resonant transmission pathway and an increase in the conductance as the contact is strained. At the same time, the coupling between the molecular orbitals and the electrodes will decrease, resulting in a decrease in the conductance. This study proposed that the alignment effect would dominate until the orbital energy level stopped changing, at which point the decrease in coupling would begin causing a decrease in the conductance. This system has also been studied using several other theoretical models, which also predict this increasing conductance behavior in BDT [26, 88, 89].

Many of the initial experimental studies on BDT did not observe this effect, though it was often noted how broad the conductance histogram is for this molecule [81, 87]. In 2012, Bruot et al. demonstrated that the difficulty in obtaining a specific conductance value for BDT was because of the orbital alignment changes as the junction is strained, which in turn causes large changes in the molecular conductance [18]. These experiments were performed at 4.2 K, and during normal break-junction measurements, many

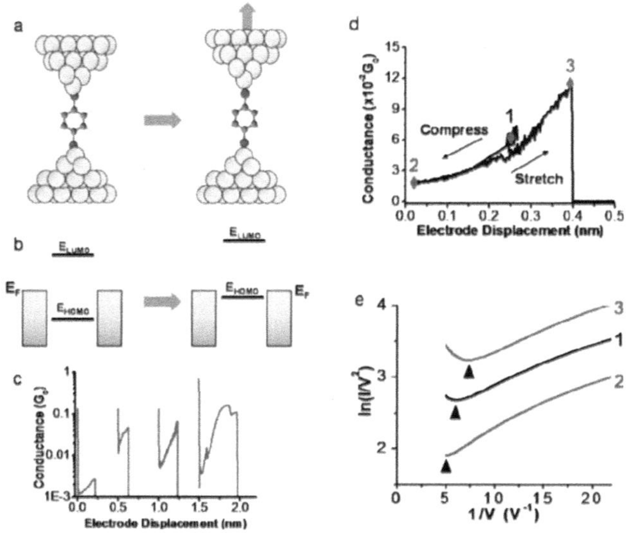

Figure 5.6 Strain-induced changes in orbital energies as described in Ref. [18]. (a) Benzendithiol molecular junction under strain conditions. (b) Corresponding energy-level diagrams when the junction is at equilibrium separation and strained. (c) Examples of several conductance vs. distance traces for BDT junctions demonstrating a large change in the conductance. (d) One example junction where the bias is swept at different electrode separations. TVS plots from the I–V characteristics showing a systematic change in the alignment between the molecular orbital energy levels and the electrode Fermi energies as sketched in (b).

traces exhibited a large increase in the conductance as the junction was strained (see Fig. 5.6). To determine if this effect was due to changes in the contact, the authors performed IETS (for a description of this technique, see Chapter 6). These measurements demonstrated that the contacts were changing as the junction was strained. Second, to demonstrate that this change in contact was accompanied by a change in the energy-level alignment, they also performed TVS studies as shown in Fig. 5.6e. These measurements clearly demonstrated that the difference between the dominant molecular orbital and the electrode Fermi energy decreased as the junction was strained, and as a result, the conductance increased. This study demonstrates that molecule–electrode linker plays an important role not only in determining the coupling between the two, but can also have an important effect on the energetics of the system. Clearly, controlling the charge transport properties of a single-molecule device will require not only careful control of the electronic structure of the molecule, but also consideration of the effects that the molecule–electrode binding may have on the molecular orbital energy levels.

5.6 Summary and Conclusions

This chapter has focused on describing the effects of the molecule–electrode contact on charge transport in single-molecule devices. A simple model for the effect of this coupling was derived, which provides a useful framework for describing model molecular systems. It has been demonstrated that accurately describing and controlling the molecule–electrode contact is important for both creating a stable mechanical system and controlling the charge transport properties. In most of the studies discussed in this chapter, gold was used as the electrode material, although this has been a useful approach for understanding coupling effects and demonstrating mechanical control of the electronic system via the molecule–electrode binding; these systems are still limited in both stability and lifetime. As such, the development of stable, long-lifetime single-molecule devices capable of being used in electronic circuits, or integrated into a larger device, will require that completely new molecule–electrode coupling paradigms emerge. This may include carbon-based electrodes such as graphene or carbon nanotubes, or new nanoscale lithographic tools that allow atomic-level control of electrode placement. Clearly, the molecule–electrode contact will continue to be an important issue in the design, development, and testing of single-molecule devices for the foreseeable future.

Acknowledgments

The author would like to acknowledge financial support from the National Science Foundation (NSF ECCS-1231915).

Problems

1. (a) Demonstrate that a one-dimensional conjugated chain of length L with a lattice spacing of a results in a conductor with a half-filled conduction band. (b) Repeat for a lattice spacing of $2a$, and demonstrate that this results in a completely filled band.
2. Starting from the matrix definition for the Hamiltonian and neglecting the self-energy terms, find the relevant Green's function term to define transport in a 3 bridge site molecular chain. Show the full inversion process of the Green's function matrix.
3. Starting from Eq. 5.15, show that Eq. 5.20 is correct, and justify any approximations or assumptions.
4. Given the conductance values for hexanedithiol, octanedithiol, and decanedithiol of $4.5 \times 10^{-4}\, G_0$, $5.9 \times 10^{-5}\, G_0$, and 6.9×10^{-6} G_0, respectively, find G_C and β for this system.

References

1. Tao, N.J., Electron transport in molecular junctions. *Nature* Nanotechnology, 2006. **1**(3): pp. 173–181.
2. Reed, M.A., and Tour, J.M., Computing with molecules. *Scientific American*, 2000. **282**: pp. 86–93.
3. Song, H., Reed, M.A., and Lee, T., Single molecule electronic devices. *Advanced Materials*, 2011. **23**(14): pp. 1583–1608.
4. Mantooth, B.A., and Weiss, P.S., Fabrication, assembly, and characterization of molecular electronic components. *Proceedings of the IEEE*, 2003. **91**(11): pp. 1785–1802.
5. Peierls, R., *More Surprises in Theoretical Physics. Princeton Series in Physics*, ed. P.W. Anderson, 1991. Princeton, New Jersey: Princeton University Press.
6. Peierls, R., *Quantum Theory of Solids*, 1955. London: Oxford University Press. p. 226.
7. Comes, R., Lambert, M., Launois, H., and Zeller, H.R., Evidence for a Peierls distortion or a Kohn anomaly in one-dimensional conductors of the type $K_2Pt(CN)_4Br_{0.30}{\cdot}xH_2O$. *Physical Review B*, 1973. **8**(2): pp. 571–575.

8. Lee, P.A., Rice, T.M., and Anderson, P.W., Fluctuation effects at a Peierls transition. *Physical Review Letters*, 1973. **31**(7): pp. 462–465.
9. Venkataraman, L., Klare, J.E., Tam, I.W., Nuckolls, C., Hybertsen, M.S., and Steigerwald, M.L., Single-molecule circuits with well-defined molecular conductance. *Nano Letters*, 2006. **6**(3): pp. 458–462.
10. Chen, F., Li, X., Hihath, J., Huang, Z., and Tao, N., Effect of anchoring groups on single-molecule conductance: Comparative study of thiol-, amine-, and carboxylic-acid-terminated molecules. *Journal of the American Chemical Society*, 2006. **128**(49): pp. 15874–15881.
11. Li, C., Pobelov, I., Wandlowski, T., Bagrets, A., Arnold, A., and Evers, F., Charge transport in single Au |alkanedithiol |Au junctions: Coordination geometries and conformational degrees of freedom. *Journal of the American Chemical Society*, 2008. **130**(1): pp. 318–326.
12. Kim, Y., Song, H., Strigl, F., Pernau, H.F., Lee, T., and Scheer, E., Conductance and vibrational states of single-molecule junctions controlled by mechanical stretching and material variation. *Physical Review Letters*, 2011. **106**(19): pp. 196804.
13. Meisner, J.S., Kamenetska, M., Krikorian, M., Steigerwald, M.L., Venkataraman, L., and Nuckolls, C., A single-molecule potentiometer. *Nano Letters*, 2011. **11**(4): pp. 1575–1579.
14. Schneebeli, S.T., Kamenetska, M., Cheng, Z., Skouta, R., Friesner, R.A., Venkataraman, L., and Breslow, R., Single-molecule conductance through multiple pi-pi-stacked benzene rings determined with direct electrode-to-benzene ring connections. *Journal of the American Chemical Society*, 2011. **133**(7): pp. 2136–2139.
15. Diez-Perez, I., Hihath, J., Hines, T., Wang, Z.S., Zhou, G., Mullen, K., and Tao, N.J., Controlling single-molecule conductance through lateral coupling of pi orbitals. *Nature Nanotechnology*, 2011. **6**(4): pp. 226–231.
16. Haiss, W., Wang, C.S., Grace, I., Batsanov, A.S., Schiffrin, D.J., Higgins, S.J., Bryce, M.R., Lambert, C.J., and Nichols, R.J., Precision control of single-molecule electrical junctions. *Nature Materials*, 2006. **5**(12): pp. 995–1002.
17. Xue, Y.Q., and Ratner, M.A., Microscopic study of electrical transport through individual molecules with metallic contacts. I. Band lineup, voltage drop, and high-field transport. *Physical Review B*, 2003. **68**: p. 115406.
18. Bruot, C., Hihath, J., and Tao, N., Mechanically controlled molecular orbital alignment in single molecule junctions. *Nature Nanotechnology*, 2011. **7**(1): pp. 35–40.

19. Ferry, D.K., *Quantum Mechanics: An Introduction for Device Physicists and Electrical Engineers*. 2nd ed., Vol. 1, 2001, Bristol BS1 6BE, United Kingdom: Institute of Physics Publishing, p. 345.
20. Mujica, V., Kemp, M., and Ratner, M.A., Electron conduction in molecular wires. I. A scattering formalism. *Journal of Chemical Physics*, 1994. **101**: pp. 6849–6864.
21. Mujica, V., Kemp, M., and Ratner, M.A., Electron conduction in molecular wires. 2. Applications to scanning-tunneling microscopy. *Journal of Chemical Physics*, 1994. **101**(8): pp. 6856–6864.
22. Nitzan, A., Electron transmission through molecules and molecular interfaces. *Annual Review of Physical Chemistry*, 2001. **52**: pp. 681–750.
23. Smit, R.H.M., Noat, Y., Untiedt, C., Lang, N.D., van Hemert, M.C., and van Ruitenbeek, J.M., Measurement of the conductance of a hydrogen molecule. *Nature* (London), 2002. **419**: pp. 906–909.
24. Tal, O., Kiguchi, M., Thijssen, W.H.A., Djukic, D., Untiedt, C., Smit, R.H.M., and van Ruitenbeek, J.M., Molecular signature of highly conductive metal-molecule-metal junctions. *Physical Review B*, 2009. **80**(8): p. 085427.
25. Datta, S., *Quantum Transport: From Atom to Transistor*. 1st ed. Vol. 1. 2005, New York: Cambridge University Press. p. 404.
26. Ke, S.H., Baranger, H.U., and Yang, W.T., Contact atomic structure and electron transport through molecules. *Journal of Chemical Physics*, 2005. **122**: p. 074704.
27. Muller, K.H., Effect of the atomic configuration of gold electrodes on the electrical conduction of alkanedithiol molecules. *Physical Review B: Condensed Matter and Materials Physics*, 2006. **73**(4): pp. 045403/1-03/6.
28. Paulsson, M., Krag, C., Frederiksen, T., and Brandbyge, M., Conductance of alkanedithiol single-molecule junctions: A molecular dynamics study. *Nano Letters*, 2009. **9**(1): pp. 117–121.
29. Li, Z. and Kosov, D.S., Nature of well-defined conductance of amine-anchored molecular junctions: Density functional calculations. *Physical Review B*, 2007. **76**: p. 035415.
30. Xue, Y., and Ratner, M.A., Microscopic study of electrical transport through individual molecules with metallic contacts. II. Effect of the interface structure. *Physical Review B*, 2003. **68**: p. 115407.
31. Ulrich, J., Esrail, D., Pontius, W., Venkataraman, L., Millar, D., and Doerrer, L.H., Variability of conductance in molecular junctions. *Journal of Physical Chemistry B*, 2006. **110**(6): pp. 2462–2466.

32. Park, Y.S., Whalley, A.C., Kamenetska, M., Steigerwald, M.L., Hybertsen, M.S., Nuckolls, C., and Venkataraman, L., Contact chemistry and single-molecule conductance: A comparison of phosphines, methyl sulfides, and amines. *Journal of the American Chemical Society*, 2007. **129**(51): p. 15768.

33. Parameswaran, R., Widawsky, J.R., Vazquez, H., Park, Y.S., Boardman, B.M., Nuckolls, C., Steigerwald, M.L., Hybertsen, M.S., and Venkataraman, L., Reliable formation of single molecule junctions with air-stable diphenylphosphine linkers. *Journal of Physical Chemistry Letters*, 2010. **1**(14): pp. 2114–2119.

34. Kim, Y., Hellmuth, T.J., Burkle, M., Pauly, F., and Scheer, E., Characteristics of amine-ended and thiol-ended alkane single-molecule junctions revealed by inelastic electron tunneling spectroscopy. *ACS Nano*, 2011. **5**(5): pp. 4104–4111.

35. Haiss, W., Martin, S., Scullion, L.E., Bouffier, L., Higgins, S.J., and Nichols, R.J., Anomalous length and voltage dependence of single molecule conductance. *Physical Chemistry Chemical Physics*, 2009. **11**(46): pp. 10831–10838.

36. Haiss, W., Nichols, R.J., van Zalinge, H., Higgins, S.J., Bethell, D., and Schiffrin, D.J., Measurement of single molecule conductivity using the spontaneous formation of molecular wires. *Physical Chemistry Chemical Physics*, 2004. **6**: pp. 4330–4337.

37. Martin, C.A., Ding, D., van der Zant, H.S.J., and van Ruitenbeek, J.M., Lithographic mechanical break junctions for single-molecule measurements in vacuum: Possibilities and limitations. *New Journal of Physics*, 2008. **10**: p. 065008.

38. Li, X., He, J., Hihath, J., Xu, B., Lindsay, S.M., and Tao, N., Conductance of single alkanedithiols. Conduction mechanism and effect of molecule-electrode contacts. *Journal of the American Chemical Society*, 2006. **128**(6): pp. 2135–2141.

39. Jang, S.Y., Reddy, P., Majumdar, A., and Segalman, R.A., Interpretation of stochastic events in single molecule conductance measurements. *Nano Letters*, 2006. **6**(10): pp. 2362–2367.

40. Xu, B., and Tao, N.J., Measurement of single-molecule resistance by repeated formation of molecular junctions. *Science*, 2003. **301**(5637): pp. 1221–1223.

41. Xu, B., Xiao, X., Yang, X., Zang, L., and Tao, N., Large gate modulation in the current of a room temperature single molecule transistor. *Journal of the American Chemical Society*, 2005. **127**: pp. 2386–2387.

42. Tomfohr, J.K., and Sankey, O.F., Complex band structure, decay lengths, and Fermi level alignment in simple molecular electronic systems. *Physical Revie B*, 2002. **65**: pp. 245105/1–05/12.
43. Segal, D., Nitzan, A., Ratner, M.A., and Davis, W.B., Activated conduction in microscopic molecular junctions. *Journal of Physical Chemistry B*, 2000. **104**: pp. 2790–2793.
44. York, R.L., Nguyen, P.T., and Slowinski, K., Long-range electron transfer through monolayers and bilayers of alkanethiols in electrochemically controlled Hg-Hg tunneling junctions. *Journal of the American Chemical Society*, 2003. **125**: pp. 5948–5953.
45. Chidsey, C.E.D., Free energy and temperature dependence of electron transfer at the metal-electrolyte interface. *Science*, 1991. **251**: pp. 919–922.
46. Holmlin, R.E., Haag, R., Chabinyc, M.L., Ismagilov, R.F., Cohen, A.E., Terfort, A., Rampi, M.A., and Whitesides, G.M., Electron transport through thin organic films in metal-insulator-metal junctions based on self-assembled monolayers. *Journal of the American Chemical Society*, 2001. **123**: pp. 5075–5085.
47. Smalley, J.F., Feldberg, S.W., Chidsey, C.E.D., Linford, M.R., Newton, M.D., and Liu, Y.-P., The kinetics of electron transfer through ferrocene-terminated alkanethiol monolayers on gold. *Journal of Physical Chemistry*, 1995. **99**: pp. 13141–13149.
48. McCreery, R.L., Molecular electronic junctions. *Chemistry of Materials*, 2004. **16**: pp. 4477–4496.
49. Beebe, J.M., Engelkes, V.B., Miller, L.L., and Frisbie, C.D., Contact resistance in metal-molecule-metal junctions based on aliphatic SAMs: Effects of surface linker and metal work function. *Journal of the American Chemical Society*, 2002. **124**(38): pp. 11268–11269.
50. Hihath, J., Arroyo, C.R., Rubio-Bollinger, G., Tao, N., and Agrait, N., Study of electron-phonon interactions in a single molecule covalently connected to two electrodes. *Nano Letters*, 2008. **8**(6): pp. 1673–1678.
51. Arroyo, C.R., Frederiksen, T., Rubio-Bollinger, G., Vélez, M., Arnau, A., Sánchez-Portal, D., and Agraït, N., Characterization of single-molecule pentanedithiol junctions by inelastic electron tunneling spectroscopy and first-principles calculations. *Physical Review B*, 2010. **81**(7): p. 075405.
52. Kim, B., Beebe, J.M., Jun, Y., Zhu, X.Y., and Frisbie, C.D., Correlation between HOMO alignment and contact resistance in molecular junctions:

Aromatic thiols versus aromatic isocyanides. *Journal of the American Chemical Society*, 2006. **128**: pp. 4970–4971.

53. Beebe, J.M., Kim, B., Frisbie, C.D., and Kushmerick, J.G., Measuring relative barrier heights in molecular electronic junctions with transition voltage spectroscopy. *ACS Nano*, 2008. **2**(5): pp. 827–832.
54. Guo, S., Hihath, J., Diez-Perez, I., and Tao, N., Measurement and statistical analysis of single-molecule current-voltage characteristics, transition voltage spectroscopy, and tunneling barrier height. *Journal of the American Chemical Society*, 2011. **133**(47): pp. 19189–19197.
55. Xia, J.L., Diez-Perez, I., and Tao, N.J., Electron transport in single molecules measured by a distance-modulation assisted break junction method. *Nano Letters*, 2008. **8**(7): pp. 1960–1964.
56. Zhou, J., and Xu, B., Determining contact potential barrier effects on electronic transport in single molecular junctions. *Applied Physics Letters*, 2011. **99**(4): p. 042104.
57. Cui, X.D., Primak, A., Zarate, X., Tomfohr, J., Sankey, O.F., Moore, A.L., Moore, T.A., Gust, D., Harris, G., and Lindsay, S.M., Reproducible measurement of single-molecule conductivity. *Science*, 2001. **294**(5542): pp. 571–574.
58. Fujihira, M., Suzuki, M., Fujii, S., and Nishikawa, A., Currents through single molecular junction of Au/hexanedithiolate/Au measured by repeated formation of break junction in STM under UHV: Effects of conformational change in an alkylene chain from gauche to trans and binding sites of thiolates on gold. *Physical Chemistry Chemical Physics*, 2006. **8**(33): pp. 3876–3884.
59. Gonzalez, M.T., Wu, S.M., Huber, R., van der Molen, S.J., Schonenberger, C., and Calame, M., Electrical conductance of molecular junctions by a robust statistical analysis. *Nano Letters*, 2006. **6**: pp. 2238–2242.
60. Haiss, W., Martin, S., Leary, E., van Zalinge, H., Higgins, S.J., Bouffier, L., and Nichols, R.J., Impact of junction formation method and surface roughness on single molecule conductance. *Journal of Physical Chemistry C*, 2009. **113**(14): pp. 5823–5833.
61. Martin, S., Haiss, W., Higgins, S., Cea, P., Lopez, M.C., and Nichols, R.J., A comprehensive study of the single molecule conductance of alpha, omega-dicarboxylic acid-terminated alkanes. *Journal of Physical Chemistry C*, 2008. **112**(10): pp. 3941–3948.
62. Patrone, L., Palacin, S., Bourgoin, J., Lagoute, J., Zambelli, T., and Gauthier, S., Direct comparison of the electronic coupling efficiency of

sulfur and selenium anchoring groups for molecules adsorbed onto gold electrodes more options. *Chemical Physics*, 2002. **281**: pp. 325–332.

63. Yasuda, S., Yoshida, S., Sasaki, J., Okutsu, Y., Nakamura, T., Taninaka, A., Takeuchi, O., and Shigekawa, H., Bond fluctuation of S/Se anchoring observed in single-molecule conductance measurements using the point contact method with scanning tunneling microscopy. *Journal of the American Chemical Society*, 2006. **128**: pp. 7746–7747.
64. Cheng, Z.L., Skouta, R., Vazquez, H., Widawsky, J.R., Schneebeli, S., Chen, W., Hybertsen, M.S., Breslow, R., and Venkataraman, L., In situ formation of highly conducting covalent Au-C contacts for single-molecule junctions. *Nature Nanotechnology*, 2011. **6**(6): pp. 353–357.
65. Hihath, J., Bruot, C., and Tao, N.J., Electron-phonon interactions in single octanedithiol molecular junctions. *ACS Nano*, 2010. **4**(7): pp. 3823–3830.
66. Xu, B., Xiao, X., and Tao, N.J., Measurements of single-molecule electromechanical properties. *Journal of the American Chemical Society*, 2003. **125**(52): pp. 16164–16165.
67. Frei, M., Aradhya, S.V., Koentopp, M., Hybertsen, M.S., and Venkataraman, L., Mechanics and chemistry: Single molecule bond rupture forces correlate with molecular backbone structure. *Nano Letters*, 2011. **11**(4): pp. 1518–1523.
68. Aradhya, S.V., Frei, M., Hybertsen, M.S., and Venkataraman, L., Van der Waals interactions at metal/organic interfaces at the single-molecule level. *Nature Materials*, 2012. **11**(10): pp. 872–876.
69. Huang, Z., Chen, F., Bennett, P.A., and Tao, N., Single molecule junctions formed via Au-thiol contact: Stability and breakdown mechanism. *Journal of the American Chemical Society*, 2007. **129**(43): pp. 13225–13231.
70. Rubio, G., Agrait, N., and Vieira, S., Atomic-sized metallic contacts: Mechanical properties and electronic transport. *Physical Review Letters*, 1996. **76**(13): pp. 2302–2305.
71. Huang, Z., Chen, F., D'Agosta, R., Bennett, P.A., Di Ventra, M., and Tao, N., Local ionic and electron heating in single-molecule junctions. *Nature Nanotechnology*, 2007. **2**(11): pp. 698–703.
72. Tsutsui, M., Shoji, K., Morimoto, K., Taniguchi, M., and Kawai, T., Thermodynamic stability of single molecule junctions. *Applied Physics Letters*, 2008. **92**(22): p. 223110.

73. Guo, X.F., Small, J.P., Klare, J.E., Wang, Y.L., Purewal, M.S., Tam, I.W., Hong, B.H., Caldwell, R., Huang, L.M., O'Brien, S., Yan, J.M., Breslow, R., Wind, S.J., Hone, J., Kim, P., and Nuckolls, C., Covalently bridging gaps in single-walled carbon nanotubes with conducting molecules. *Science*, 2006. **311**: pp. 356–359.

74. Yan, H.J., Bergren, A.J., and McCreery, R.L., All-carbon molecular tunnel junctions. *Journal of the American Chemical Society*, 2011. **133**(47): pp. 19168–19177.

75. Prins, F., Barreiro, A., Ruitenberg, J.W., Seldenthuis, J.S., Aliaga-Alcalde, N., Vandersypen, L.M.K., and van der Zant, H.S.J., Room-temperature gating of molecular junctions using few-layer graphene nanogap electrodes. *Nano Letters*, 2011. **11**(11): pp. 4607–4611.

76. Roy, S., Vedala, H., Roy, A.D., Kim, D.-H., Doud, M., Mathee, K., Shin, H.-K., Shimamoto, N., Prasad, V., and Choi, W., Direct electrical measurements on single-molecule genomic dna using single-walled carbon nanotubes. *Nano Letters*, 2007. **8**(1): pp. 26–30.

77. Kornilovitch, P.E., and Bratkovsky, A.M., Orientational dependence of current through molecular films. *Physical Review B: Condensed Matter*, 2001. **64**(19): p. 195413.

78. Basch, H., Cohen, R., and Ratner, M.A., Interface geometry and molecular junction conductance: Geometric fluctuation and stochastic switching. *Nano Letters*, 2005. **5**: pp. 1668–1675.

79. Toyoda, K., Morimoto, K., and Morita, K., First-principles study on current through a single pi conjugate molecule for analysis of carrier injection through an organic/metal interface. *Surface Science*, 2006. **600**(23): pp. 5080–5083.

80. Meisner, J.S., Kamenetska, M., Krikorian, M., Steigerwald, M.L., Venkataraman, L., and Nuckolls, C., A single-molecule potentiometer. *Nano Letters*, 2011. **11**(4): pp. 1575–1579.

81. Xiao, X.Y., Xu, B.Q., and Tao, N.J., Measurement of single molecule conductance: Benzenedithiol and benzenedimethanethiol. *Nano Letters*, 2004. **4**(2): pp. 267–271.

82. Reed, M.A., Zhou, C., Muller, C.J., Burgin, T.P., and Tour, J.M., Conductance of a molecular junction. *Science*, 1997. **278**: pp. 252–254.

83. Song, H., Kim, Y., Jang, Y.H., Jeong, H., Reed, M.A., and Lee, T., Observation of molecular orbital gating. *Nature* (London), 2009. **462**(7276): pp. 1039–1043.

84. Tsutsui, M., Teramae, Y., Kurokawa, S., and Sakai, A., High-conductance states of single benzenedithiol molecules. *Applied Physics Letters*, 2006. **89**(16): p. 163111.

85. Taniguchi, M., Tsutsui, M., Yokota, K., and Kawai, T., Inelastic electron tunneling spectroscopy of single-molecule junctions using a mechanically controllable break junction. *Nanotechnology*, 2009. **20**(43): p. 434008.

86. Romaner, L., Heimel, G., Gruber, M., Bredas, J.L., and Zojer, E., Stretching and breaking of a molecular junction. *Small*, 2006. **2**(12): pp. 1468–1475.

87. Kim, Y., Pietsch, T., Erbe, A., Belzig, W., and Scheer, E., Benzenedithiol: A broad-range single-channel molecular conductor. *Nano Letters*, 2011. **11**(9): pp. 3734–3738.

88. Toher, C., and Sanvito, S., Efficient atomic self-interaction correction scheme for nonequilibrium quantum transport. *Physical Review Letters*, 2007. **99**(5): p. 056801.

89. Pontes, R., Rocha, A., Sanvito, S., Fazzio, A., and da Silva, A.J.R., Ab initio calculations of structural evolution and conductance of benzene-1,4-dithiol on gold leads. *ACS Nano*, 2011. **5**: pp. 795–804.

Chapter 6

Vibrational Excitations in Single-Molecule Junctions

Joseph S. Seldenthuis, Herre S. J. van der Zant, and Johannes M. Thijssen
Kavli Institute of Nanoscience, Delft University of Technology, Lorentzweg 1, 2628 CJ Delft, The Netherlands
h.s.j.vanderzant@tudelft.nl

Vibrational excitations have an important effect on the transport properties of molecular junctions. Which modes are important depends strongly on the coupling of the molecule to the leads. In this chapter we analyze these effects for both strongly and weakly coupled molecules and give approximate selection rules for the vibrational modes. For weakly coupled molecules the low-frequency modes tend to be dominant. Their relative intensities are determined by the so-called Franck–Condon factors. For strongly coupled conjugated molecules, on the other hand, the modes involving stretching of the π-bonds determine the transport properties. This chapter shows how to calculate vibrational effects in both regimes for the full vibrational spectrum of a molecule.

Handbook of Single-Molecule Electronics
Edited by Kasper Moth-Poulsen

ISBN 978-981-4463-38-6 (Hardcover), 978-981-4463-39-3 (eBook)
www.panstanford.com

6.1 Introduction

In this chapter we consider the effects of vibrational excitations on the transport properties of single-molecule junctions. Vibrational excitations have been measured in molecular junctions using scanning tunneling microscopes (STMs) [Stipe et al. (1998); Qiu et al. (2003); Dong et al. (2004); Ćavar et al. (2005); Wu et al. (2008)], in mechanical break-junctions (MBJs) [Smit et al. (2002); Parks et al. (2007)], and electromigrated break-junctions (EMBJs) [Park et al. (2000); Yu et al. (2004); Osorio et al. (2007, 2010)].

The effect of vibrational excitations on transport strongly depends on the coupling between the molecule and the electrodes. In the weak-coupling regime, transport is incoherent and the current is dominated by sequential tunneling processes. Electrons hop onto the molecule one at a time, are localized there for a while, and dephase before tunneling off again. While current flows, the charge of the molecule therefore fluctuates between integer multiples of the electron charge e. It is this continuous charging and discharging that excites the vibrational modes: the equilibrium geometry of the nuclei differs for different charge states, and since they move much more slowly than the electrons, the tunneling of an electron leaves the nuclei in an excited state.

In the opposite limit, the strong-coupling regime, transport is coherent. Electrons are delocalized over the molecule and the electrodes, and the charge does not fluctuate between integer multiples of e during transport. The current is generally dominated by elastic tunneling processes, which do not excite vibrational modes. It is however possible for electrons to tunnel inelastically by emitting or absorbing a phonon. Such a process opens a new transmission channel which can lead to either an increase or a reduction of the conductance.

In this chapter we will investigate the effect of vibrational excitations in both regimes. There are many sophisticated theoretical methods for addressing these effects. A comprehensive review can be found in Ref. [Galperin et al. (2007)]. However, most of those methods are only feasible for small 'toy model' systems with only a few vibrational modes. We will focus on two first-order approaches: one for the weak and one for the strong-coupling regime. Both have the important advantage that their computational cost scales favorably; they are applicable to systems with hundreds of vibrational modes, and can be used to examine the full vibrational spectrum of the molecules under investigation. We describe both approaches in sufficient detail to allow practical implementation, but the more technical sections can be skipped on first reading.

To test the merits of the methods, we apply them to two π-conjugated molecules, both examples of the prototypical molecular wires: in the weak-coupling regime the oligo(phenylene vinylene) derivative OPV-5 [Kubatkin et al. (2003); Osorio et al. (2007)], and in the strong-coupling regime the oligo(phenylene ethynylene) derivative OPE-3 [Bumm et al. (1996); Kushmerick et al. (2004)]. Comparison of the calculations with the measured spectra shows good agreement.

For molecules in bulk or gas phase, selection rules exists to determine which vibrational modes are active in, e.g., infrared or Raman spectra. For electrical transport through molecules in a junction, however, no such selection rules exists. The binding of a molecule to the electrodes generally breaks all symmetries and changes the vibrational modes by restricting the movements of the nuclei. These effects can be taken into account in a calculation, which can therefore be used to predict the vibrational spectrum in transport.

6.2 Vibrational Modes

In order to investigate the effects of vibrational excitation, we first have to calculate the frequencies and normal modes of a molecule. This is generally done with quantum chemistry methods such as Hartree–Fock or density functional theory (DFT) and by assuming that the molecule behaves as a harmonic oscillator. The calculated vibrational spectra tend to match the measured spectra very well, provided the frequencies are corrected with the appropriate scaling factor [Scott and Radom (1996)]. For most molecules, the energies of the vibrational modes range from several milli-electron volts (meV)—note that 1 meV $\approx$ 8 cm^{-1} $\approx$ 250 GHz—for the lowest modes, such as bending or twisting of the entire molecule, to several hundred meV for the highest modes; the C–H stretching mode, e.g., is found between 355–375 meV.

6.2.1 *Born–Oppenheimer Approximation*

The Hamiltonian of a molecule consisting of n electrons and N nuclei is given by

$$\begin{aligned}\hat{H} =& \sum_{i=1}^{n} \frac{\hat{\mathbf{p}}_i^2}{2m_\mathrm{e}} + \frac{e^2}{4\pi\epsilon_0} \sum_{i=1}^{n} \sum_{i'=i+1}^{n} \frac{1}{|\mathbf{r}_i - \mathbf{r}_{i'}|} \\ &+ \sum_{j=1}^{N} \frac{\hat{\mathbf{P}}_j^2}{2M_j} + \frac{e^2}{4\pi\epsilon_0} \sum_{j=1}^{N} \sum_{j'=j+1}^{N} \frac{Z_j Z_{j'}}{|\mathbf{R}_j - \mathbf{R}_{j'}|} \\ &- \frac{e^2}{4\pi\epsilon_0} \sum_{i=1}^{n} \sum_{j=1}^{N} \frac{Z_j}{|\mathbf{r}_i - \mathbf{R}_j|},\end{aligned}$$

where i numbers the electrons with mass m_e and charge $-e$, and j numbers the nuclei with mass M_j and charge eZ_j. $\hat{\mathbf{p}}_i = -\mathrm{i}\hbar\frac{\partial}{\partial \mathbf{r}_i}$ and $\hat{\mathbf{P}}_j = -\mathrm{i}\hbar\frac{\partial}{\partial \mathbf{R}_j}$ are the momentum operators of the electrons and nuclei, respectively. The first line in the Hamiltonian describes the electrons, the second the nuclei and the third the Coulomb interaction between the electrons and nuclei. The number of degrees of freedom in this Hamiltonian is far too large for it to be tractable for all but the smallest molecules. Therefore a number of approximations has to be made.

The first is the Born–Oppenheimer approximation, where we assume that we can separate the $3n$ electronic from the $3N$ nuclear degrees of freedom, i.e.,

$$\Psi(\mathbf{r}, \mathbf{R}) = \psi_\mathrm{e}(\mathbf{r}, \mathbf{R})\psi_\mathrm{n}(\mathbf{R}),$$

where $\psi_\mathrm{e}(\mathbf{r}, \mathbf{R})$ is the electronic and $\psi_\mathrm{n}(\mathbf{R})$ the nuclear wave-function. Note that this neglects vibronic coupling effects, which are crucial to the understanding of non-adiabatic processes. The approximation is motivated by the fact that the nuclei are much heavier than the electrons (the mass of a proton is 1836 times that of an electron) and therefore move much slower. The electrons only feel the instantaneous nuclear coordinates, while the nuclei only feel the average electron distribution.

The electrons are described by the Hamiltonian

$$\hat{H}_\text{e} = \sum_{i=1}^{n} \frac{\hat{\mathbf{p}}_i^2}{2m_\text{e}} + V_\text{e}(\mathbf{r}, \mathbf{R}), \tag{6.1}$$

with eigenstates $\psi_\text{e}(\mathbf{r}, \mathbf{R})$: $\hat{H}_\text{e}\psi_\text{e}(\mathbf{r}, \mathbf{R}) = E_\text{e}\psi_\text{e}(\mathbf{r}, \mathbf{R})$. This equation is generally solved with quantum-chemistry methods such as Hartree–Fock or density functional theory (DFT), while keeping $\mathbf{R}$ fixed. By varying $\mathbf{R}$ and repeatedly solving this equation we can obtain the effective potential for the nuclei $V_\text{n}(\mathbf{R})$. For slow nuclear motion (the adiabatic limit), the Hamiltonian

$$\hat{H}_\text{n} = \sum_{j=1}^{N} \frac{\hat{\mathbf{P}}_j^2}{2M_j} + V_\text{n}(\mathbf{R}) \tag{6.2}$$

is then a good approximation. The eigenstates $\psi_\text{n}(\mathbf{R})$ of this Hamiltonian satisfy $\hat{H}_\text{n}\psi_\text{n}(\mathbf{R}) = E_\text{n}\psi_\text{n}(\mathbf{R})$. The equilibrium geometry of the nuclei ($\mathbf{R}_0$) is obtained by minimizing E_n.

6.2.2 *Harmonic Oscillator*

At geometric equilibrium ($\mathbf{R} = \mathbf{R}_0$), the gradient of the nuclear potential vanishes. The lowest-order term in $V_\text{n}(\mathbf{R})$ is therefore quadratic in $\mathbf{R}$. If we discard all higher-order terms, we are left with a harmonic potential:

$$V_\text{n}(\mathbf{R}) \approx \tfrac{1}{2}\left(\mathbf{R} - \mathbf{R}_0\right)^T \mathbf{K}\left(\mathbf{R} - \mathbf{R}_0\right). \tag{6.3}$$

K is the Hessian matrix:

$$K_{ij} = \left.\frac{\partial^2 V_\text{n}(\mathbf{R})}{\partial R_i \partial R_j}\right|_{\mathbf{R}=\mathbf{R}_0},$$

where i and j run over all $3N$ nuclear coordinates. In the case of a single oscillator, **K** equals the spring constant. Most quantum-chemistry packages are able to calculate the Hessian directly.

The potential in Eq. 6.3 describes a system of coupled harmonic oscillators. The oscillators can be decoupled by diagonalizing the Hessian [Wilson et al. (1955)]. First, however, we need to get rid of the mass-term in Eq. 6.2. We can do this by switching to the mass-weighted coordinates $\tilde{\mathbf{R}} = \sqrt{\mathbf{M}}\left(\mathbf{R} - \mathbf{R}_0\right)$, where $\mathbf{M}$ is a diagonal matrix containing the mass of the atoms. The Hamiltonian then becomes

$$\hat{H}_{\mathrm{n}} = \sum_{j=1}^{N} \tfrac{1}{2}\hat{\tilde{\mathbf{P}}}_j + \tfrac{1}{2}\tilde{\mathbf{R}}^T \sqrt{\mathbf{M}^{-1}}\mathbf{K}\sqrt{\mathbf{M}^{-1}}\tilde{\mathbf{R}}.$$

Now we can diagonalize the mass-weighted Hessian:

$$\sqrt{\mathbf{M}^{-1}}\mathbf{K}\sqrt{\mathbf{M}^{-1}} = \boldsymbol{\Omega}\boldsymbol{\omega}^2\boldsymbol{\Omega}^T.$$

$\boldsymbol{\omega}$ is a diagonal matrix, the elements of which are the frequencies of the normal modes. At equilibrium, the Hessian is positive definite and so are the eigenvalues, hence the frequencies are real. Imaginary frequencies sometimes result from numerical calculations, which indicates that the molecule is not in its equilibrium geometry. The columns of $\boldsymbol{\Omega}$ are the mass-weighted normal-mode displacements.

An isolated molecule has three translational and three rotational degrees of freedom. These six degrees of freedom (or five in the case of a linear molecule) appear in the results of a vibrational-mode calculation as modes with zero frequency, since motion in any of these directions is not associated with a restoring force. We therefore have only $N_q = 3N - 6$ actual vibrational modes. Note that for a molecule in a transport junction, we generally fix the atoms in the electrodes and only look at the vibrational modes of the molecule. However, since the molecule is attached to the electrodes it no longer has any rotational or translational degrees of freedom. In this case we have $N_q = 3N$.

We now define the vector of (dimensionless) normal coordinates

$$\mathbf{q} = \sqrt{\frac{\boldsymbol{\omega}}{2\hbar}}\boldsymbol{\Omega}^T\sqrt{\mathbf{M}}\,(\mathbf{R} - \mathbf{R}_0)\,. \tag{6.4}$$

For a single oscillator this equals $q = \frac{R-R_0}{2\Delta R}$, where $\Delta R = \sqrt{\frac{\hbar}{2m\omega}}$ is the zero-point motion; a measure of the uncertainty in the position in the ground state of the harmonic oscillator. In terms of the normal coordinates the nuclear potential (Eq. 6.3) becomes

$$V_{\mathrm{n}}(\mathbf{q}) = \mathbf{q}^T\hbar\boldsymbol{\omega}\mathbf{q},$$

and the nuclear Hamiltonian simplifies to

$$\hat{H}_{\mathrm{n}} = \sum_{i=1}^{N_q} \hbar\omega_i \left(q_i^2 - \frac{1}{4}\frac{\partial^2}{\partial q_i^2}\right) = \sum_{i=1}^{N_q} \hbar\omega_i \left(\hat{a}_i^\dagger \hat{a}_i + \tfrac{1}{2}\right), \tag{6.5}$$

where

$$\hat{a}_i^\dagger = q_i - \frac{\mathrm{i}}{2\hbar}\hat{p}_i = q_i - \frac{1}{2}\frac{\partial}{\partial q_i}, \tag{6.6a}$$

$$\hat{a}_i = q_i + \frac{\mathrm{i}}{2\hbar}\hat{p}_i = q_i + \frac{1}{2}\frac{\partial}{\partial q_i}, \tag{6.6b}$$

are the ladder operators. In this context, $\hat{p}_i = -\mathrm{i}\hbar\frac{\partial}{\partial q_i}$. Using the fact that $[q_i, \hat{p}_i] = \mathrm{i}\hbar$, it is easily seen that the ladder operators satisfy the commutation relations for bosonic creation and annihilation operators:

$$\left[\hat{a}_i, \hat{a}_j^\dagger\right] = \delta_{ij}, \tag{6.7a}$$

$$\left[\hat{a}_i, \hat{a}_j\right] = \left[\hat{a}_i^\dagger, \hat{a}_j^\dagger\right] = 0. \tag{6.7b}$$

$\hat{a}_i^\dagger$ creates a phonon in (or adds a vibrational quantum to) mode i, while $\hat{a}_i$ destroys a phonon. The combination $\hat{n}_i = \hat{a}_i^\dagger \hat{a}_i$ counts the number of quanta in mode i. The total nuclear energy is therefore given by

$$E_\mathrm{n} = \sum_i \hbar\omega_i \left(n_i + \tfrac{1}{2}\right).$$

The nuclear wave-functions for arbitrary vibrational quanta can be obtained by starting from the vibrational ground-state ($n = 0$):

$$\psi_i^0(q_i) = \left(\frac{\omega_i}{\pi\hbar}\right)^{\frac{1}{4}} e^{-q_i^2},$$

and acting on it with the required number of creation operators:

$$\psi_\mathbf{n}(\mathbf{q}) = \prod_{i=1}^{N_q} \frac{1}{\sqrt{n_i!}} \left(\hat{a}_i^\dagger\right)^{n_i} \psi_i^0(q_i). \tag{6.8}$$

6.2.3 *Morse Potential*

An example of a harmonic oscillator potential with the first few wave-functions is shown in Fig. 6.1a. The energy spectrum is evenly spaced by $\hbar\omega$ and, in principle, contains an infinite number of modes. In practice, this is only a good approximation for the first few vibrational quanta. Once the vibrational energy is larger than the dissociation energy, bond-breaking will occur. The Morse potential is

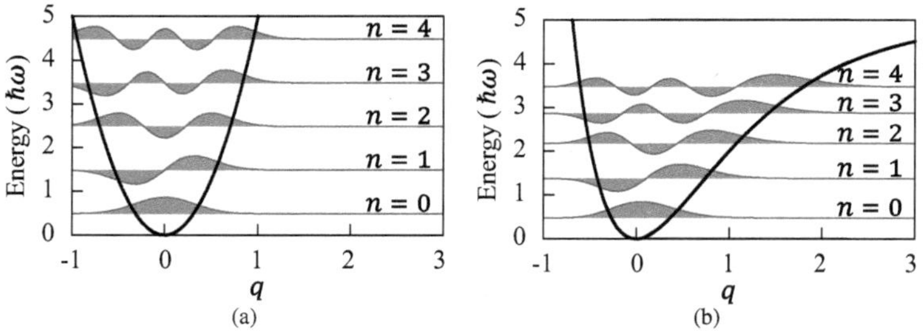

Figure 6.1 (a) Harmonic oscillator and (b) Morse potential with the first five wave functions.

an improvement on the harmonic oscillator potential that explicitly includes this effect.

For a single mode. the Morse potential is given by [Morse (1929)]

$$V_{\mathrm{n}}(R) = D\left(1 - e^{-a(R-R_0)}\right)^2,$$

where D is the dissociation energy, and $a = \sqrt{\frac{1}{2}K/D}$. Close to equilibrium we recover the harmonic oscillator:

$$V_{\mathrm{n}}(R) \approx \tfrac{1}{2}K\left(R - R_0\right)^2 - \tfrac{1}{2}Ka\left(R - R_0\right)^3 + \ldots,$$

but for large deviations the potential differs (see Fig. 6.1b). The energy spectrum is no longer evenly spaced [Morse (1929)]:

$$E_{\mathrm{n}} = \hbar\omega\left(n + \tfrac{1}{2}\right)\left(1 - \frac{\hbar\omega\left(n + \frac{1}{2}\right)}{4D}\right). \qquad (6.9)$$

More importantly, the number of bound states is now finite. Eq. 6.9 is only valid for $n + \frac{1}{2} \leq \frac{D}{\hbar\omega}$. If the number of vibrational quanta is larger, the bond dissociates and the spectrum becomes continuous.

There exist analytical expressions for the wave-functions of the Morse potential [Morse (1929)], and also for the ladder operators [Dong et al. (2002)], but for the remainder of this chapter we will stick to the harmonic oscillator. It is however useful to keep in mind when and how this approximation breaks down.

6.3 Franck–Condon Principle

Charge transfer and electronic transitions, due to, e.g., the absorption or emission of a photon, change the electron density of the

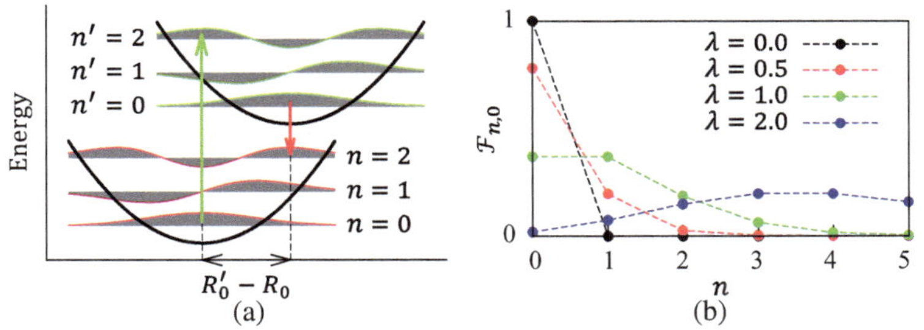

Figure 6.2 (a) Illustration of the Franck–Condon principle. Electronic transitions are vertical in this picture, since they are instantaneous on the timescale of the nuclei. If the equilibrium geometries are different, this can leave the molecule in a vibrationally excited state. (b) Ground-state Franck–Condon factors of a single harmonic oscillator (with $\omega = \omega'$) for different values of the electron–phonon coupling λ.

molecule and therefore the potential landscape felt by the nuclei. Within the Born–Oppenheimer approximation we take the nuclei to be stationary on the timescale of the dynamics of the electrons. This means that an electronic transition is most likely to occur without changes in the positions of the nuclei. This is called the Franck–Condon principle, and the electronic transition is a vertical transition (see Fig. 6.2a). If the molecule is in the vibrational ground state before the transition, it is in a vibrationally excited state after the transition, since the nuclei are still in the equilibrium geometry of the initial electronic state, which is not necessarily the equilibrium geometry of the final state. Several vibrational modes may be excited during the transition. A vibrational excitation is more likely to happen if the initial and final vibrational wave-functions overlap more significantly. The probability of such an excitation by the Franck–Condon factor; the square of the overlap between the two vibrational wave-functions that are involved in the transition.

The Franck–Condon factor can be derived from Fermi's golden rule, which states that the probability (per unit time) of an electronic transition from state Ψ to state Ψ' due to a perturbation described by the Hamiltonian $\hat{H}'$ is given by

$$T = \frac{2\pi}{\hbar} \left|\langle \Psi | \hat{H}' | \Psi' \rangle\right|^2 \rho, \tag{6.10}$$

where ρ is the density of final states. Using the Born–Oppenheimer approximation, we can separate the electronic and nuclear wave-functions. Since $\hat{H}'$ only acts on the electronic wave-function, we can write the transition matrix element as

$$\left|\langle\Psi|\hat{H}'|\Psi'\rangle\right|^2 = \left|\langle\psi_{\mathbf{n}}\psi_{\mathrm{e}}|\hat{H}'|\psi'_{\mathrm{e}}\psi'_{\mathbf{n}'}\rangle\right|^2 = \mathcal{F}_{\mathbf{n},\mathbf{n}'}\left|\langle\psi_{\mathrm{e}}|\hat{H}'|\psi'_{\mathrm{e}}\rangle\right|^2,$$

where

$$\mathcal{F}_{\mathbf{n},\mathbf{n}'} = \left|\langle\psi_{\mathbf{n}}|\psi'_{\mathbf{n}'}\rangle\right|^2 = \left|\mathcal{I}_{\mathbf{n},\mathbf{n}'}\right|^2$$

is the Franck–Condon factor, with

$$\mathcal{I}_{\mathbf{n},\mathbf{n}'} = \langle\psi_{\mathbf{n}}|\psi'_{\mathbf{n}'}\rangle \equiv \int d\mathbf{R}\,\psi^{\dagger}_{\mathbf{n}}(\mathbf{R})\psi'_{\mathbf{n}'}(\mathbf{R})$$

the overlap integral of the nuclear wave functions. When the wave functions are normalized, the Franck–Condon factors sum to 1:

$$\sum_{\mathbf{n}}\mathcal{F}_{\mathbf{n},\mathbf{n}'} = \sum_{\mathbf{n}}\langle\psi'_{\mathbf{n}'}\underbrace{|\psi_{\mathbf{n}}\rangle\langle\psi_{\mathbf{n}}|}_{\text{unit operator}}\psi'_{\mathbf{n}'}\rangle = \langle\psi'_{\mathbf{n}'}|\psi'_{\mathbf{n}'}\rangle = 1,$$

and similarly for $\sum_{\mathbf{n}'}\mathcal{F}_{\mathbf{n},\mathbf{n}'}$. This is to be expected: given an electronic transition, Franck–Condon factors represent the probability of an accompanying vibrational transition. The sum of all Franck–Condon factors, i.e., all probabilities, must equal 1.

6.3.1 *Electron–Phonon Coupling*

As we will show below for the harmonic oscillator, the Franck–Condon factors are determined by the electron–phonon coupling λ (Eq. 6.15). The electron–phonon coupling is a dimensionless measure of the displacement of the equilibrium positions of the nuclei in the basis of the normal coordinates. Since it measures the displacement in each normal mode due to the electronic transition, it determines which modes will be excited. Certain selection rules apply—they reflect the symmetry of the vibrational mode and the geometry. As a result if these, it is possible for certain modes never to be excited by a particular electronic transition. In this case the electron–phonon coupling for that mode is zero.

In the case of a single harmonic oscillator, where the frequency in both states is equal ($\omega = \omega'$), a simple expression for ground-state to excited-state Franck–Condon factors can be derived (Eq. 6.13):

$$\mathcal{F}_{n,0} = \left|\mathcal{I}_{n,0}\right|^2 = \frac{\lambda^{2n}}{n!}e^{-\lambda^2}.$$

The Franck-Condon factors are shown in Fig. 6.2b for different values of λ. From this expression it is clear that in the case where $\lambda = 0$, $\mathcal{F}_{0,0} = 1$ and all other factors are zero; no vibrational modes are excited. When $\lambda = 1$, $\mathcal{F}_{0,0} = \mathcal{F}_{1,0}$, while for larger values $\mathcal{F}_{0,0}$ is no longer the dominant contribution. The case $\lambda = 1$ corresponds to the situation where the displacement of the nuclei equals the sum of the quantum uncertainty in the position of both states. For larger displacements there is only little overlap between the vibrational ground states and not exciting a vibrational mode becomes increasingly unlikely.

When $\lambda > 1$ we speak of Franck–Condon blockade [Koch et al. (2006)], since the purely electronic transition, without a vibrational excitation, becomes a blocked process. This can be seen in Fig. 6.2b, where $\mathcal{F}_{0,0}$ nearly vanishes for $\lambda = 2$. Below, we shall investigate how Franck–Condon blockade manifests itself in the case of electrical transport. First we will describe how Franck–Condon factors can be calculated for a molecule.

6.3.2 *Recursion Relations*

In the case of the harmonic oscillator, it is possible to obtain analytical expressions for the Franck-Condon factors in the form of recursion relations. We will derive this explicitly for the single-mode oscillator and then show how the approach can be generalized for multiple modes.

6.3.2.1 Single harmonic oscillator

For the initial electronic state, with frequency ω and equilibrium position R_0, the normal coordinate is given by Eq. 6.4. The normal coordinate of the final electronic state, with a possibly different frequency ω' and equilibrium position R_0', is related to the first via

$$q' = \sqrt{\frac{m\omega'}{2\hbar}}\left(R - R_0'\right) = \sqrt{\frac{\omega'}{\omega}}q + \lambda, \tag{6.11}$$

where

$$\lambda = \sqrt{\frac{m\omega'}{2\hbar}}\left(R_0 - R_0'\right)$$

is the electron–phonon coupling; it measures the displacement of the nuclei in units of the normal coordinates. In the literature, the electron–phonon coupling is related to the Huang–Rhys factor $g = \lambda^2$.

The overlap integral of the vibrational ground states is

$$\mathcal{I}_{0,0} = \int_{-\infty}^{\infty} \mathrm{d}R \psi_0^\dagger(R)\psi_0'(R) = \sqrt{\sqrt{\frac{\omega'}{\omega}\frac{2}{\pi}}} \int_{-\infty}^{\infty} \mathrm{d}q e^{-q^2-q'^2}.$$

For a single Gaussian integral, we have $\int_{-\infty}^{\infty} \mathrm{d}x e^{-x^2} = \sqrt{\pi}$. Variable substitution gives

$$\int_{-\infty}^{\infty} \mathrm{d}q e^{-q^2-q'^2} = \int_{-\infty}^{\infty} \mathrm{d}q e^{-A(q+b)^2-c} = \sqrt{\frac{\pi}{A}} e^{-c},$$

where

$$A = \frac{\omega+\omega'}{\omega}, \qquad b = \frac{\sqrt{\omega\omega'}}{\omega+\omega'}\lambda, \qquad c = \lambda^2 \frac{\omega}{\omega+\omega'}.$$

The ground-state overlap integral therefore becomes

$$\mathcal{I}_{0,0} = \sqrt{\frac{\sqrt{4\omega\omega'}}{\omega+\omega'}} e^{-\lambda^2 \frac{\omega}{\omega+\omega'}}$$

In the case of $\omega = \omega'$, we get for the ground state–to–ground state Franck–Condon factor: $\mathcal{F}_{0,0} = \left|\mathcal{I}_{0,0}\right|^2 = e^{-\lambda^2}$. If $\lambda = 0$, $\mathcal{F}_{0,0} = 1$; the equilibrium positions overlap and no vibrational transitions take place.

In the case of the single harmonic oscillator there exists a direct expression for the overlap integral between two arbitrary vibrational wave-functions [Smith (1969)]. However, here we will use a method based on the ladder operators as this approach can easily be generalized to multiple oscillators [Palma and Morales (1983); Sandoval et al. (1989)]. In terms of the ladder operators, the overlap integral is given by (see Eq. 6.8)

$$\mathcal{I}_{n,n'} = \int_{-\infty}^{\infty} \mathrm{d}R \psi_n^\dagger(R)\psi_{n'}'(R) = \int_{-\infty}^{\infty} \mathrm{d}R \psi_0^\dagger(R) \frac{(\hat{a})^n}{\sqrt{n!}} \frac{\left(\hat{a}'^\dagger\right)^{n'}}{\sqrt{n'!}} \psi_0'(R).$$

The ladder operators $\hat{a}$ and $\hat{a}'^\dagger$ are defined in terms of q and q' (see Eq. 6.7), which are related via Eq. 6.11. This allows us to write $\hat{a}$ as a linear combination of $\hat{a}^\dagger$ and $\hat{a}'$, which reduces the number of creation and annihilation operators in the overlap integral. Similarly, we can write $\hat{a}'^{\dagger}$ as a linear combination of $\hat{a}'$ and $\hat{a}^{\dagger}$:

$$\hat{a} = \left(\frac{2\omega}{\omega+\omega'} - 1\right)\hat{a}^{\dagger} + \frac{\sqrt{4\omega\omega'}}{\omega+\omega'}\hat{a}' - \frac{\sqrt{4\omega\omega'}}{\omega+\omega'}\lambda,$$
$$\hat{a}'^{\dagger} = \left(1 - \frac{2\omega}{\omega+\omega'}\right)\hat{a}' + \frac{\sqrt{4\omega\omega'}}{\omega+\omega'}\hat{a}^{\dagger} + \frac{2\omega}{\omega+\omega'}\lambda.$$

In the case of $\omega = \omega'$, this reduces to $\hat{a} = \hat{a}' - \lambda$ and $\hat{a}'^{\dagger} = \hat{a}^{\dagger} + \lambda$. Together, these relations enable us to write the overlap integrals of higher vibrational excitations in terms of those of lower excitations, leading to the following two recursion relations:

$$\begin{aligned}\mathcal{I}_{n,n'} =& \sqrt{\frac{n-1}{n}}\left(\frac{2\omega}{\omega+\omega'} - 1\right)\mathcal{I}_{n-2,n'} + \sqrt{\frac{n'}{n}}\frac{\sqrt{4\omega\omega'}}{\omega+\omega'}\mathcal{I}_{n-1,n'-1} \\ &- \frac{\lambda}{\sqrt{n}}\frac{\sqrt{4\omega\omega'}}{\omega+\omega'}\mathcal{I}_{n-1,n'} \\ =& \sqrt{\frac{n'-1}{n'}}\left(1 - \frac{2\omega}{\omega+\omega'}\right)\mathcal{I}_{n,n'-2} + \sqrt{\frac{n}{n'}}\frac{\sqrt{4\omega\omega'}}{\omega+\omega'}\mathcal{I}_{n-1,n'-1} \\ &+ \frac{\lambda}{\sqrt{n'}}\frac{2\omega}{\omega+\omega'}\mathcal{I}_{n,n'-1}.\end{aligned}$$

When $\omega = \omega'$, repeatedly applying these relations yields the following expression for the overlap integral:

$$\mathcal{I}_{n,n'} = e^{-\frac{1}{2}\lambda^2}\sqrt{n!n'!}\sum_{k=0}^{[n,n']}\frac{(-\lambda)^{n-k}\lambda^{n'-k}}{k!(n-k)!\,(n'-k)!}, \tag{6.12}$$

where $[n, n']$ is the smaller of n and n'. The ground-state Franck–Condon factor then simply becomes

$$\mathcal{F}_{n,0} = \left|\mathcal{I}_{n,0}\right|^2 = \frac{\lambda^{2n}}{n!}e^{-\lambda^2}, \tag{6.13}$$

and similarly for $\mathcal{F}_{0,n'}$.

6.3.2.2 Multiple harmonic oscillators

For a molecule with multiple vibrational modes, the normal mode coordinates $\mathbf{q}$ and $\mathbf{q}'$ of the first and second electronic state, respectively, are related via (see Eqs. 6.4 and 6.11)

$$\mathbf{q}' = \sqrt{\frac{\omega'}{2\hbar}}\boldsymbol{\Omega}'^{T}\sqrt{\mathbf{M}}\left(\mathbf{R} - \mathbf{R}'_0\right) = \mathbf{L}\mathbf{q} + \boldsymbol{\lambda}, \tag{6.14}$$

where

$$\boldsymbol{\lambda} = \sqrt{\frac{\boldsymbol{\omega}'}{2\hbar}}\boldsymbol{\Omega}'^{T}\sqrt{\mathbf{M}}\left(\mathbf{R}_0 - \mathbf{R}'_0\right) \tag{6.15}$$

is the electron–phonon coupling. $\mathbf{L} = \sqrt{\boldsymbol{\omega}'}\mathbf{J}\sqrt{\boldsymbol{\omega}^{-1}}$ is the transformation matrix from the $\mathbf{q}$-basis to the $\mathbf{q}'$-basis, with $\mathbf{J} = \boldsymbol{\Omega}'^{T}\mathbf{M}\boldsymbol{\Omega}$ the so-called Duschinsky mixing matrix [Duschinsky (1937)]. Due to the three translational and three rotational degrees of freedom of an isolated molecule, the $\mathbf{R}$-basis has a larger dimension ($3N$) than the $\mathbf{q}$ and $\mathbf{q}'$-basis ($N_q = 3N - 6$). The translational degrees of freedom can be eliminated by shifting the origin of both geometries to the center of mass. It is not possible to eliminate the rotational degrees of freedom, but their mixing with the normal coordinates can be minimized by rotating one of the geometries to minimize the root-mean-square distance between the nuclei of the different states [Zhixing (1989); Sando and Spears (2001); Coutsias et al. (2004)].

The ground-state overlap integral is given by

$$\mathcal{I}_{0,0} = \int_{-\infty}^{\infty} d\mathbf{R}\psi_0(\mathbf{R})\psi'_0(\mathbf{R}) = \sqrt{\sqrt{\frac{\det(\boldsymbol{\omega}')}{\det(\boldsymbol{\omega})}}\left(\frac{2}{\pi}\right)^{N_q}\int_{-\infty}^{\infty} d\mathbf{q}e^{-\mathbf{q}^T\mathbf{q}-\mathbf{q}'^T\mathbf{q}'}}.$$

Using variable substitution we can evaluate the N_q-dimensional Gaussian integral:

$$\begin{aligned}\int_{-\infty}^{\infty} d\mathbf{q}e^{-\mathbf{q}^T\mathbf{q}-\mathbf{q}'^T\mathbf{q}'} &= \int_{-\infty}^{\infty} d\mathbf{q}e^{-\mathbf{q}^T\mathbf{q}-(\mathbf{L}\mathbf{q}+\lambda)^T(\mathbf{L}\mathbf{q}+\lambda)} \\ &= \int_{-\infty}^{\infty} d\mathbf{q}e^{-(\mathbf{q}+\mathbf{b})^T\mathbf{A}(\mathbf{q}+\mathbf{b})-c} \\ &= \sqrt{\frac{\pi^{N_q}}{\det(\mathbf{A})}}e^{-c},\end{aligned}$$

where

$$\mathbf{A} = \mathbf{1}+\mathbf{L}^T\mathbf{L}, \qquad \mathbf{b} = \left(\mathbf{1}+\mathbf{L}^T\mathbf{L}\right)^{-1}\mathbf{L}^T\boldsymbol{\lambda}, \qquad c = \boldsymbol{\lambda}^T\left(\mathbf{1}+\mathbf{L}\mathbf{L}^T\right)^{-1}\boldsymbol{\lambda},$$

hence the ground-state overlap integral becomes

$$\mathcal{I}_{0,0} = \sqrt{\frac{\sqrt{\det(4\boldsymbol{\omega}\boldsymbol{\omega}')}}{\det\left(\mathbf{J}^T\boldsymbol{\omega}'\mathbf{J}+\boldsymbol{\omega}\right)}}e^{-\boldsymbol{\lambda}^T\left(\mathbf{1}+\mathbf{L}\mathbf{L}^T\right)^{-1}\boldsymbol{\lambda}}.$$

For the overlap integral between arbitrary vibrational wavefunctions, most derivations are based on the generating function approach of Sharp and Rosenstock [Sharp and Rosenstock (1964)] or the coherent state approach of Doktorov et al. [Doktorov et al. (1977)]. Here, we again use the alternative approach based on the ladder operators (Eq. 6.7). The resulting recursion relations are equivalent, but the derivation is more straightforward. The overlap integral is given by

$$
\begin{aligned}
\mathcal{I}_{\mathbf{n},\mathbf{n}'} &= \int_{-\infty}^{\infty} \mathrm{d}\mathbf{R}\psi_{\mathbf{n}}^{\dagger}(\mathbf{R})\psi_{\mathbf{n}'}'(\mathbf{R}) \\
&= \int_{-\infty}^{\infty} \mathrm{d}\mathbf{R}\psi_{0}^{\dagger}(\mathbf{R}) \prod_{i=1}^{N_q} \frac{(\hat{a}_i)^{n_i}}{\sqrt{n_i!}} \prod_{j=1}^{N_q} \frac{\left(\hat{a}_i'^{\dagger}\right)^{n_j'}}{\sqrt{n_j'!}} \psi_0'(\mathbf{R}).
\end{aligned}
$$

Using Eq. 6.14 we can write

$$
\begin{aligned}
\hat{\mathbf{a}} &= -\mathbf{A}\hat{\mathbf{a}}^{\dagger} + \mathbf{B}\hat{\mathbf{a}}' + \mathbf{c}, \\
\hat{\mathbf{a}}'^{\dagger} &= \mathbf{A}\hat{\mathbf{a}}' + \mathbf{B}\hat{\mathbf{a}}^{\dagger} + \mathbf{d},
\end{aligned}
$$

where we have now defined

$$
\begin{aligned}
\mathbf{A} &= \left(\mathbf{L}+\mathbf{L}^{-1}\right)^{-1}\left(\mathbf{L}-\mathbf{L}^{-1}\right), & \mathbf{B} &= 2\left(\mathbf{L}+\mathbf{L}^{-1}\right)^{-1}, \\
\mathbf{c} &= -2\left(\mathbf{L}+\mathbf{L}^{-1}\right)^{-1}\boldsymbol{\lambda}, & \mathbf{d} &= 2\left(\mathbf{L}+\mathbf{L}^{-1}\right)^{-1}\mathbf{L}^{-1}\boldsymbol{\lambda}.
\end{aligned}
$$

We then get the recursion relations

$$
\begin{aligned}
\mathcal{I}_{\mathbf{n},\mathbf{n}'} = &-\sum_i \sqrt{\frac{n_i-1}{n_i}} A_{ii}\mathcal{I}_{n_i-2} - \sum_{i,j\neq i} \sqrt{\frac{n_j}{n_i}} A_{ij}\mathcal{I}_{n_i-1,n_j-1} \\
&+\sum_{i,j} \sqrt{\frac{n_j'}{n_i}} B_{ij}\mathcal{I}_{n_i-1,n_j'-1} + \sum_i \frac{c_i}{\sqrt{n_i}}\mathcal{I}_{n_i-1},
\end{aligned}
$$

and

$$
\begin{aligned}
\mathcal{I}_{\mathbf{n},\mathbf{n}'} = &\sum_i \sqrt{\frac{n_i'-1}{n_i'}} A_{ii}\mathcal{I}_{n_i'-2} + \sum_{i,j\neq i} \sqrt{\frac{n_j'}{n_i'}} A_{ij}\mathcal{I}_{n_i'-1,n_j'-1} \\
&+\sum_{i,j} \sqrt{\frac{n_i}{n_j'}} B_{ij}\mathcal{I}_{n_i-1,n_j'-1} + \sum_i \frac{d_i}{\sqrt{n_i'}}\mathcal{I}_{n_i'-1},
\end{aligned}
$$

where, on the right-hand side, we have only indicated the occupation numbers that deviate from those on the left-hand side.

6.3.3 *Numerical Evaluation*

We will briefly discuss some practical aspects concerning the numerical evaluation of Franck–Condon factors. Implementations exist in several quantum chemistry programs such as Gaussian [Santoro et al. (2007a,b, 2008); Barone et al. (2009)] and Amsterdam Density Functional (ADF) [Seldenthuis et al. (2008)].

Although there are in principle infinitely many Franck–Condon factors, in practice we can only evaluate a finite number. We can of course keep calculating factors until the sum is sufficiently close to one, but then we do not know in advance how many factors we need. An easier way is to simply specify a maximum number of vibrational quanta that can be distributed over the normal modes at any given time [Ruhoff (1994); Ruhoff and Ratner (2000)]. In this case we can determine the number of factors in advance: when we have at most n vibrational quanta distributed over N_q normal modes, there are

$$N_n = \binom{N_q + n}{n}$$

different vibrationally excited states (including the ground state). For the same number of vibrational quanta in the initial and final state, this yields an $N_n \times N_n$ matrix of Franck–Condon factors. Note that the size of this matrix scales rapidly with both n and N_q.

6.3.3.1 Example: emission spectrum of Pt(4,6-dFppy)(acac)

As an example of a numerical calculation we will look at the emission spectrum of the triplet-to-singlet transition of the organic light-emitting diode Pt(4,6-dFppy)(acac). This spectrum was measured by Rausch et al. [Rausch et al. (2009)] at both room temperature and 20 K (Fig. 6.3a). The molecule has $N = 35$ atoms and therefore $N_q = 3N - 6 = 99$ vibrational modes. Since we expect the molecule to be in the vibrational ground state before emission, we only calculate the ground-state to excited-state Franck–Condon factors. Taking up to five vibrational quanta into account results in almost 92 million

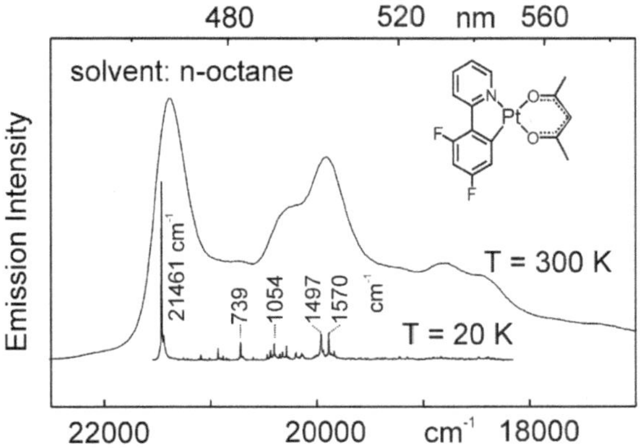

(a)

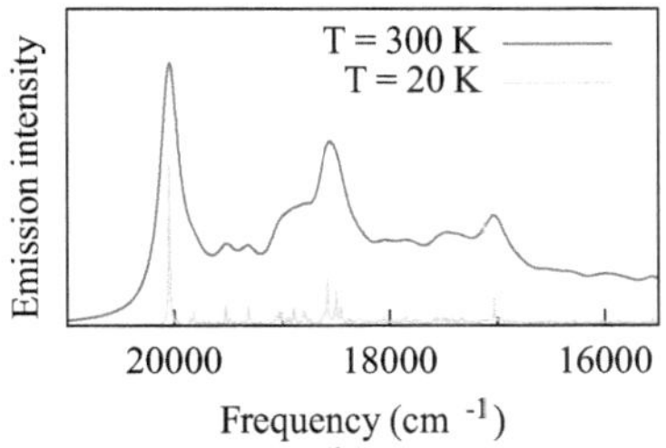

(b)

Figure 6.3 (a) Measured [Rausch et al. (2009)] and (b) calculated emission spectrum of Pt(4,6-dFppy)(acac). The calculation has been performed with ADF [Fonseca Guerra et al. (1998); te Velde et al. (2001); ADF (2013)] using the BP86 exchange-correlation potential and a double-ζ singly polarized basis set. The location of the main peak, or zero-phonon line, in the calculation (20048 cm^{-1}) is underestimated by 7% with respect to the measurement (21461 cm^{-1}). This is most likely caused by the inability of DFT to accurately calculate excited states. We thank Kento Mori for providing this example.

Franck–Condon factors. Figure 6.3b shows the calculated spectrum, where the effect of temperature has been simulated by broadening the peaks by $k_B T$. The calculated spectrum shows good agreement with the measurement (up to a constant shift in energy). Note that the zero-phonon peak is dominant as none of the electron–phonon couplings is larger than 1.

6.4 Vibrational Modes in Transport: Weak-Coupling Regime

When analyzing the effects of vibrational excitations on a molecule in a junction, we distinguish between the weak-coupling and the strong-coupling regime. In the weak-coupling regime, the electronic coupling between the molecule and the electrodes (determined by the overlap of the electronic wave-functions) is too small for electrons to coherently travel from one lead to the other. Instead, they tunnel onto the molecule, are localized there for a while, and dephase before tunneling off again. Due to the small size of a molecule, quantum level splitting and Coulomb repulsion prevent multiple electrons from entering the molecule simultaneously, hence the dominant transport mechanism is sequential tunneling.

Transport through weakly coupled junctions is generally modeled with the master equation approach [Beenakker (1991); Bonet et al. (2002); Timm (2008)]. We will describe this method below, including implementation details for junctions with both single [Boese and Schoeller (2001); McCarthy et al. (2003); Braig and Flensberg (2003); Mitra et al. (2004); Koch et al. (2004); Wegewijs and Nowack (2005); Koch et al. (2006)] and multiple vibrational modes [Chang et al. (2007); Seldenthuis et al. (2008, 2010)].

6.4.1 *Master Equation*

The central quantity in the master equation formalism is the occupation probability of the many-body states of the system. This quantity, denoted by P_i for a certain many-body state i, corresponds to the diagonal of the density matrix and is a measure of the probability that the system will be in that state at time t. Here, i is a general index combining all relevant electronic and vibrational quantum numbers. The time evolution of the occupation probability is governed by the master equation [Beenakker (1991); Bonet et al. (2002); Timm (2008)]:

$$\frac{\mathrm{d}P_i}{\mathrm{d}t} = \sum_{i' \neq i} \left(W_{i' \to i} P_{i'} - W_{i \to i'} P_i \right), \tag{6.16}$$

where $W_{i' \to i}$ is the transition rate from state i' to state i. This can be written in matrix-vector form, where the indices of the initial and final states (i' and i) correspond to the columns and rows, respectively:

$$\dot{\mathbf{P}} = \mathbf{W}\mathbf{P}.$$

The off-diagonal elements are the transition rates, while the diagonal elements (the second term on the right-hand side in Eq. 6.16) contain minus the sum of the corresponding columns. The steady-state situation corresponds to $\dot{\mathbf{P}} = \mathbf{0}$, in which case $\mathbf{P}$ is the null space of $\mathbf{W}$. Since the sum of the columns of $\mathbf{W}$ is zero, the null space is guaranteed to exist.

6.4.1.1 Transition rates

In the weak-coupling regime, the transition rates can be obtained from Fermi's golden rule (see Eq. 6.10). In the case of charging the molecule from the leads, the transition rate is determined by the position of the chemical potential of the charging process with respect to the chemical potential of the leads. This chemical potential is defined as the discrete derivative $\mu = \frac{\mathrm{d}E}{\mathrm{d}N}$, and is equal for both the charging and discharging process:

$$\mu_{i \to i'} = \frac{E_{i'} - E_i}{N_{i'} - N_i} = \frac{E_i - E_{i'}}{N_i - N_{i'}} = \mu_{i' \to i}.$$

If the junction is symmetrically biased, the chemical potential of the source and drain electrode is given by $\mu_{\mathrm{S,D}} = E_{\mathrm{F}} \mp \frac{1}{2}eV_b$, where E_{F} is the Fermi energy of the leads and V_b is the bias voltage. In an asymmetric junction, the chemical potential of the charging process can shift along with the bias voltage. Additionally, a gate electrode, with gate voltage V_g, can shift the chemical potential up and down. In general, the effective chemical potential is given by [Thijssen and van der Zant (2008)]

$$\mu_{i \to i'} = \frac{E_{i'} - E_i}{N_{i'} - N_i} + \alpha V_b + \beta V_g,$$

where α and β are the coupling to the bias and gate voltage, respectively. If there is no chemical potential corresponding to a charging process inside the bias window, no current can flow. This effect is called Coulomb blockade. It can be lifted by applying a gate voltage aligning the chemical potential with that of the leads.

For the charging process (from the source electrode), the transition rate is

$$W^{\mathrm{S}}_{i\to i'} = \mathcal{F}_{i,i'} \frac{\Gamma_{\mathrm{S}}}{\hbar} f_{\mathrm{S}} \left(\mu_{i\to i'}\right),$$

where $\mathcal{F}_{i,i'}$ is the Franck–Condon factor and

$$f(\epsilon) = \frac{1}{e^{\frac{\epsilon-\mu}{k_{\mathrm{B}}T}} + 1}$$

is the Fermi distribution on the lead. The latter ensures that the transition can only occur if there is an electron available on the lead. Γ_{S} is the coupling to the lead in units of energy, while $\Gamma_{\mathrm{S}}/\hbar$ is the maximum possible transition rate. Similarly, the rate for the discharging process (to the drain electrode) is given by

$$W^{\mathrm{D}}_{i\to i'} = \mathcal{F}_{i,i'} \frac{\Gamma_{\mathrm{D}}}{\hbar} \left(1 - f_{\mathrm{D}} \left(\mu_{i\to i'}\right)\right).$$

It can only occur when the destination for the electron on the lead is unoccupied. Note that each lead can act as a source and a drain depending on whether the chemical potential is above or below the Fermi energy. Also, at finite temperatures, the electrodes are partially occupied around the Fermi energy, and they will act as both source and drain for the same chemical potential.

In addition to charging and discharging, neutral excitations can occur. The transition rate for a process involving the absorption or emission of a photon is

$$W^{\mathrm{P}}_{i\to i'} = \mathcal{F}_{i,i'} \frac{\omega^3}{3\pi\epsilon_0\hbar c^3} \left|\langle i|\hat{\mu}|i'\rangle\right|^2,$$

where ω is the frequency of the photon and $\hat{\mu}$ is the transition dipole moment. This rate can be used to describe, e.g., photoconductance or electroluminescence [Seldenthuis et al. (2008)].

Finally, due to the coupling of the vibrational modes of the molecule with the phonon bath of the leads, the vibrational excitations decay over time. Splitting the general index i into an electronic quantum number n and a vibrational quantum number ν, we have the following relaxation process (see Eq. 6.16) [Koch et al. (2004); Wegewijs and Nowack (2005)]:

$$\frac{\mathrm{d}P_{n\nu}}{\mathrm{d}t} = \ldots - \frac{1}{\tau_{n\nu}} \left(P_{n\nu} - P^{\mathrm{eq}}_{n\nu} \sum_{\nu'} P_{n\nu'} \right),$$

where $\tau_{n\nu}$ is the relaxation time and

$$P_{n\nu}^{\text{eq}} = \frac{e^{-\frac{E_{n\nu}}{k_\text{B}T}}}{\sum_{\nu'} e^{-\frac{E_{n\nu'}}{k_\text{B}T}}}$$

is the equilibrium occupation of the vibrational excitations according to the Boltzmann distribution. Since $\sum_{\nu} P_{n\nu}^{\text{eq}} = 1$, this process can be described by the transition rate

$$W_{n\nu\to n\nu'}^{\text{R}} = \frac{1}{\tau_{n\nu}} P_{n\nu'}^{\text{eq}}.$$

The limiting cases of $\tau_{n\nu} \to 0$ and $\tau_{n\nu} \to \infty$ describe equilibrated and unequilibrated phonons, respectively [Koch et al. (2004)].

Note that all transition rates except relaxation are proportional to the Franck–Condon factors. The rate matrix therefore consists of blocks: each off-diagonal block corresponds to an electronic transition, while the elements inside that block are the vibrational transition rates given by the Franck–Condon factors. Within a particular electronic state the only transition process is vibrational relaxation, the rates of which make up the diagonal blocks of the rate matrix. Finally, the actual diagonal of the rate matrix contains the sum of the columns.

6.4.1.2 Calculating the properties of interest

Once the transition rates have been determined from Fermi's golden rule, the steady-state occupation probabilities can be obtained by calculating the null space of the rate matrix (see next section for details). The properties of interest can then be obtained from the occupation probabilities. The average charge or spin state can be calculated simply by performing a weighted sum of the occupation probabilities. Properties involving the transition rates, such as the current, or the amount of absorbed or emitted light, are obtained from the rate matrices. For example, the electroluminescence intensity (measured as the number of emitted photons), can be calculated with

$$L = \sum_{i,i'} W_{i\to i'}^{\text{E}} P_i,$$

where $W_{i\to i'}^{\text{E}}$ is the photo-emission rate.

Calculating the current is slightly more involved, since we have to distinguish between electrons hopping from the lead onto the molecule and *vice versa*. At steady state, the current from the left lead to the molecule is equal to the current from the molecule to the right lead (Kirchhoff's law). We will therefore focus on the left lead. Since electrons are negatively charged, the current *from* the left lead to the molecule is equal to the rate of electrons hopping from the molecule *to* the left lead minus the rate of those hopping onto the molecule. If we combine the charging and discharging rates ($\mathbf{W}^{\mathrm{S}}$ and $\mathbf{W}^{\mathrm{D}}$) of the left lead into a single matrix $\mathbf{W}^{\mathrm{L}}$, and number the states (with index i) in order of increasing number of electrons on the molecule, then the upper-diagonal blocks of $\mathbf{W}^{\mathrm{L}}$ corresponds to electrons hopping from the molecule to the lead, and the lower-diagonal blocks to electrons hopping from the lead to the molecule. In matrix-vector form, the current from the left lead can then be calculated with

$$I^{\mathrm{L}} = e\,\mathrm{sum}\left(\left[\mathrm{tril}\left(\mathbf{W}^{\mathrm{L}}\right) - \mathrm{triu}\left(\mathbf{W}^{\mathrm{L}}\right)\right]\mathbf{P}\right),$$

where $\mathrm{tril}\left(\mathbf{W}^{\mathrm{L}}\right)$ and $\mathrm{triu}\left(\mathbf{W}^{\mathrm{L}}\right)$ are the lower- ad upper-diagonal parts of $\mathbf{W}^{\mathrm{L}}$, respectively, and we take e to be positive.

6.4.1.3 Numerical evaluation

Most molecules measured in a junction will have dozens, if not hundreds, of vibrational modes. Due to the combinatorial scaling of the number of Franck–Condon factors, the dimension of the rate matrices will be much larger than that. Most of the Franck–Condon factors, especially those corresponding to different vibrational modes, are close to zero. This suggest the use of a sparse storage scheme for the rate matrix. A particularly suitable storage scheme is the modified compressed sparse column (MCSC) scheme. Here, only the non-zero (i.e., larger than a certain cutoff value) rates are stored column-wise, while the diagonal is stored separately (since it can be used as a preconditioner, see below). The row indices of each element are stored in a separate array. For N_{nz} non-zero elements, $N_{\mathrm{nz}} + 1$ values and indices need to be stored. Apart from saving memory, using a sparse matrix storage scheme also speeds up the calculation, since only operations involving non-zero elements have to be performed.

Due to the large dimension of the rate matrix, direct methods for solving the null space, such as singular value decomposition (SVD), quickly become unfeasible. Also, direct methods cannot generally be performed on sparse matrices. We therefore use iterative methods. A particularly efficient method is the bi-conjugate gradient stabilized (Bi-CGSTAB) method [van der Vorst (1992)]. Like the bi-conjugate gradient (Bi-CG) and the conjugate gradient squared (CGS) methods, it is applicable to non-symmetric linear systems, but has faster and smoother convergence. Its implementation also uses less memory and tends to be simpler than the more general generalized minimal residual (GMRES) method. Like many iterative methods, the convergence of the Bi-CGSTAB method can be accelerated considerably by the use of a preconditioner. Since **W** is diagonally dominant (the diagonal elements contain minus the sum of the corresponding columns), the inverse of the diagonal is a good approximation of $\mathbf{W}^{-1}$. This is the Jacobi preconditioner, and it is one of the main motivations for using the MCSC storage scheme.

One of the most important factors determining fast convergence to a correct solution is the choice of the initial guess of the occupation probabilities of the many-body states (including both electrons and phonons). Since the system is described in terms of many-body states with different energies, we expect the equilibrium occupation probabilities of the *closed* system to be governed by the Boltzmann distribution:

$$P_i^0 = \frac{e^{-\frac{E_i}{k_B T}}}{\sum_{i'} e^{-\frac{E_{i'}}{k_B T}}}.$$

As the main motivation for the master-equation approach is weak coupling, this should be a good initial guess for the coupled system. When we need the steady-state occupation probabilities for many closely related systems, e.g., when calculating the current–voltage characteristics, the best choice for the initial guess is usually the converged result of the previous calculation. If the relevant parameter, such as the voltage, changes slowly between calculations, we expect the new occupation probabilities to differ only slightly from the old ones. In practice, we therefore use the Boltzmann distribution as the initial guess at the start of a calculation, and subsequently use the previously converged probabilities for the following points.

We noted before that the null space is guaranteed to exist. It is not, however, guaranteed to be unique. If the transition rates to and from a particular state are (close to) zero, any possible occupation of that state is stable, even though in reality it would never be occupied. This is especially problematic for highly excited vibrational states, as their Franck–Condon factors are generally small. This can partially be solved by introducing vibrational relaxation, as the vibrationally excited states then acquire a finite lifetime. In general, the problem can be solved by choosing a physically plausible initial guess. At low temperatures, e.g., the Boltzmann distribution ensures that the highly excited vibrational states will never be occupied and that the resulting steady-state occupation probabilities will be physically acceptable.

6.4.2 *Selection Rules*

6.4.2.1 Single-level model

To illustrate the effect of vibrational excitations on the current–voltage (I–V) characteristics of a weakly coupled molecular junction, we first take a look at a simple system consisting of a single electronic level coupled to a single vibrational mode. Figure 6.4 shows the stability diagrams—plots of the differential conductance ($\mathrm{d}I/\mathrm{d}V_b$) as a function of the bias (V_b) and gate (V_g) voltage—and the Franck–Condon factors of such a system for different values of the electron–phonon coupling λ.

The case $\lambda = 0$ corresponds to the situation where there is no shift in the nuclear coordinates upon charging the molecule, hence no vibrational modes are excited. This is essentially the case of a single-level model without vibrational excitations. At low bias, the current through such a junction is suppressed until the chemical potential of the charging process enters the bias window. This Coulomb blockade can be lifted with a gate by tuning the gate voltage in such a way as to make the chemical potential of the molecule resonant with the Fermi energy of the leads ($V_g = 0$ in Fig. 6.4). In the absence of vibrational excitations, there is only a single

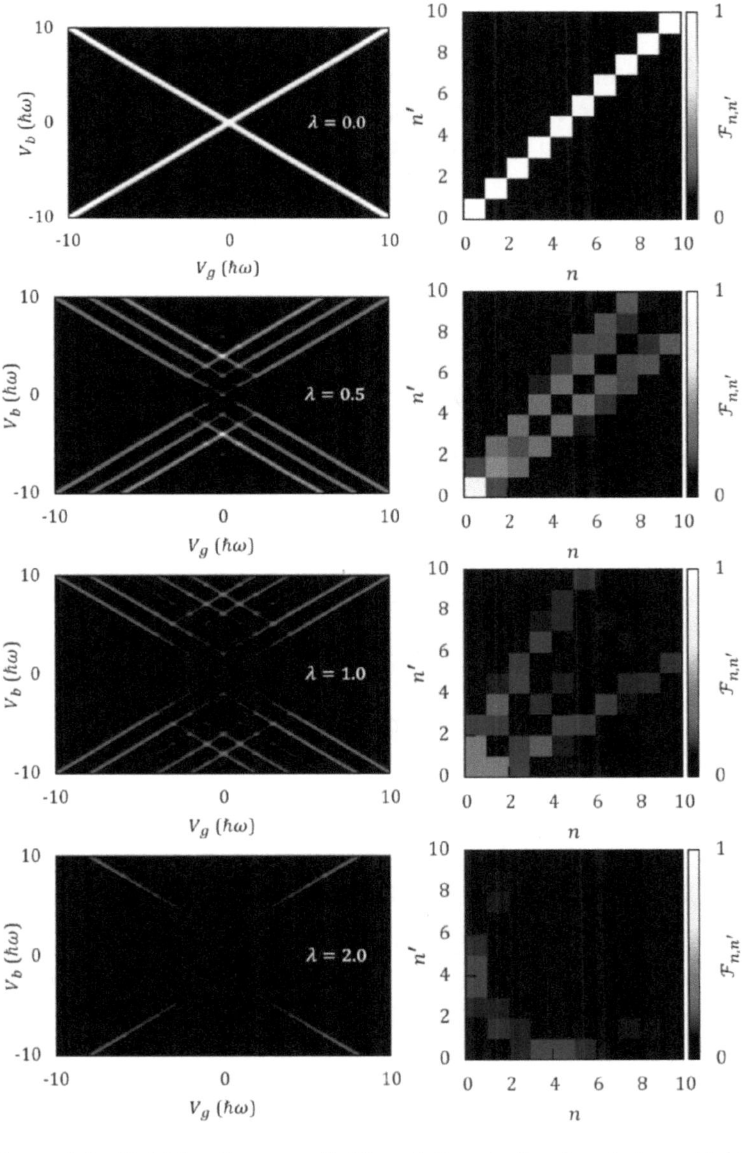

Figure 6.4 Stability diagrams (left) and Franck–Condon factors (right) of a single-molecule junction with a single vibrational mode ($\omega = \omega'$) for different values of the electron–phonon coupling λ. No vibrational relaxation is taken into account, hence the phonons are unequilibrated.

chemical potential, corresponding to the charging and discharging of the molecule in the vibrational ground state. This gives rise to two lines in the stability diagram, the so-called Coulomb diamond edges, each of which corresponds to the bias and gate voltages at which the chemical potential comes into resonance with one of the leads [Thijssen and van der Zant (2008)].

For non-zero values of λ, charging the molecule leads to a shift in the equilibrium positions of the nuclei and to the excitation of the vibrational mode. There are now several chemical potentials, and therefore several lines in the stability diagram, since the electronic transitions are now accompanied by different vibrational transitions. As long as $\lambda < 1$, the ground-state to ground-state transition still has the largest probability, and the zero-phonon line in the stability diagram (the Coulomb diamond edge) still has the highest intensity. However, for $\lambda > 1$ this is no longer the case. Electronic transitions without a vibrational excitation become exceedingly unlikely and the zero-phonon line is suppressed, a phenomenon known as Franck–Condon blockade [Koch et al. (2006)]. A striking characteristic of this type of blockade is that it cannot be lifted with a gate. This can be seen in Fig. 6.4 in the case of $\lambda = 2$, where it is shown that the current is suppressed at low bias for all gate voltages.

For a more detailed analysis of this system, see Ref. [Wegewijs and Nowack (2005)].

6.4.2.2 Example: weakly coupled OPV-5 junction

A clear measurement of the effect of vibrational modes on the I-V characteristics of a weakly coupled molecule was obtained by Osorio et al. [Osorio et al. (2007)] in the case of the oligo(phenylene vinylene) derivative OPV-5 [Kubatkin et al. (2003)], a π-conjugated molecule. The measured stability diagram is shown in Fig. 6.5a. A background conductance makes it difficult to resolve all excitation lines at a single color scale, but close inspection reveals 17 modes in the energy range below 125 meV (see Tab. 1 in Ref. [Osorio et al. (2007)]). The energies of the excitations can be determined from the bias voltage at which the excitation lines cross the diamond edge.

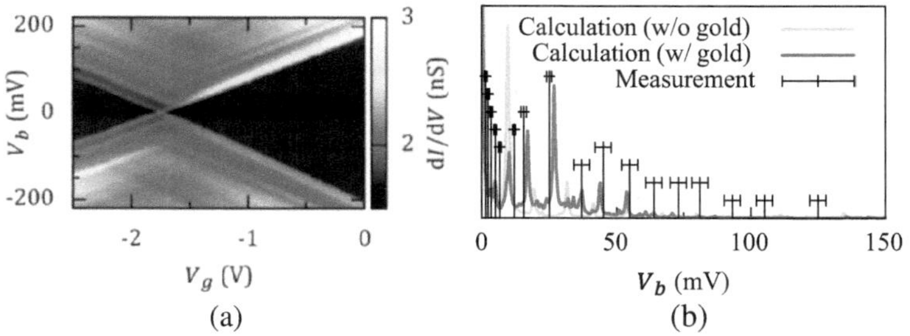

Figure 6.5 (a) Measured stability diagram of a single OPV-5 molecule [Osorio et al. (2007)]. (b) Calculated vibrational spectrum of OPV-5 with (red line) and without (green line) gold atoms on either side of the molecule (see Fig. 6.6). The calculation has been performed with ADF [Fonseca Guerra et al. (1998); te Velde et al. (2001); ADF (2013)]. The black lines are the measured excitations with the horizontal bars indicating the uncertainty in the measurement.

The calculated vibrational spectrum of the neutral charge state is shown in Fig. 6.5b (see also Ref. [Seldenthuis et al. (2008)]). For comparison, black lines show the measured excitation energies. The uncertainties in the measured values, indicated by the horizontal bars, are caused by the broadening of the excitation lines due to temperature and the coupling to the leads. The green line in Fig. 6.5b is the calculated spectrum of an isolated molecule, without gold atoms. Comparison with the measurement shows a large discrepancy for excitations below 50 meV. As these excitations correspond to bending and stretching of the entire molecule (see Fig. 6.6), they are greatly influenced by the coupling to the contacts. Adding two gold atoms to either side of the molecule can account for most of this influence on modes above 10 meV (red line in Fig. 6.5b). The omission of the non-conjugated sidearms from the calculation lowers the mass of the molecule, which might explain the discrepancy between the calculation and the measurement for the modes below 10 meV, which involve motions of the entire molecule. Also, the contact geometry in the measurement is unknown, so any mode involving a significant distortion of the gold-sulfur bond is expected to be inaccurate.

Most of the vibrational modes in the calculation have electron–phonon couplings below 0.1 and are not expected to give rise to

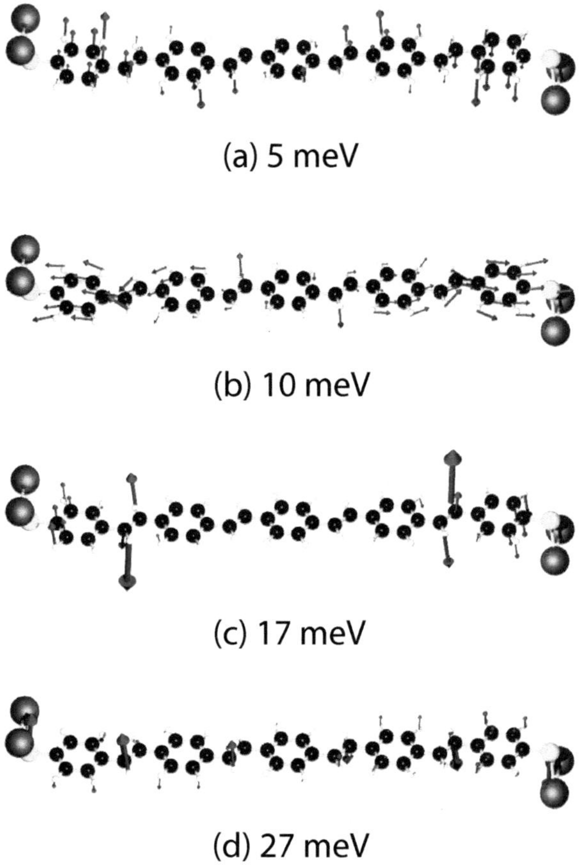

Figure 6.6 The four dominant low-energy vibrational modes in the calculated spectrum of OPV-5 (see Fig. 6.5b). The dodecane side-arms of the measured molecule (see Fig. 1 in Ref. [Osorio et al. (2007)]) are omitted in the calculation and two gold atoms have been added to either side of the molecule to simulate the presence of the leads.

extra excitation lines corresponding to higher harmonics. However, the modes at 17 and 27 meV (Fig. 6.6c and d), with electron–phonon couplings of respectively 0.6 and 0.7, are expected to give rise to excitation lines at 34, 51–54 and possibly 81 meV. These lines are indeed observed in the measurement and the calculation (see Fig. 6.5b).

In both the calculation and the measurement, only excitations below 125 meV are visible, with the dominant modes occurring below 30 meV. This can be understood by looking at the nature of the electronic transition. OPV-5 is a conjugated molecule, which means that the highest occupied molecular orbital (HOMO) and the lowest unoccupied molecular orbital (LUMO) are both delocalized over the entire molecule. Charging the molecule, which entails removing an electron from the HOMO or adding one to the LUMO, will therefore change the charge density everywhere, leading to an expansion or a contraction or bending of the entire molecule. Vibrational modes involving such a motion will therefore have the largest electron–phonon couplings (see Eq. 6.11). Since these modes involve the combined motion of (almost) all the nuclei, the relative displacements are small while the effective mass of the mode is large, hence the frequencies will be low. In the weak-coupling regime, it is therefore the low-frequency modes which will have the largest effect on the I–V characteristics. This is in marked contrast with the strong-coupling regime, as we shall see below.

It should be emphasized that in Fig. 6.5b all visible vibrational excitations, for both the calculation and the measurement, are shown. Comparing the spectrum to Raman and IR spectroscopy data reveals a close match [Osorio et al. (2007)], but the optical spectra predict many more modes observed in neither the measurement nor the calculation. The calculation, on the other hand, predicts only a handful visible excitations out of a total of a 109 vibrational modes under 125 meV. A Franck–Condon calculation is thus able to provide what we might call 'selection rules' for vibrational excitations in weakly coupled single-molecule junctions.

6.5 Vibrational Modes in Transport: Strong-Coupling Regime

In the strong coupling regime, orbitals of the molecule hybridize with those of the contacts, and electrons can travel coherently from one lead to the other. The eigenstates of the isolated molecule therefore no longer present a suitable basis for the description of transport, like in the master equation approach.

6.5.1 *Nonequilibrium Green's Function Formalism*

In this section we introduce the nonequilibrium Green's function (NEGF) approach [Meir and Wingreen (1992); Jauho et al. (1994); Haug and Jauho (1997)], which is particularly suitable for describing transport through single-molecule junctions [Datta (1995); Taylor et al. (2001); Derosa and Seminario (2001); Xue et al. (2002)] in the strong-coupling regime. As this formalism is most conveniently expressed in terms of second quantization, we will first briefly discuss this notation. We then give an overview of the NEGF approach to elastic transport, i.e., in the absence of vibrational excitations, before showing one of the ways to incorporate inelastic tunneling.

Although the NEGF approach is commonly applied to strongly coupled systems, it is a general formalism for describing many-body processes and therefore in principle also applicable to weakly coupled systems. At the end of this section we show how, by taking the appropriate weak-coupling limits, it is possible to rederive the Franck–Condon factors within second quantization.

6.5.1.1 Second quantization

Using second quantization [Mahan (1981)], we can write the electronic Hamiltonian (Eq. 6.1) of an isolated molecule as

$$\hat{H}_m = \sum_i \epsilon_i \hat{d}_i^\dagger \hat{d}_i + \sum_{i,j} \tau_{ij} \hat{d}_i^\dagger \hat{d}_j + \tfrac{1}{2} \sum_{i,j,k,l} U_{ijkl} \hat{d}_i^\dagger \hat{d}_j^\dagger \hat{d}_k \hat{d}_l. \tag{6.17}$$

$\hat{d}_i^\dagger$ and $\hat{d}_i$ are the creation and annihilation operators for electrons in the single-particle states, or molecular orbitals, with energy ϵ_i. They obey the fermionic anti-commutation relations

$$\left\{\hat{d}_i, \hat{d}_j^\dagger\right\} = \delta_{ij},$$
$$\left\{\hat{d}_i, \hat{d}_j\right\} = \left\{\hat{d}_i^\dagger, \hat{d}_j^\dagger\right\} = 0.$$

The single-particle and two-particle interactions are described by τ_{ij} and U_{ijkl}, respectively. In a junction, the molecule is coupled to contacts via the Hamiltonian

$$\hat{H}_c = \sum_k \epsilon_k \hat{c}_k^\dagger \hat{c}_k + \sum_{ik} V_{ki} \hat{c}_k^\dagger \hat{d}_i + \text{h.c.} \tag{6.18}$$

$\hat{c}_k^\dagger$ and $\hat{c}_k$ are the creation and annihilation operators of the *non-interacting* electrons in the contacts, while V_{ik} describes the coupling to the molecule.

In the weak-coupling regime, vibrational excitations facilitate transitions between different electronic states. In the strong-coupling regime, on the other hand, transport is dominated by coherent processes. Vibrational excitations are inelastic perturbations, caused by the dependence of the parameters of the electronic Hamiltonian (ϵ_i, τ_{ij}, *etc.*) on the nuclear coordinates. When these perturbations are taken into account as first-order corrections to elastic transport, this leads to a markedly different effect on the I-V characteristics than in the weak-coupling regime.

6.5.1.2 Elastic transport

In the weak-coupling regime, transport is dominated by sequential tunneling, a first-order process that can be described by Fermi's golden rule. In the strong-coupling regime, first-order perturbation theory in the coupling to the leads is no longer sufficient and we need a different approach. A general method to take higher-order processes into account is the nonequilibrium Green's function (NEGF) formalism [Meir and Wingreen (1992); Jauho et al. (1994); Haug and Jauho (1997)]. We shall briefly introduce this method without including inelastic processes.

In the absence of capacitive interactions, i.e., $U_{ijkl} = 0$ in Eq. 6.17, the elastic current through a strongly coupled junction is given by the Landauer formula:

$$I = \frac{2e}{h} \int_{-\infty}^{\infty} \mathrm{d}\epsilon \, [f_{\mathrm{R}}(\epsilon) - f_{\mathrm{L}}(\epsilon)] \, \mathcal{T}(\epsilon), \tag{6.19}$$

where $f_{\mathrm{L}}(\epsilon)$ and $f_{\mathrm{R}}(\epsilon)$ are the Fermi distribution on the left and right lead, respectively, and $\mathcal{T}(\epsilon)$ is the transmission probability of an electron with energy ϵ. In the low-temperature limit, the Fermi distribution becomes a step function and the integral reduces to

$$I = \frac{2e}{h} \int_{\mu_{\mathrm{L}}}^{\mu_{\mathrm{R}}} \mathrm{d}\epsilon \, \mathcal{T}(\epsilon).$$

From this we see that only electrons within the bias window contribute to the current.

The transmission of a junction can be obtained from the NEGF formalism via

$$\mathcal{T}(\epsilon) = \mathrm{Tr}\left\{\mathbf{\Gamma}^{\mathrm{L}}(\epsilon)\mathbf{G}^{r}(\epsilon)\mathbf{\Gamma}^{\mathrm{R}}(\epsilon)\mathbf{G}^{a}(\epsilon)\right\}, \tag{6.20}$$

where $\mathbf{\Gamma}^{\mathrm{L,R}}(\epsilon)$ describes the coupling to the leads (see below), and $\mathbf{G}^{r}(\epsilon)$ and $\mathbf{G}^{a}(\epsilon)$ are, respectively, the retarded and advanced Green's function of the transport region. The former is given by

$$\mathbf{G}^{r}(\epsilon) = \left[\epsilon\mathbf{1} - \mathbf{H} - \mathbf{\Sigma}^{\mathrm{L}}(\epsilon) - \mathbf{\Sigma}^{\mathrm{R}}(\epsilon)\right]^{-1},$$

while $\mathbf{G}^{a}(\epsilon) = [\mathbf{G}^{r}(\epsilon)]^{\dagger}$. $\mathbf{H}$ is the single-particle Hamiltonian of the molecule (Eq. 6.17 with $U_{ijkl} = 0$):

$$H_{ij} = \epsilon_i\delta_{ij} + \tau_{ij},$$

The effect of the (non-interacting) leads is described by the (retarded) self-energy (see Eq. 6.18)

$$\Sigma_{ij}^{\mathrm{L,R}}(\epsilon) = \lim_{\eta\to 0}\sum_{k}\frac{V_{ki}^{\mathrm{L,R}*}V_{kj}^{\mathrm{L,R}}}{\epsilon - \epsilon_k + \mathrm{i}\eta}.$$

The self-energy can be split into a Hermitian part $\mathbf{\Lambda}^{\mathrm{L,R}}(\epsilon)$ and an anti-Hermitian part $\mathbf{\Gamma}^{\mathrm{L,R}}(\epsilon)$, which we encountered in Eq. 6.20:

$$\mathbf{\Sigma}^{\mathrm{L,R}}(\epsilon) = \mathbf{\Lambda}^{\mathrm{L,R}}(\epsilon) - \frac{\mathrm{i}}{2}\mathbf{\Gamma}^{\mathrm{L,R}}(\epsilon).$$

$\mathbf{\Lambda}^{\mathrm{L,R}}(\epsilon)$ shifts the molecular resonances while $\mathbf{\Gamma}^{\mathrm{L,R}}(\epsilon)$ induces a broadening. The latter is given by

$$\Gamma_{ij}^{\mathrm{L,R}}(\epsilon) = 2\pi\sum_{k}V_{ki}^{\mathrm{L,R}*}V_{kj}^{\mathrm{L,R}}\delta\left(\epsilon - \epsilon_k\right).$$

The sum of delta functions shows that $\mathbf{\Gamma}^{\mathrm{L,R}}(\epsilon)$ is proportional to the density of states (DOS) in the leads. For metals such as gold, the DOS is a slowly varying function near the Fermi energy. It is therefore common to take $\mathbf{\Gamma}^{\mathrm{L,R}}$ and $\mathbf{\Lambda}^{\mathrm{L,R}}$ to be constant, which greatly simplifies the calculations. This approximation is known as the wide-band limit (WBL).

It seems natural to identify the transport region with the molecule. However, in a strongly coupled junction, the molecular orbitals hybridize with those of the leads, and the self-energy at the metal-molecule interface will depend strongly on the contact geometry. It is therefore common to include part of the electrodes into a so-called "extended molecule". The self-energies then describe the metal-metal coupling deeper inside the leads.

6.5.1.3 Inelastic transport

So far we have only considered elastic transport processes; electrons enter the drain electrode with the same energy with which they left the source. It is, of course, also possible for electrons to gain or lose energy on the molecule by coupling to phonons. We therefore have to generalize Eqs. 6.19 and 6.20 to include inelastic processes. There are several ways to do this [Galperin et al. (2007)], the most common of which is the self-consistent Born approximation (SCBA) [Frederiksen et al. (2007)]. Here, however, we will use the scattering, or Tersoff–Haman [Tersoff and Hamann (1983, 1985)], approach which is primarily applicable in the off-resonance regime. Its main advantage is that it scales well and can easily be applied to molecules with a large number of vibrational modes.

First, we note that Eq. 6.19 can be written as

$$I = \frac{2e}{h} \int_{-\infty}^{\infty} \mathrm{d}\epsilon\, \mathcal{T}(\epsilon)\, [f_{\mathrm{R}}(\epsilon)\,(1 - f_{\mathrm{L}}(\epsilon)) - f_{\mathrm{L}}(\epsilon)\,(1 - f_{\mathrm{R}}(\epsilon))]\,.$$

Since electrons carry negative charge, positive current from left to right equals electrons traveling from right to left. The first term in brackets says that the current is proportional to the availability of electrons on the right lead, given by $f_{\mathrm{R}}(\epsilon)$, and the availability of holes for the electrons to disappear into on the left electrode, given by $1 - f_{\mathrm{L}}(\epsilon)$. The second term says that the current is reduced by electrons traveling in the opposite direction.

This expression is readily generalized to include inelastic processes, in which case electrons can enter the molecule with a different energy (ϵ) than with which they leave it (ϵ'). The transmission now depends on both ϵ and ϵ' and the current is given by

$$\begin{aligned} I = \frac{2e}{h} \int_{-\infty}^{\infty} \mathrm{d}\epsilon \int_{-\infty}^{\infty} \mathrm{d}\epsilon' \big[& \mathcal{T}^{\mathrm{R}\to\mathrm{L}}\left(\epsilon, \epsilon'\right) f_{\mathrm{R}}(\epsilon)\left(1 - f_{\mathrm{L}}\left(\epsilon'\right)\right) \\ & - \mathcal{T}^{\mathrm{L}\to\mathrm{R}}\left(\epsilon, \epsilon'\right) f_{\mathrm{L}}(\epsilon)\left(1 - f_{\mathrm{R}}\left(\epsilon'\right)\right) \big]. \end{aligned}$$

For elastic processes: $\mathcal{T}_{\mathrm{el}}^{\mathrm{R}\to\mathrm{L}}(\epsilon, \epsilon') = \mathcal{T}_{\mathrm{el}}^{\mathrm{L}\to\mathrm{R}}(\epsilon, \epsilon') = \mathcal{T}(\epsilon)\delta(\epsilon - \epsilon')$. For simple models it is possible to calculate the inelastic transmission exactly [Wingreen et al. (1988); Jonson (1989); Wingreen et al. (1989)], but for real molecules with many vibrational modes we have to resort to perturbation theory. Here we follow the approach of Troisi et al. [Troisi et al. (2003)]. First, using the cyclic property of the trace we write the elastic transmission (Eq. 6.20) as [Solomon et al. (2008)]

$$\mathcal{T}(\epsilon) = \mathrm{Tr}\left\{\mathbf{T}(\epsilon)\mathbf{T}^{\dagger}(\epsilon)\right\},$$

where

$$\mathbf{T}(\epsilon) = \sqrt{\mathbf{\Gamma}^{\mathrm{L}}(\epsilon)}\mathbf{G}^{r}(\epsilon)\sqrt{\mathbf{\Gamma}^{\mathrm{R}}(\epsilon)}.$$

$T_{ij}(\epsilon)$ can be interpreted as the (complex) transmission from atomic orbital i on the left interface to orbital j on the right. This is the quantity we generalize to include inelastic processes. The inelastic transmission of an electron is accompanied by a vibrational transition on the molecule. If $|\mathbf{n}\rangle$ and $|\mathbf{n}'\rangle$ are the initial and final vibrational states, conservation of energy requires that

$$\epsilon' - \epsilon = \sum_i \hbar\omega_i \left(n_i' - n_i\right).$$

For an electron traveling from left to right, the complex inelastic transmission is given by

$$\mathbf{T}_{\mathrm{inel}}^{\mathrm{L}\to\mathrm{R}}\left(\epsilon, \epsilon'\right) = \sqrt{\mathbf{\Gamma}^{\mathrm{L}}(\epsilon)}\left\langle \mathbf{n}\middle|\hat{\mathbf{G}}_{\mathrm{inel}}^{r}\left(\epsilon''\right)\middle|\mathbf{n}'\right\rangle\sqrt{\mathbf{\Gamma}^{\mathrm{R}}\left(\epsilon'\right)},$$

where

$$\epsilon'' = \epsilon + \sum_i \hbar\omega_i n_i = \epsilon' + \sum_i \hbar\omega_i n_i'.$$

Note that in the wide-band limit, both $\mathbf{\Gamma}^{\mathrm{L}}$ and $\mathbf{\Gamma}^{\mathrm{R}}$ are independent of ϵ, hence $\mathbf{T}_{\mathrm{inel}}^{\mathrm{R}\to\mathrm{L}}\left(\epsilon, \epsilon'\right) = \mathbf{T}_{\mathrm{inel}}^{\mathrm{L}\to\mathrm{R}}\left(\epsilon, \epsilon'\right)$.

To first order in the nuclear displacements, the inelastic retarded Green's function is

$$\begin{aligned}\hat{\mathbf{G}}_{\mathrm{inel}}^{r}(\epsilon) &= \sum_i \left.\frac{\partial \mathbf{G}^{r}(\epsilon, \mathbf{q})}{\partial q_i}\right|_{\mathbf{q}=\mathbf{0}} q_i \\ &\approx \mathbf{G}_{\mathrm{el}}^{r}(\epsilon)\sum_i \left[\left.\frac{\partial \mathbf{H}(\mathbf{q})}{\partial q_i}\right|_{\mathbf{q}=\mathbf{0}} q_i\right]\mathbf{G}_{\mathrm{el}}^{r}(\epsilon) \\ &= \mathbf{G}_{\mathrm{el}}^{r}(\epsilon)\left[\sum_i \boldsymbol{\lambda}^{i}\hbar\omega_i\left(\hat{a}_i + \hat{a}_i^{\dagger}\right)\right]\mathbf{G}_{\mathrm{el}}^{r}(\epsilon), \end{aligned} \tag{6.21}$$

where

$$\lambda^i_{jk} = \frac{1}{2\hbar\omega_i} \frac{\partial H_{jk}(\mathbf{q})}{\partial q_i}\bigg|_{\mathbf{q}=\mathbf{0}} \tag{6.22}$$

the electron–phonon coupling in the strong-coupling regime. This definition of the electron–phonon coupling is very different from the λ used in the calculation of Franck–Condon factors (Eq. 6.15), which is proportional to the nuclear displacement due to an electronic transition. Note that although the first derivative of the total energy with respect to the normal coordinates vanishes at the equilibrium geometry, the first derivatives of the elements of the electronic Hamiltonian are generally non-zero.

Since q_i appears to first order in $\hat{\mathbf{G}}^r_{\text{inel}}(\epsilon)$, only initial and final states differing by a single vibrational quantum contribute to the inelastic transmission:

$$\left\langle \mathbf{n} \middle| \hat{a}_i + \hat{a}_i^\dagger \middle| \mathbf{n}' \right\rangle = \left(\delta_{n_i,n_i'-1}\sqrt{n_i+1} + \delta_{n_i,n_i'+1}\sqrt{n_i} \right) \prod_{j\neq i} \delta_{n_j,n_j'}.$$

If we now assume the molecule to be in the vibrational ground state before each tunneling event, i.e., $|\mathbf{n}\rangle = |\mathbf{0}\rangle$, the inelastic transmission acquires the following simple form:

$$\mathcal{T}^{\text{L}\to\text{R}}_{\text{inel}}\left(\epsilon, \epsilon'\right) = \sum_i \delta\left(\epsilon - \epsilon' - \hbar\omega_i\right) \mathcal{T}^{\text{L}\to\text{R}}_i\left(\epsilon\right),$$

where

$$\mathcal{T}^{\text{L}\to\text{R}}_i\left(\epsilon\right) = \hbar^2\omega_i^2 \,\text{Tr}\left\{ \mathbf{G}^r_{\text{el}}(\epsilon)\boldsymbol{\lambda}^i\mathbf{G}^r_{\text{el}}(\epsilon)\boldsymbol{\Gamma}^{\text{L}}(\epsilon)\mathbf{G}^a_{\text{el}}(\epsilon)\boldsymbol{\lambda}^i\mathbf{G}^a_{\text{el}}(\epsilon)\boldsymbol{\Gamma}^{\text{R}}\left(\epsilon - \hbar\omega_i\right) \right\} \tag{6.23}$$

is the contribution from mode i. The relative importance of each vibrational mode can be expressed with the ratio

$$\mathcal{R}^{\text{L}\to\text{R}}_i(\epsilon) = \frac{\mathcal{T}^{\text{L}\to\text{R}}_i(\epsilon)}{\mathcal{T}^{\text{L}\to\text{R}}_{\text{el}}(\epsilon)}. \tag{6.24}$$

In the off-resonance regime, inelastic processes only serve to increase the total transmission. They therefore appear as positive steps in the differential conductance $\text{d}I/\text{d}V_b$ at voltages where the bias window equals $eV_b = \hbar\omega_i$, and as peaks in the second derivative $\text{d}^2I/\text{d}V_b^2$. These peaks are used in inelastic tunneling spectroscopy (IETS) to identify molecules in a junction. Eq. 6.24 can be used to estimate the relative intensity of the peaks. Close to an electronic resonance, on the other hand, it is known that inelastic processes can induce a renormalization of the elastic transmission, leading to a dip in the IETS signal [Galperin et al. (2004)]. This is, however, beyond the scope of this chapter.

6.5.2 *Selection Rules*

While in the weak-coupling regime vibrational excitations lead to peaks in the differential conductance during resonant transport, in the strong-coupling regime they lead to steps in the off-resonance conductance, and therefore peaks in the $\mathrm{d}^2I/\mathrm{d}V^2$. Moreover, the selection rules that determine which modes appear in the spectrum are different. In the weak-coupling regime, modes are excited by nuclear displacements due to charging. As we showed in the previous section, this means that the low-frequency modes are generally dominant. In the strong-coupling regime, on the other hand, the peak intensities are largest for modes that modify the pathway of the electrons, e.g., by stretching the carbon–carbon bonds in a π-conjugated molecule [Troisi and Ratner (2006)].

6.5.2.1 Example: strongly coupled OPE-3 junction

To illustrate the difference with the weak-coupling regime, we will look in detail at the IETS spectrum of a typical π-conjugated molecule: the oligo(phenylene ethynylene) derivative OPE-3 (see Fig. 6.7). For a strongly coupled molecule we expect the contact geometry to have a significant influence on the I–V characteristics. We therefore study two binding configurations: the bridge-site configuration, where the thiol groups bind to two gold atoms and the molecule is nearly parallel to the surface (Fig. 6.7c), and the hollow-site configuration, where the molecule is perpendicular to the surface and the thiol groups bind to three gold atoms in a tetrahedral configuration (Fig. 6.7d). For both configurations, the extended molecule used for the transport calculations includes 27 gold atoms of the leads on either side. This corresponds to three atomic layers, which should provide sufficient screening from the bulk electrodes.

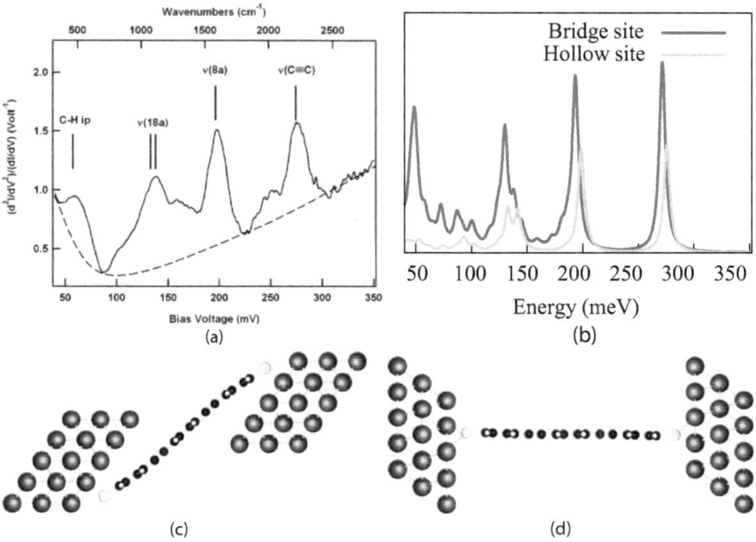

Figure 6.7 (a) Measured [Kushmerick et al. (2004)] and (b) calculated IETS spectrum of OPE-3. The calculation has been performed with ADF [Fonseca Guerra et al. (1998); te Velde et al. (2001); ADF (2013)]. The peaks in (b) have been calculated with Eq. 6.24 and broadened by 77 K for clarity. The red and green lines correspond to a bridge-site (c) and hollow-site (d) binding configuration, respectively. The four dominant modes in the bridge-site configuration are shown in Fig. 6.8.

The calculated IETS spectrum for both configurations is shown in Fig. 6.7b. The spectrum is constructed by broadening the peaks obtained from Eq. 6.24 at the Fermi energy by 77 K for clarity. Both spectra show several peaks below 50 meV, but only the bridge-site configuration (red line) has a peak at 49 meV. Above 100 meV the two spectra are similar, each showing three peaks around 135, 200 and 275 meV, although the peak around 135 meV is weaker in the hollow-site configuration (green line). The vibrational modes corresponding to the four largest peaks in the bridge-site configuration are shown in Fig. 6.8. With the exception of the mode at 49 meV, these are also the modes responsible for the peaks in the hollow-site configuration.

A close look at the nuclear displacements in Fig. 6.8 reveals that the dominant vibrational modes all involve the stretching of

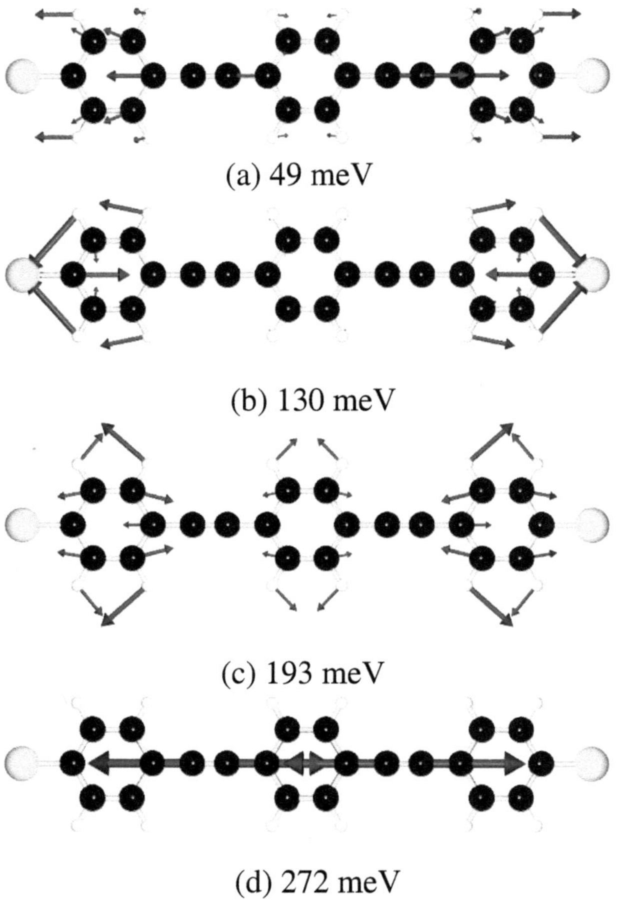

Figure 6.8 The four dominant vibrational modes in the IETS spectrum of OPE-3 in the bridge-site configuration (Fig. 6.7c). The modes involve the stretching of (a) the ethynyl C–C bond, (b) the C–S bond, (c) the phenyl C=C bond, and (d) the ethynyl C≡C bond.

π-bonds: the ethynyl C–C bond at 49 meV, the C–S bond at 130 meV, the phenyl C=C bond at 193, and the ethynyl C≡C bond at 272 meV. Note that modes involving stronger bonds appear at higher frequencies, as would be expected. As the mode at 130 meV involves the thiols, we would expect this mode to be primarily affected by the contact geometry, and indeed it is more dominant in the bridge-site than in the hollow-site configuration. However, this is a little misleading, as Fig. 6.7b shows the ratio of the inelastic and elastic transmissions (Eq. 6.24). The actual contribution of the mode at 130 meV to the inelastic transmission is nearly the same for both configurations, but the elastic transmission is three times larger for the hollow-site configuration than for the bridge-site, leading to a lower ratio for the former.

Comparison of the calculated spectrum to the measurement by Kushmerick et al. (2004) (see Fig. 6.7a) and previous calculations [Troisi and Ratner (2005); Paulsson et al. (2006)] shows good agreement. Both the calculation and measurement show three dominant peaks above 100 meV with the same relative intensity. Kushmerick et al. assign the peak at 274 meV to the C≡C bond stretch (Fig. 6.8d) and the peaks at 138 and 196 meV to phenyl breathing modes (Fig. 6.8b and c). These are the same modes obtained from the calculation, although we ascribe the peaks in the spectrum to the C–S and C=C bond stretches resulting from the phenyl breathing mode rather than to the breathing mode itself.

The measured and calculated IETS spectra of OPE-3 are markedly different from the measured and calculated spectra of the similar OPV-5 molecule in the weak-coupling regime (*cf.* Fig. 6.5). For comparison, we also calculate the Franck–Condon spectrum of OPE-3 in the weak-coupling regime (see Fig. 6.9). The result bears no resemblance to the IETS spectra in Fig. 6.7. Like OPV-5, only the modes below 50 meV give rise to a large electron–phonon coupling, while almost nothing is visible above 50 meV. This clearly demonstrates the difference between the effects of vibrational excitations in both regimes.

6.5.3 *Franck–Condon Factors Revisited*

As noted above, second quantization and the NEGF formalism are not limited to strongly coupled systems. We will demonstrate this here by rederiving the Franck–Condon factors for a simple system within second quantization. This will also illustrate the link between the definitions of the electron–phonon coupling in both regimes (Eqs. 6.15 and 6.22).

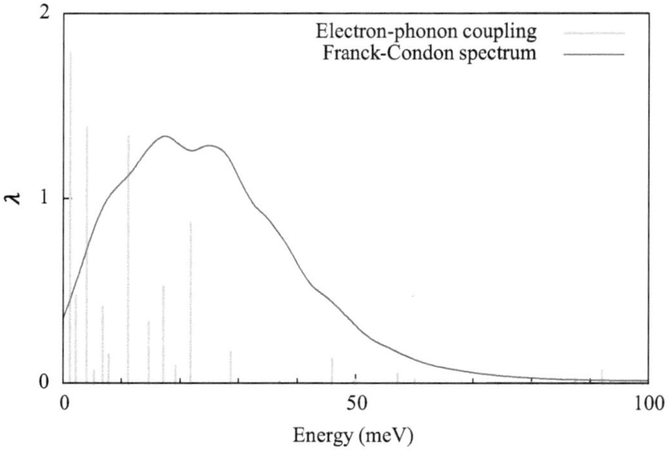

Figure 6.9 Calculated electron–phonon couplings (green lines) and Franck–Condon spectrum (red line) of a single OPE-3 molecule in the weak-coupling regime. As in the OPV-5 calculation (Figs. 6.5 and 6.6), two gold atoms are added to either side of the molecule to simulate the presence of the leads. The Franck–Condon spectrum takes up to four vibrational quanta into account and has been broadened by 77 K for clarity.

To first order, the dependence of the electronic Hamiltonian on the nuclear coordinates q_i (Eq. 6.4) is linear. Recalling the fact that $q_i = \left(\hat{a}_i + \hat{a}_i^\dagger\right)/2$ (see Eq. 6.6), the Hamiltonian of a single orbital coupled to a single vibrational mode is given by

$$\hat{H}_m = \epsilon_0 \hat{d}^\dagger \hat{d} + \hbar\omega \left(\hat{a}^\dagger \hat{a} + \tfrac{1}{2}\right) + \lambda\hbar\omega \left(\hat{a} + \hat{a}^\dagger\right) \hat{d}^\dagger \hat{d}. \qquad (6.25)$$

The first two terms are simply the electronic (Eq. 6.17) and nuclear (Eq. 6.5) Hamiltonians, while the third term describes the electron–phonon interactions. Since there is only one electron, there are no electronic interactions. The dimensionless electron–phonon coupling λ is given by the first derivative of the electronic Hamiltonian with respect to the normal coordinate (*cf.* Eq. 6.22):

$$\lambda = \frac{1}{2\hbar\omega} \frac{\partial \epsilon_0}{\partial q}\bigg|_{q=0}.$$

Although this expression does not look similar to the electron–phonon coupling in strongly coupled systems (Eq. 6.15) it can actually be interpreted as a shift in the equilibrium positions of the nuclei (see Problem 2). However, the correspondence is not exact.

In order to proceed we need to decouple the electrons and phonons, which can be achieved by the Lang–Firsov, or polaron transformation: we transform all operators according the Campbell–Baker–Hausdorff formula

$$\hat{O}' \equiv e^{\hat{S}}\hat{O}e^{-\hat{S}} = \hat{O} + [\hat{S}, \hat{O}] + \frac{1}{2!}[\hat{S}, [\hat{S}, \hat{O}]] + \frac{1}{3!}[\hat{S}, [\hat{S}, [\hat{S}, \hat{O}]]] + \ldots,$$

with

$$\hat{S} = \lambda\left(\hat{a} - \hat{a}^\dagger\right)\hat{d}^\dagger\hat{d} = \lambda\frac{\mathrm{i}}{\hbar}\hat{p}\hat{d}^\dagger\hat{d}.$$

For the creation and annihilation operators we get

$$\begin{aligned} \hat{a}' &= \hat{a} + \lambda\hat{d}^\dagger\hat{d}, & \hat{d}' &= \hat{d}\hat{D}^\dagger, \\ \hat{a}'^\dagger &= \hat{a}^\dagger + \lambda\hat{d}^\dagger\hat{d}, & \hat{d}'^\dagger &= \hat{d}^\dagger\hat{D}, \end{aligned}$$

where

$$\hat{D} = e^{\lambda\left(\hat{a} - \hat{a}^\dagger\right)} \tag{6.26}$$

is the displacement operator, so named because (*cf.* Eq. 6.11)

$$q' = \hat{D}q\hat{D}^\dagger = \hat{D}\frac{\hat{a} + \hat{a}^\dagger}{2}\hat{D}^\dagger = \frac{\hat{a} + \hat{a}^\dagger + 2\lambda}{2} = q + \lambda.$$

Applying the polaron transformation to the Hamiltonian (Eq. 6.25) yields

$$\hat{H}_m' = \left(\epsilon_0 - \hbar\omega\lambda^2\right)\hat{d}^\dagger\hat{d} + \hbar\omega\left(\hat{a}^\dagger\hat{a} + \tfrac{1}{2}\right),$$

which indeed decouples the electrons and phonons, and shifts the energy of the electrons by the reorganization energy $\hbar\omega\lambda^2$ (see Problems 1 and 2). Eigenstates of this Hamiltonian are products of electronic and vibrational states with energy $E = \left(\epsilon_0 - \hbar\omega\lambda^2\right)n + \hbar\omega\left(m + \frac{1}{2}\right)$, for n electrons and m phonons. However, this reduction in the complexity of the molecular Hamiltonian comes at the cost of an increased complexity in the Hamiltonian describing the coupling to the leads (Eq. 6.18). Even if we assume the electrons in the leads and their coupling to the molecule to be independent of the normal coordinates, the polaron transformation still transforms the creation and annihilation operators for the electrons on the molecule [Ryndyk et al. (2009)]:

$$\hat{H}_c' = \sum_k \epsilon_k\hat{c}_k^\dagger\hat{c}_k + \sum_k V_k\hat{c}_k^\dagger\hat{d}\hat{D}^\dagger + \text{h.c.}$$

Since $\hat{D}^\dagger$ contains $\hat{a}$ and $\hat{a}^\dagger$ in the exponent (Eq. 6.26), this equation is infinite-order in $\hat{a}$ and $\hat{a}^\dagger$. Although it is possible to derive the transport characteristics from this equation [Wingreen et al. (1988); Jonson (1989); Wingreen et al. (1989)], we will take a different approach here: we show the correspondence to the Franck–Condon picture by taking the weak-coupling limit.

In the weak-coupling regime, we treat the coupling to the leads as a perturbation to the device Hamiltonian:

$$\hat{H}' = \sum_k V_k \hat{c}_k^\dagger \hat{d} \hat{D}^\dagger + \text{h.c.}$$

The transition rates between different states can then be obtained from Fermi's golden rule (Eq. 6.10), with transition matrix elements $\langle nm | \hat{H}' | n'm' \rangle$. Since the eigenstates of the device Hamiltonian are products of electronic and vibrational states, and since $\hat{D}$ only couples to vibrational states, the vibrational transition matrix element is simply $\mathcal{I}_{m,m'} = \langle m | \hat{D} | m' \rangle$. For the vibrational ground state–to-ground state transition we get

$$\begin{aligned}\mathcal{I}_{0,0} \equiv \langle 0 | \hat{D} | 0 \rangle &= \sum_{n=0}^{\infty} \frac{\lambda^n}{n!} \left\langle 0 \middle| \left(\hat{a}^\dagger - \hat{a}\right)^n \middle| 0 \right\rangle \\ &= \sum_{n=0}^{\infty} \frac{\left(-\frac{1}{2}\lambda^2\right)^n}{n!^2} \left\langle 0 \middle| \hat{a}^n \left(\hat{a}^\dagger\right)^n \middle| 0 \right\rangle \\ &= \sum_{n=0}^{\infty} \frac{\left(-\frac{1}{2}\lambda^2\right)^n}{n!} = e^{-\frac{1}{2}\lambda^2}.\end{aligned}$$

Using the fact that $\hat{a}\hat{D} = \hat{D}(\hat{a} - \lambda)$, we get the following recursion relation for the higher transitions:

$$\begin{aligned}\mathcal{I}_{m,m'} \equiv \langle m | \hat{D} | m' \rangle &= \left\langle 0 \middle| \frac{\hat{a}^m}{\sqrt{m!}} \hat{D} \frac{\left(\hat{a}^\dagger\right)^{m'}}{\sqrt{m'!}} \middle| 0 \right\rangle \\ &= \left\langle 0 \middle| \frac{\hat{a}^{m-1}}{\sqrt{(m-1)!}} \hat{D} \frac{\hat{a} - \lambda}{\sqrt{m}} \frac{\left(\hat{a}^\dagger\right)^{m'}}{\sqrt{m'!}} \middle| 0 \right\rangle \\ &= \sqrt{\frac{m'}{m}} \mathcal{I}_{m-1,m'-1} - \frac{\lambda}{\sqrt{n}} \mathcal{I}_{m-1,m'}.\end{aligned}$$

Note that these two equations are exactly equal to the overlap integral of two vibrational wave functions displaced by λ (with

$\omega = \omega'$). We have therefore rederived the Franck–Condon factors in the weak-coupling regime, but starting from the electron–phonon coupling in second quantization, instead of the nuclear wave function overlap.

Problems

1. Reorganization Energy: After an electronic transition, the nuclei are in a vibrationally excited state due to the displacement of the equilibrium positions. Derive an expression for the total vibrational excitation energy (see Eq. 6.3) in terms of the electron–phonon coupling (Eq. 6.15). This energy is called the vibrational reorganization energy. Note the similarity to the energy shift in the next exercise.

2. Electron–Phonon Coupling: This exercise will show the relationship between the electron–phonon coupling from the weak coupling and the one from second quantization (Eqs. 6.15 and 6.22). Take the single-level Hamiltonian in second quantization (Eq. 6.25). If the molecule is weakly coupled to the electrodes, the electron is localized on the molecule and the occupation of the level is (close to) an integer, i.e., $n = \langle \hat{d}^\dagger \hat{d} \rangle$ is either 0 or 1. When $n = 0$, Eq. 6.25 reduces to the harmonic oscillator Hamiltonian (Eq. 6.5):

$$\hat{H}_m = \hbar\omega \left(-\frac{1}{4}\frac{\partial^2}{\partial q^2} + q^2 \right).$$

Cast Eq. 6.25 into this form in the case of $n = 1$ by defining a suitable normal coordinate q'.

3. IETS in a Hydrogen Molecule: (a) An isolated hydrogen molecule consists of only two atoms and therefore has a single vibrational mode: a stretching of the H–H bond. By modeling the molecule as a two-level system, with two sites at energy ϵ_0 coupled by τ, i.e.,

$$\mathbf{H} = \begin{bmatrix} \epsilon_0 & \tau \\ \tau & \epsilon_0 \end{bmatrix},$$

calculate the elastic transmission using Eq. 6.20. Use the wide-band limit for the electrodes and couple the molecule symmetrically, i.e., $\Gamma^L = \Gamma^R$.

(b) We now assume that the vibrational mode only changes the coupling between the atoms, not the energies, i.e.,

$$\boldsymbol{\lambda} = \begin{bmatrix} 0 & \lambda \\ \lambda & 0 \end{bmatrix}.$$

Derive an expression for the inelastic retarded Green's function (Eq. 6.21) with this electron–phonon coupling.

(c) Calculate the inelastic contribution to the transmission (Eq. 6.23).

(d) Finally, calculate the IETS intensity of the mode with Eq. 6.24 and take the off-resonance limit.

References

ADF2013 (2013). SCM, Theoretical Chemistry, Vrije Universiteit: Amsterdam, The Netherlands, http://www.scm.com.

Barone, V., Bloino, J., Biczysko, M. and Santoro, F. (2009). Fully integrated approach to compute vibrationally resolved optical spectra: From small molecules to macrosystems, *J. Chem. Theory Comput.* **5**, 3, pp. 540–554, doi:10.1021/ct8004744.

Beenakker, C. W. J. (1991). Theory of coulomb-blockade oscillations in the conductance of a quantum dot, *Phys. Rev. B* **44**, 4, pp. 1646–1656, doi: 10.1103/PhysRevB.44.1646.

Boese, D. and Schoeller, H. (2001). Influence of nanomechanical properties on single-electron tunneling: A vibrating single-electron transistor, *Europhys. Lett.* **54**, 5, pp. 668–674, doi:10.1209/epl/i2001-00367-8.

Bonet, E., Deshmukh, M. M. and Ralph, D. C. (2002). Solving rate equations for electron tunneling via discrete quantum states, *Phys. Rev. B* **65**, 4, p. 045317, doi:10.1103/PhysRevB.65.045317.

Braig, S. and Flensberg, K. (2003). Vibrational sidebands and dissipative tunneling in molecular transistors, *Phys. Rev. B* **68**, 20, p. 205324, doi: 10.1103/PhysRevB.68.205324.

Bumm, L. A., Arnold, J. J., Cygan, M. T., Dunbar, T. D., Burgin, T. P., Jones, L., Allara, D. L., Tour, J. M. and Weiss, P. S. (1996). Are single molecular wires conducting? *Science* **271**, 5256, pp. 1705–1707, doi:10.1126/science.271.5256.1705.

Ćavar, E., Blüm, M.-C., Pivetta, M., Patthey, F., Chergui, M. and Schneider, W.-D. (2005). Fluorescence and phosphorescence from individual C_{60}

molecules excited by local electron tunneling, *Phys. Rev. Lett.* **95**, 19, p. 196102, doi:10.1103/PhysRevLett.95.196102.

Chang, C. T.-C., Sethna, J. P., Pasupathy, A. N., Park, J., Ralph, D. C. and McEuen, P. L. (2007). Phonons and conduction in molecular quantum dots: Density functional calculations of Franck–Condon emission rates for bifullerenes in external fields, *Phys. Rev. B* **76**, 4, p. 045435, doi: 10.1103/PhysRevB.76.045435.

Coutsias, E. A., Seok, C. and Dill, K. A. (2004). Using quaternions to calculate rmsd, *J. Comput. Chem.* **25**, 15, pp. 1849–1857, doi:10.1002/jcc.20110.

Datta, S. (1995). *Electron Transport in Mesoscopic Systems* (Cambridge University Press, Cambridge).

Derosa, P. A. and Seminario, J. M. (2001). Electron transport through single molecules: Scattering treatment using density functional and green function theories, *J. Phys. Chem. B* **105**, 2, pp. 471–481, doi:10.1021/jp003033+.

Doktorov, E. V., Malkin, I. A. and Man'ko, V. I. (1977). Dynamical symmetry of vibronic transitions in polyatomic molecules and the Franck–Condon principle, *J. Mol. Spectrosc.* **64**, 2, pp. 302–326, doi:10.1016/0022-2852(77)90269-7.

Dong, S.-H., Lemus, R. and Frank, A. (2002). Ladder operators for the morse potential, *Int. J. Quantum Chem.* **86**, 5, pp. 433–439, doi:10.1002/qua.10038.

Dong, Z.-C., Guo, X.-L., Trifonov, A. S., Dorozhkin, P. S., Miki, K., Kimura, K., Yokoyama, S. and Mashiko, S. (2004). Vibrationally resolved fluorescence from organic molecules near metal surfaces in a scanning tunneling microscope, *Phys. Rev. Lett.* **92**, 8, p. 086801, doi:10.1103/PhysRevLett.92.086801.

Duschinsky, F. (1937). Meaning of the electronic spectrum of polyatomic molecules. i. the Franck–Condon principle, *Acta Physicochim. URSS* **7**, pp. 551–566.

Fonseca Guerra, C., Snijders, J., te Velde, G. and Baerends, E. (1998). Towards an order-*n* DFT method, *Theor. Chem. Acc.* **99**, pp. 391–403, doi:10.1007/s002140050353.

Frederiksen, T., Paulsson, M., Brandbyge, M. and Jauho, A.-P. (2007). Inelastic transport theory from first principles: Methodology and application to nanoscale devices, *Phys. Rev. B* **75**, 20, p. 205413, doi:10.1103/PhysRevB.75.205413.

Galperin, M., Ratner, M. A. and Nitzan, A. (2004). Inelastic electron tunneling spectroscopy in molecular junctions: Peaks and dips, *J. Chem. Phys.* **121**, 23, pp. 11965–11979, doi:10.1063/1.1814076.

Galperin, M., Ratner, M. A. and Nitzan, A. (2007). Molecular transport junctions: Vibrational effects, *J. Phys.: Condens. Matter* **19**, 10, p. 103201, doi:10.1088/0953-8984/19/10/103201.

Haug, H. and Jauho, A.-P. (1997). *Quantum Kinetics in Transport and Optics of Semiconductors* (Springer, Berlin, Heidelberg), doi:10.1007/978-3-540-73564-9.

Jauho, A.-P., Wingreen, N. S. and Meir, Y. (1994). Time-dependent transport in interacting and noninteracting resonant-tunneling systems, *Phys. Rev. B* **50**, 8, pp. 5528–5544, doi:10.1103/PhysRevB.50.5528.

Jonson, M. (1989). Quantum-mechanical resonant tunneling in the presence of a boson field, *Phys. Rev. B* **39**, 9, pp. 5924–5933, doi:10.1103/PhysRevB.39.5924.

Koch, J., von Oppen, F. and Andreev, A. V. (2006). Theory of the Franck–Condon blockade regime, *Phys. Rev. B* **74**, 20, p. 205438, doi:10.1103/PhysRevB.74.205438.

Koch, J., von Oppen, F., Oreg, Y. and Sela, E. (2004). Thermopower of single-molecule devices, *Phys. Rev. B* **70**, 19, p. 195107, doi:10.1103/PhysRevB.70.195107.

Kubatkin, S., Danilov, A., Hjort, M., Cornil, J., Bredas, J. L., Stuhr-Hansen, N., Hedegård, P. and Bjørnholm, T. (2003). Single-electron transistor of a single organic molecule with access to several redox states, *Nature* **425**, 6959, pp. 698–701, doi:10.1038/nature02010.

Kushmerick, J. G., Lazorcik, J., Patterson, C. H., Shashidhar, R., Seferos, D. S. and Bazan, G. C. (2004). Vibronic contributions to charge transport across molecular junctions, *Nano Lett.* **4**, 4, pp. 639–642, doi:10.1021/nl049871n.

Mahan, G. D. (1981). *Many-Particle Physics* (Plenum Press, New York).

McCarthy, K. D., Prokof'ev, N. and Tuominen, M. T. (2003). Incoherent dynamics of vibrating single-molecule transistors, *Phys. Rev. B* **67**, 24, p. 245415, doi:10.1103/PhysRevB.67.245415.

Meir, Y. and Wingreen, N. S. (1992). Landauer formula for the current through an interacting electron region, *Phys. Rev. Lett.* **68**, 16, pp. 2512–2515, doi:10.1103/PhysRevLett.68.2512.

Mitra, A., Aleiner, I. and Millis, A. J. (2004). Phonon effects in molecular transistors: Quantal and classical treatment, *Phys. Rev. B* **69**, 24, p. 245302, doi:10.1103/PhysRevB.69.245302.

Morse, P. M. (1929). Diatomic molecules according to the wave mechanics. ii. vibrational levels, *Phys. Rev.* **34**, 1, pp. 57–64, doi:10.1103/PhysRev.34.57.

Osorio, E. A., O'Neill, K., Stuhr-Hansen, N., Nielsen, O. F., Bjørnholm, T. and van der Zant, H. S. J. (2007). Addition energies and vibrational fine structure measured in electromigrated single-molecule junctions based on an oligophenylenevinylene derivative, *Adv. Mater.* **19**, 2, pp. 281–285, doi:10.1002/adma.200601876.

Osorio, E. A., Ruben, M., Seldenthuis, J. S., Lehn, J. M. and van der Zant, H. S. J. (2010). Conductance switching and vibrational fine structure of a [2 × 2] Co_4^{II} gridlike single molecule measured in a three-terminal device, *Small* **6**, 2, pp. 174–178, doi:10.1002/smll.200901559.

Palma, A. and Morales, J. (1983). Franck–Condon factors and ladder operators. i. harmonic oscillator, *Int. J. Quantum Chem.* **24**, S17, pp. 393–400, doi:10.1002/qua.560240843.

Park, H., Park, J., Lim, A. K. L., Anderson, E. H., Alivisatos, A. P. and McEuen, P. L. (2000). Nanomechanical oscillations in a single-C_{60} transistor, *Nature* **407**, 6800, pp. 57–60, doi:10.1038/35024031.

Parks, J. J., Champagne, A. R., Hutchison, G. R., Flores-Torres, S., Abruna, H. D. and Ralph, D. C. (2007). Tuning the Kondo effect with a mechanically controllable break junction, *Phys. Rev. Lett.* **99**, 2, p. 026601, doi:10.1103/PhysRevLett.99.026601.

Paulsson, M., Frederiksen, T. and Brandbyge, M. (2006). Inelastic transport through molecules: Comparing first-principles calculations to experiments, *Nano Lett.* **6**, 2, pp. 258–262, doi:10.1021/nl052224r.

Qiu, X. H., Nazin, G. V. and Ho, W. (2003). Vibrationally resolved fluorescence excited with submolecular precision, *Science* **299**, 5606, pp. 542–546, doi:10.1126/science.1078675.

Rausch, A. F., Thompson, M. E. and Yersin, H. (2009). Triplet state relaxation processes of the OLED emitter pt(4,6-dfppy)(acac), *Chem. Phys. Lett.* **468**, 1-3, pp. 46–51, doi:10.1016/j.cplett.2008.11.075.

Ruhoff, P. T. (1994). Recursion-relations for multidimensional Franck–Condon overlap integrals, *Chem. Phys.* **186**, 2-3, pp. 355–374, doi:10.1016/0301-0104(94)00173-1.

Ruhoff, P. T. and Ratner, M. A. (2000). Algorithms for computing Franck–Condon overlap integrals, *Int. J. Quantum Chem.* **77**, 1, pp. 383–392, doi:10.1002/(SICI)1097-461X(2000)77:1⟨383::AID-QUA38⟩3.0.CO;2-0.

Ryndyk, D. A., Gutiérrez, R., Song, B. and Cuniberti, G. (2009). Green function techniques in the treatment of quantum transport at the molecular scale, in I. Burghardt, V. May, D. A. Micha and E. R. Bittner (eds.), *Springer Series in Chemical Physics*, Vol. 93 (Springer, Berlin, Heidelberg), pp. 213–335, doi:10.1007/978-3-642-02306-4_9.

Sando, G. M. and Spears, K. G. (2001). *Ab Initio* computation of the duschinsky mixing of vibrations and nonlinear effects, *J. Phys. Chem. A* **105**, 22, pp. 5326–5333, doi:10.1021/jp004230b.

Sandoval, L., Palma, A. and Rivas-Silva, F. (1989). Operator algebra and recurrence relations for the Franck–Condon factors, *Int. J. Quantum Chem.* **36**, S23, pp. 183–186, doi:10.1002/qua.560360822.

Santoro, F., Improta, R., Lami, A., Bloino, J. and Barone, V. (2007a). Effective method to compute Franck–Condon integrals for optical spectra of large molecules in solution, *J. Chem. Phys.* **126**, 8, p. 084509, doi:10.1063/1.2437197.

Santoro, F., Lami, A., Improta, R. and Barone, V. (2007b). Effective method to compute vibrationally resolved optical spectra of large molecules at finite temperature in the gas phase and in solution, *J. Chem. Phys.* **126**, 18, p. 184102, doi:10.1063/1.2721539.

Santoro, F., Lami, A., Improta, R., Bloino, J. and Barone, V. (2008). Effective method for the computation of optical spectra of large molecules at finite temperature including the duschinsky and Herzberg–Teller effect: The q_x band of porphyrin as a case study, *J. Chem. Phys.* **128**, 22, p. 224311, doi:10.1063/1.2929846.

Scott, A. P. and Radom, L. (1996). Harmonic vibrational frequencies: An evaluation of Hartree–Fock, møller-plesset, quadratic configuration interaction, density functional theory, and semiempirical scale factors, *J. Phys. Chem.* **100**, 41, pp. 16502–16513, doi:10.1021/jp960976r.

Seldenthuis, J. S., van der Zant, H. S. J., Ratner, M. A. and Thijssen, J. M. (2008). Vibrational excitations in weakly coupled single-molecule junctions: A computational analysis, *ACS Nano* **2**, 7, pp. 1445–1451, doi:10.1021/nn800170h.

Seldenthuis, J. S., van der Zant, H. S. J., Ratner, M. A. and Thijssen, J. M. (2010). Electroluminescence spectra in weakly coupled single-molecule junctions, *Phys. Rev. B* **81**, 20, p. 205430, doi:10.1103/PhysRevB.81.205430.

Sharp, T. E. and Rosenstock, H. M. (1964). Franck–Condon factors for polyatomic molecules, *J. Chem. Phys.* **41**, 11, pp. 3453–3463, doi:10.1063/1.1725748.

Smit, R. H. M., Noat, Y., Untiedt, C., Lang, N. D., van Hemert, M. C. and van Ruitenbeek, J. M. (2002). Measurement of the conductance of a hydrogen molecule, *Nature* **419**, 6910, pp. 906–909, doi:10.1038/nature01103.

Smith, W. L. (1969). The overlap integral of two harmonic-oscillator wave functions, *J. Phys. B* **2**, 1, p. 1, doi:10.1088/0022-3700/2/1/301.

Solomon, G. C., Andrews, D. Q., Hansen, T., Goldsmith, R. H., Wasielewski, M. R., Van Duyne, R. P. and Ratner, M. A. (2008). Understanding quantum interference in coherent molecular conduction, *J. Chem. Phys.* **129**, 5, p. 054701, doi:10.1063/1.2958275.

Stipe, B. C., Rezaei, M. A. and Ho, W. (1998). Single-molecule vibrational spectroscopy and microscopy, *Science* **280**, 5370, pp. 1732–1735, doi: 10.1126/science.280.5370.1732.

Taylor, J., Guo, H. and Wang, J. (2001). *Ab Initio* modeling of quantum transport properties of molecular electronic devices, *Phys. Rev. B* **63**, 24, p. 245407, doi:10.1103/PhysRevB.63.245407.

te Velde, G., Bickelhaupt, F., van Gisbergen, S., Fonseca Guerra, C., Baerends, E., Snijders, J. and Ziegler, T. (2001). Chemistry with adf, *J. Comput. Chem.* **22**, pp. 931–967, doi:10.1002/jcc.1056.

Tersoff, J. and Hamann, D. R. (1983). Theory and application for the scanning tunneling microscope, *Phys. Rev. Lett.* **50**, 25, pp. 1998–2001, doi:10.1103/PhysRevLett.50.1998.

Tersoff, J. and Hamann, D. R. (1985). Theory of the scanning tunneling microscope, *Phys. Rev. B* **31**, 2, pp. 805–813, doi:10.1103/PhysRevB.31.805.

Thijssen, J. M. and van der Zant, H. S. J. (2008). Charge transport and single-electron effects in nanoscale systems, *Phys. Status Solidi B* **245**, 8, pp. 1455–1470, doi:10.1002/pssb.200743470.

Timm, C. (2008). Tunneling through molecules and quantum dots: Master-equation approaches, *Phys. Rev. B* **77**, 19, p. 195416, doi:10.1103/PhysRevB.77.195416.

Troisi, A. and Ratner, M. A. (2005). Modeling the inelastic electron tunneling spectra of molecular wire junctions, *Phys. Rev. B* **72**, 3, p. 033408, doi: 10.1103/PhysRevB.72.033408.

Troisi, A. and Ratner, M. A. (2006). Molecular transport junctions: Propensity rules for inelastic electron tunneling spectra, *Nano Lett.* **6**, 8, pp. 1784–1788, doi:10.1021/nl0609394.

Troisi, A., Ratner, M. A. and Nitzan, A. (2003). Vibronic effects in off-resonant molecular wire conduction, *J. Chem. Phys.* **118**, 13, pp. 6072–6082, doi: 10.1063/1.1556854.

van der Vorst, H. (1992). Bi-cgstab: A fast and smoothly converging variant of bi-cg for the solution of nonsymmetric linear systems, *SIAM J. Sci. and Stat. Comput.* **13**, 2, pp. 631–644, doi:10.1137/0913035.

Wegewijs, M. R. and Nowack, K. C. (2005). Nuclear wavefunction interference in single-molecule electron transport, *New J. Phys.* **7**, p. 239, doi: 10.1088/1367-2630/7/1/239.

Wilson, E. B., Decius, J. C. and Cross, P. C. (1955). *Molecular Vibrations: The Theory of Infrared and Raman Vibrational Spectra* (McGraw-Hill, New York).

Wingreen, N. S., Jacobsen, K. W. and Wilkins, J. W. (1988). Resonant tunneling with electron-phonon interaction: An exactly solvable model, *Phys. Rev. Lett.* **61**, 12, pp. 1396–1399, doi:10.1103/PhysRevLett.61.1396.

Wingreen, N. S., Jacobsen, K. W. and Wilkins, J. W. (1989). Inelastic scattering in resonant tunneling, *Phys. Rev. B* **40**, 17, pp. 11834–11850, doi:10.1103/PhysRevB.40.11834.

Wu, S. W., Nazin, G. V. and Ho, W. (2008). Intramolecular photon emission from a single molecule in a scanning tunneling microscope, *Phys. Rev. B* **77**, 20, p. 205430, doi:10.1103/PhysRevB.77.205430.

Xue, Y., Datta, S. and Ratner, M. A. (2002). First-principles based matrix green's function approach to molecular electronic devices: General formalism, *Chem. Phys.* **281**, 2–3, pp. 151–170, doi:10.1016/S0301-0104(02)00446-9.

Yu, L. H., Keane, Z. K., Ciszek, J. W., Cheng, L., Stewart, M. P., Tour, J. M. and Natelson, D. (2004). Inelastic electron tunneling *via* molecular vibrations in single-molecule transistors, *Phys. Rev. Lett.* **93**, 26, p. 266802, doi:10.1002/qua.20653.

Zhixing, C. (1989). Rotation procedure in intrinsic reaction coordinate calculations, *Theor. Chim. Acta* **75**, 6, pp. 481–484, doi:10.1007/BF00527679.

Chapter 7

Self-Assembly at Interfaces

Tina A. Gschneidtner and Kasper Moth-Poulsen

Department of Chemistry and Chemical Engineering,
Chalmers University of Technology,
Gothenburg, 41296 Sweden
kasper.moth-poulsen@chalmers.se

Due to the microscopic size of single-molecule components, it is impractical to assemble a large number of single-molecule components via direct top-down manipulation. Instead, self-assembly methods, meaning spontaneous ordering and organization of molecules without direct human intervention, are proposed as the most feasible way of building up multiple single-molecule devices [1]. The self-assembly process is directed by weak interactions. In this chapter, we will introduce the basic concepts of self-assembly and put it into a context of single-molecule electronic devices. We will discuss mechanisms of formation of self-assembled monolayers and how typical single-molecule components interact with surfaces. Finally, we will present some recent developments in electrode materials using single molecules.

Handbook of Single-Molecule Electronics
Edited by Kasper Moth-Poulsen

ISBN 978-981-4463-38-6 (Hardcover), 978-981-4463-39-3 (eBook)
www.panstanford.com

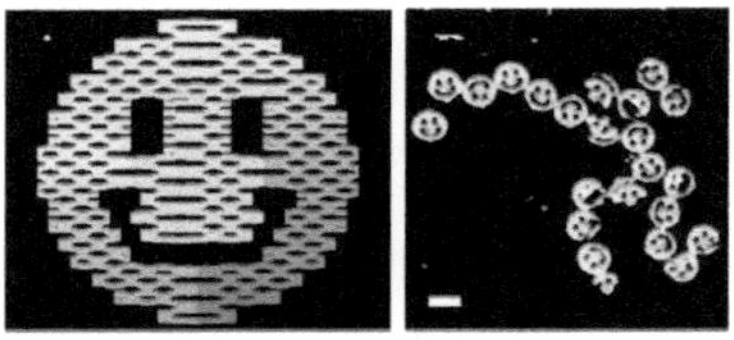

Figure 7.1 (a) Sketch of a folding path for DNA origami; DNA smiley out of a number of double helices. (b) Atomic force microscopy image of several DNA smileys (scale bar 100 nm). Reprinted with permission from Ref. [2]. Copyright (2006) Nature Publishing Group.

7.1 Introduction

Self-assembly is ubiquitous in nature and responsible for the structure and function of all biological systems. From the structure of DNA to the activity of enzymes, all are governed by weak interactions to form advanced three-dimensional structures.

During recent decades, the use of self-assembly in chemistry and materials science has evolved. Together with an increased understanding on how these weak forces function, scientists have developed ways to use natural or synthetic self-assembling units for new materials and nanoscale structures with advanced shapes and functions.

An example of the use of DNA self-assembly in a creative and programmable way is DNA origami [2]. DNA origami is a folding of DNA to construct nanoscale structures. The choice of the DNA-base sequence allows for design of two- and three-dimensional objects in a predictable way, due to complementary base-pair interactions (Fig. 7.1).

Synthetic examples of self-assembled supramolecules include the so-called rotaxane and catenanes by Stoddart et al. [3, 4] (Fig. 7.2a,b). These are interlocked molecules (interlocked macrocyclic rings and macrocyclic rings trapped onto a linear unit, respectively) consisting of two or more separate components, which are not connected by chemical (i.e., covalent) bonds. The synthesis of such molecules is possible due to the clever use of noncovalent interactions, and both catenanes (Fig. 7.2a) and rotaxanes (Fig. 7.2b) have been used as active components in molecular electronic devices. Lehn et al. have used metal–ligand interactions to form

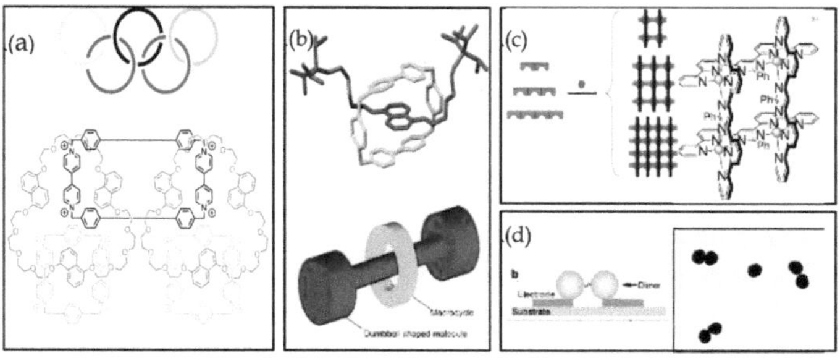

Figure 7.2 (a) Stoddart Olympics: linear array of five interlocked rings, catenanes, out of molecular compounds; reprinted with permission from Ref. [3], copyright (1994) Wiley-VCH Verlag GmbH & Co. KGaA; (b) dumbbell-shaped compounds, rotaxanes, threaded through a macrocycle; reprinted with permission from Ref. [4], copyright (1998) John Wiley & Sons; (c) metallogrids from planar, polytopic organic ligands coordinating metals; reprinted with permission from Ref. [5], copyright (2011) NZIC; and (d) gold dimers with aromatic linker molecules assembled between them; reprinted with permission from Ref. [7], copyright (2005) Nature Publishing Group.

two-dimensional "grids" with a well-defined positioning of metal ions, only a fraction of a nanometer from each other (Fig. 7.2c) [5]. Israel et al. have assembled nanoparticle dimers, linked together by a single molecule, and thus bridging the length scale of single molecules (1–2 nm) with a length scale readily accessible by lithographic techniques (30 nm) [6] (Fig. 7.2d).

Spontaneously formed self-assembled monolayers (SAMs) on well-defined solid surfaces are widely used for fabrication of devices by the so-called **bottom-up** approach (Fig. 7.3a). The approach

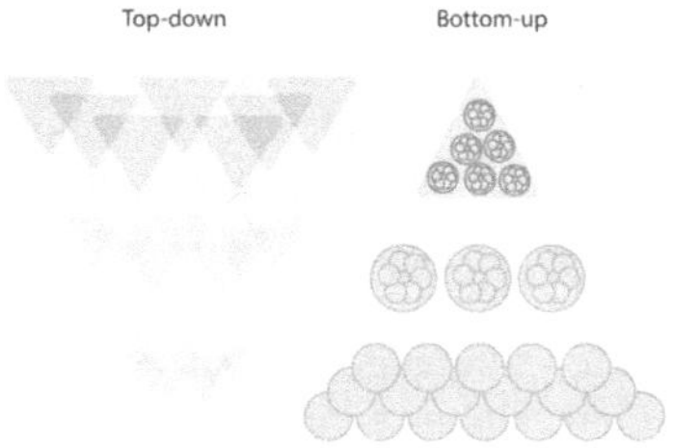

Figure 7.3 Schematic picture of a bottom-up and a top-down approach to obtain nanosized structures Ref. [8].

is called bottom-up because one starts from the fundamental parts—molecules and atoms—and assembles them to the desired structures—nanoparticles or monolayers. With this concept, a single molecule can be placed between two nanoclusters or between two electrodes for single-molecule electronics. In contrast, the **top-down** approach breaks down a macroscopic system, e.g., a metal block, into several subsystems (Fig. 7.3b). A nanogap between two electrodes for use in single-molecule electronics is typically fabricated by elaborative nano- and microfabrication techniques with built-in challenges in terms of resolution.

All single-molecule electronics experiments rely on self-assembly processes on one level or another. This chapter will focus on self-assembly on the molecular length scale and introduce the reader to the basics of forces governing self-assembly of molecules on surfaces in general. To limit the scope, the chapter is focused on self-assembled systems relevant in single-molecule electronics. Self-assembly on the device scale is discussed in Chapter 12, whereas a general introduction to top-down approaches to fabrication of single-molecule devices was presented in Chapter 2.

Several self-assembly systems have been explored in the context of single-molecule electronics, but the most successfully explored combination is thiol-functionalized molecules on gold surfaces [Au(111)]. An example is shown in Fig. 7.4a, where the tunneling current through a single organic, conducting molecule is measured

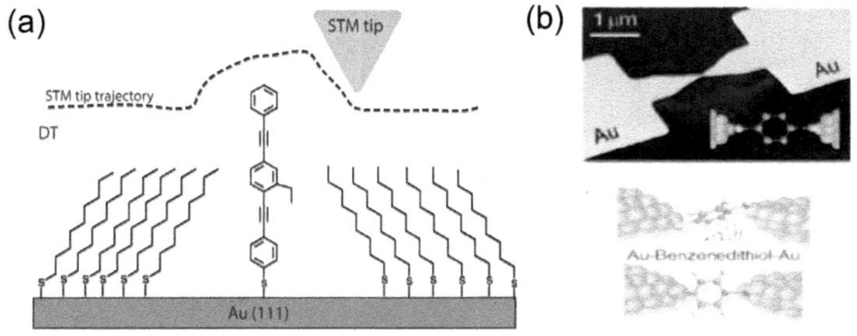

Figure 7.4 (a) Scanning tunneling microscopy can measure a single organic, conducting molecule in a mixed SAM layer on a gold surface [9]. (b) Benzenedithiol incorporated into a mechanically controllable break junction—a broad-range single-channel molecule; reprinted with permission from Ref. [11], copyright (2011) American Chemical Society.

by an STM tip. The conducting molecule is here embedded in an isolating self-assembled matrix of nonconducting alkane thiols [9]. Since the STM tip is operated in constant current mode, the apparent height of the conducting molecule is determined relatively higher than the nonconducting molecules making it possible to extract information about the tunneling barrier of the conducting molecule relative to the nonconducting one [10]. Using self-assembly in terms of single molecular electronics, Scheer et al. [11] showed that the molecular conformation and thus the transport properties can be tuned, due to the displacement of benzenedithiol in a nanogap electrode. High and low conductances, depending on the conformation, can be obtained (Fig. 7.4b).

The following sections introduces the basic process of self-assembly (Section 7.2.1), including the importance of the chemisorption of thiol compounds on gold surfaces (Sections 7.2.2–7.2.4). Further, we will discuss alternative chemical anchor groups and electrode materials, such as silicon oxide and graphene (Section 7.3).

7.2 Self-Assembled Monolayers

7.2.1 *Inter- and Intramolecular Interactions between the Molecule and the Surface*

Formed between metal surfaces and tailored molecules, SAMs are today used for a broad range of applications, including biosensors and therapeutic nanoparticles, due to their tunable physical and chemical properties [1, 12, 13]. In single-molecule electronics, SAMs have been used to place single molecules on metal surfaces in many different device types, such as molecular break junctions, scanning probe methods, and nanoparticle based systems.

Inter- and intramolecular forces play a key role in the self-assembled arrangement of molecules adsorbed on a solid surface. In the following, the different interactions and forces between organic molecules itself and with the surface are introduced to understand the basics in the formation of SAMs.

Organic molecules that form SAMs need to contain three structural motifs: a head group (linking unit to the surface), the

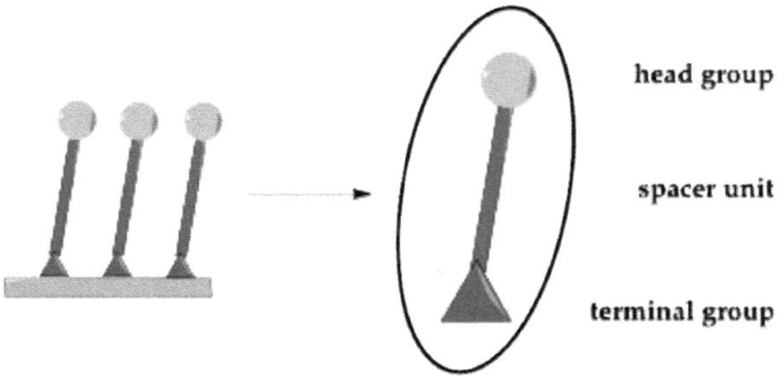

Figure 7.5 Principle of self-assembly of an organic molecule on a surface: the **terminal group** to bind the molecule to the surface, the spacer unit for the assembly to a monolayer, and the **head group** for the functionality of the surface.

spacer unit (backbone/main chain), and the terminal functional group (active) (Fig. 7.5) [1, 14, 15].

The **head group** should have a specific affinity for the substrate, like sulfur to gold. In general, the head group can be physisorbed or chemisorbed to the metal surface. For example, thiol molecules are chemisorbed on Au by a strong thiolate-Au bond (40–50 kcal/mol). This group guides the process of self-assembly since it is linking the molecule to the metal surface. This is usually a strong interaction and a main driving force in the process of self-assembly. Multidentate[a] adsorbates can give an even stronger interaction between the ligand and the metal [16]. The enhanced stability is due to the chelate effect of ligands with more than one site to attach to the same substrate.

The head group plays an important role in the self-assembly process due to the affinity between the ligand and the metal, and due to the fact that adsorbates lower the free-surface energy of the metal [1] (discussed in detail in Section 7.2.2).

The second part of the molecule, the **spacer unit**, is the main driving force to form two-dimensional horizontal order in SAMs. The spacer unit aligns laterally to each other to maximize the attractive interaction between them. This attraction can be van der Waals (VdW) forces or hydrogen bonds. These intermolecular forces are caused by permanent or induced dipole moments in the

[a]Multident: more than one head group can bind to the surface.

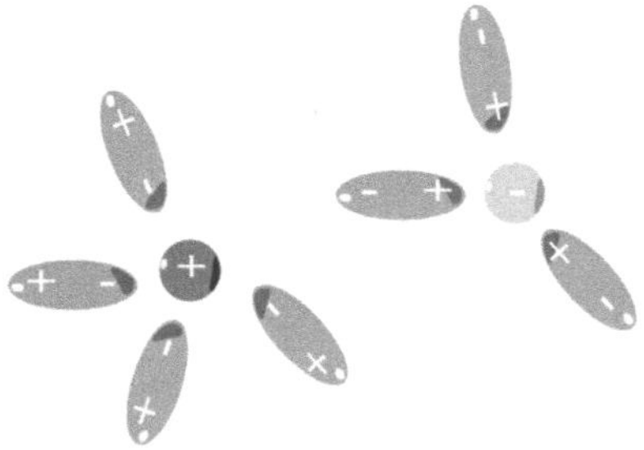

Figure 7.6 Ion–dipole forces.

molecule. VdW force is a synonym for all kind of attractive and repulsive intermolecular forces, which are relatively weak compared to covalent, ionic, or metallic bonds. VdW forces are divided into ion–dipole, dipole–dipole, and induced dipole forces and will be explained in the following.

Ion–dipole and **ion-induced dipole force** (12–40 kcal/mol) are two types of strong intermolecular interactions. The forces are found between ions and dipole molecules or molecules that can become a dipole due to the presence of a strong ionic molecule in its vicinity. These interactions are stronger than dipole–dipole interactions due to the involvement of the stronger electrostatics of ions. Polar molecules align to allow the maximum attraction, as shown in Fig. 7.6.

Another type of VdW force is the **dipole–dipole** interaction (0.5–2 kcal/mol) (Fig. 7.7). Polar molecules such as HCl show a permanent dipole due to a difference in the electronegativity of the hydrogen and chlorine atoms in the molecule. The more electronegative atom (here Cl) tends to attract the bonded electron pair between hydrogen and chlorine. It becomes negatively charged ($\delta-$) due to the slight increase in electron density. The less electronegative atom

Figure 7.7 Dipole–dipole interactions.

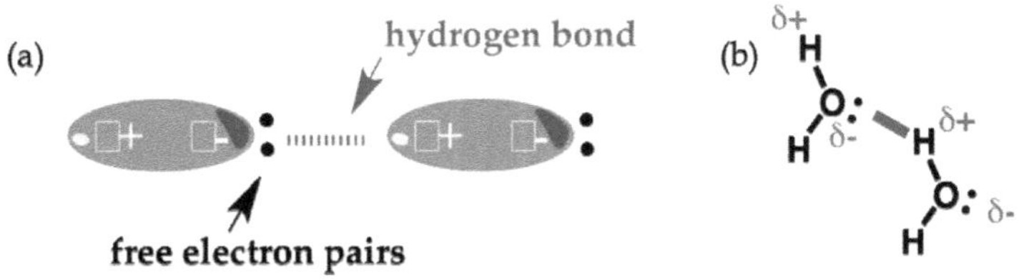

Figure 7.8 (a) Schematic picture of a hydrogen bond between two different atoms in a molecule with different electronegativities. The electronegative one is depicted with lone-pair electrons, which attract the electropositve one. (b) Example for hydrogen bonding between water molecules: hydrogen bond between the electronegative oxygen and the electropositve hydrogen.

(here H) becomes electron deficient and positively charged ($\delta+$). These permanent dipoles induce an electrostatic attraction between the electronegative chlorine and the positively charged hydrogen. Since opposite charges attract, when polar molecules approach each other, they orient themselves in a head-to-tail manner, as shown in Fig. 7.7.

Hydrogen bond (12–16 kcal/mol) is a special type of dipole-dipole interaction in which a hydrogen atom is situated between two small, electronegative atoms—such as nitrogen, oxygen, or fluoride—one of which has a lone pair of electrons (Fig. 7.8). The electrons associated with the hydrogen are displaced toward the electronegative atom. As a result, the electron density of the hydrogen atom is reduced. Thus, the proton of the hydrogen atom will be more exposed to the attraction of lone-pair electrons from other molecules (Fig. 7.8). The most prominent example is water, in which bridges between several molecules are made through hydrogen bonding.

Another type of VdW force, which is weaker than the dipole-dipole force, is the **induced dipole** force ($\sim$1 kcal/mol) (Fig. 7.9). If a particle with a permanent, instantaneous, or induced dipole approaches another particle (with or without a dipole), the dipole of the first molecule induces a dipole in the second one.

The weakest but most important force for SAMs is the dispersion force or **induced-dipole–induced-dipole** interaction in nonpolar, electrically symmetrical molecules such as alkane chains in alkane thiols for SAMs on gold substrates ($<$1 kcal/mol). Due to a fluctuation in the electron cloud in a nonpolar molecule, a displacement

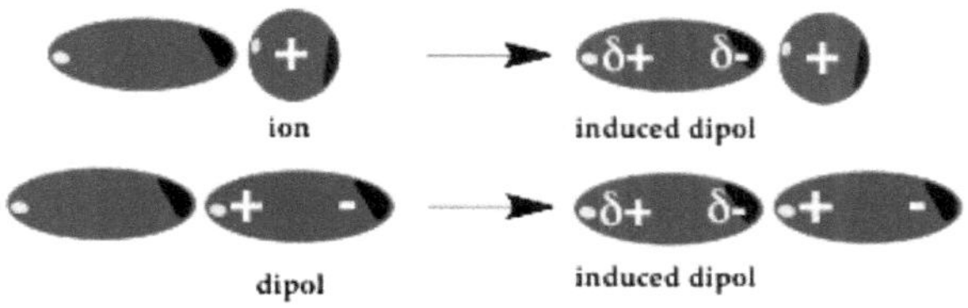

Figure 7.9 Induced forces.

between nuclei and electrons occurs; hence there may be more negative charge on one side of the molecule than on the other side at a certain time. The temporary polarity in the molecule induces dipoles in the surrounding media, such as neighboring molecules (Fig. 7.10).

This correlated movement of electrons is the source of the dispersion energy and is always present in all types of molecules. Dispersion interactions are always attractive. If the electron clouds start to overlap, higher energy orbitals for the electrons are occupied (Pauli principle), which causes an exchange repulsion.

The strength of these VdW forces, also called London forces, depends on several factors besides the molecular **size**, such as **polarizability** of molecules, molecular **shapes**, **temperature**, and **pressure**.

With longer alkane sections in the spacer unit, the total size of the electron cloud is increasing; therefore, it is more diffuse and polarizable [17]. The displacement of the electrons to form an uneven distribution of charge and a temporary dipole is easier, which results in stronger VdW forces, and thereby stronger lateral interaction in the spacer unit. Therefore, small molecules are harder to polarize than larger ones. As a consequence, an increase in the number of CH_2 units in the hydrocarbon chain backbone increases SAM order and coverage [17].

The more contact molecules display to their neighbors, the stronger their interaction is, meaning the greater is the operating VdW forces. Therefore, branched alkanes have lower boiling points than straight chains and will introduce disorder in the two-dimensional lattice of the SAMs. Also high pressure and low temperature favor the VdW force strength due to less movement of the molecules and closer contact of the molecules to each other.

The molecular backbone stabilizes the SAM via intermolecular forces with 6 kcal/mol for hexanethiol or 10 kcal/mol for dodecanethiol [17]. The resulting binding energies increase with approximately 1 kcal/mol per CH_2 group, for chains longer than three carbon atoms. Therefore, longer alkane chains stabilize the SAMs [18], and eight methylene (CH_2)-groups are usually required for highly ordered SAMs [19].

In summary, each part of the adsorbate contributes a small part to the attractive interaction between the organic molecule and the metal surface. The largest contribution originates from the head group and the substrate with around 50 kcal/mol (for S-Au). About 1–2 kcal/mol per methylene unit via VdW interaction arises from the spacer unit and a few kcal/mol due to the terminal group [14, 15].

To come back to Fig. 7.5 and the build-up of a typical molecule for self-assembly, the last part, the **terminal group**, has to be mentioned. This group determines surface properties, such as hydrophilicity or hydrophobicity, depending on the functional group [1, 15]. Furthermore, it can be used to anchor different types of molecules, nanostructures, or biomolecules to the SAM [1, 20].

7.2.2 *Chemisorption vs Physisorption*

When a molecule is placed on a surface, two different modes of interaction have to be considered: physisorption and chemisorption. **Physisorption** is the accumulation of molecules on a surface, which leaves the surface as well as the chemical species of the adsorbate intact. The main driving force in this process is the VdW force (see Section 7.2). Chemisorption is nonspecific, and the adsorbate can typically desorb from the surface reversibly. Typical physisorption energies are 2–10 kcal/mol. In **chemisorption**, covalent or ionic bonds are formed between the molecule and the surface, which can be hundred times stronger than the VdW force. Thereby a chemical reaction occurs between the adsorbate and the surface; e.g., thiols (SH) adsorb on a gold (Au) surface forming a new strong bond, a thiolate (Au–S), and releasing H_2 (Fig. 7.10). Formally, a sulfenyl radical (RS*) is created by the deprotonation of the thiol (SH) group, and the exact fate of the H_2 is debated [21–23]. The protonated thiol group can, likewise, interact by a weaker coordination bond through

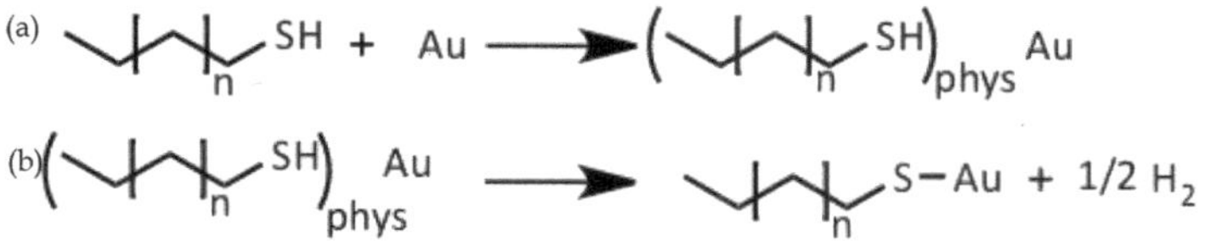

Figure 7.10 (a) Physisorption and (b) chemisorption.

the sulfur lone-pair electrons. The newly formed thiolate (Au–S) is almost as strong as the Au–Au bond and can, therefore, modify the gold–gold interaction at the gold–sulfur interface (see Section 7.2.4) [24]. Typical chemisorption energies are 15–100 kcal/mol.

The self-assembly process of organic molecules on metal surfaces has been extensively studied for alkane thiol–gold systems. Especially the kinetics of adsorption and desorption was studied on Au(111) surfaces, which acts as a model system for the assembly of thiols and dithiols [25, 26]. Chemisorption of thiols or dithiols on gold can occur in the gas phase as well as in solution.

During the self-assembly process, both chemisorption and physisorption occur. A simple picture of chemisorption via physisorption during the gas-phase deposition of alkyl thiols onto a gold surface is shown in Fig. 7.11 [27–29].

The first step in assembly on a surface is often physisorption (see Fig. 7.10a and Fig. 7.11a), which is a gas-like state, were alkane thiols are adsorbed via VdW forces. Subsequently, some molecules start chemisorbing by the sulfur head group forming a covalent bond, a

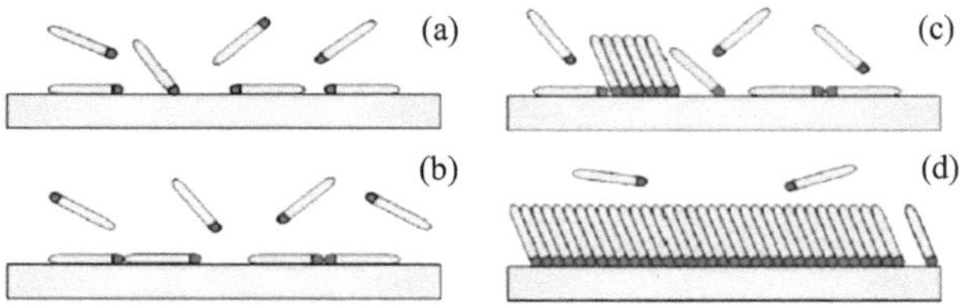

Figure 7.11 Scheme of the process of self-assembly of molecules to a monolayer: (a) physisorption, (b) lying-down phase formation, (c) nucleation of the standing-up phase, and (d) completion of the standing-up phase. Reprinted with permission from Ref. [15]. Copyright (2010) Royal Society of Chemistry.

thiolate (Figs. 7.10b, 7.11b, and 7.11c). At a certain concentration, more molecules can chemisorb as well that physisorbed molecule can desorb from the surface [28].

The last step in the self-assembly process is the slow formation of ordered domains up to highly ordered and oriented monolayers. Highly ordered structures can just be formed, if the concentration of the molecules is high enough to form fully covering monolayers.

To form a gold–thiol bond, the thiol group (S–H, or S–S in disulfides) has to be activated. The bonding energy or strength of the newly formed gold–thiol (Au–S) bond is estimated to be around 50 kcal/mol and is, therefore, a driving force in the activation of the thiol–hydrogen (S–H) bond [30]. Hydrogen (H_2) is supposed to be a byproduct in a gas-phase adsorption process [1, 31]. In solution, hydrogen can form water with oxygen in the reaction medium, called oxidative adsorption [29], followed by a dissociative S–H adsorption.

To monitor the processes of physisorption and chemisorption, X-ray photoelectron spectroscopy (XPS) can be used. In a recently published review by Salvarezza et al. [15], data of XPS studies was used to show the differences between thiolate on gold (C2), free thiol (C3), and atomic sulfur (C1). XPS can be used to measure the elemental composition, the electronic or chemical state. It is used for surface chemical analysis for the top 1–10 nm of a material. XPS can reveal the detailed chemical states of carbon atoms such as alcohols (C–OH), hydrocarbons (C–C), or trifluorocarbons (CF_3). Organosulfur compounds can be identified as alkane thiols (R–CSH), dialkyl disulfides (R–C–SS–C–R) or dialkyl sulfides (R–C–S–C–R) [16,

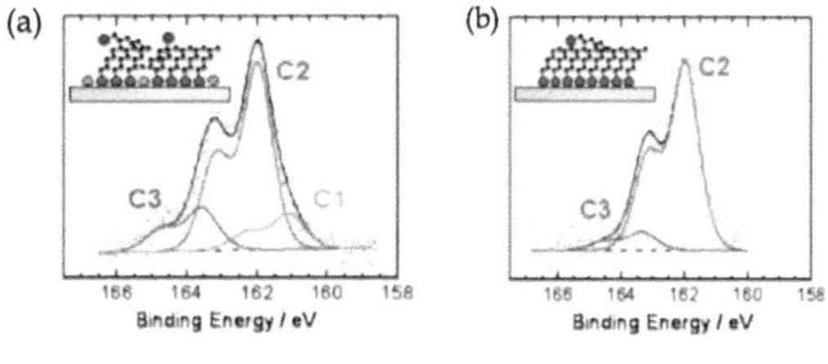

Figure 7.12 XPS spectra (S 2p electrons) of the standing-up phases on Au(111). (a) Propanethilate SAM. C1: atomic sulfur (green), C2: thiolate (red), C3: physorbed molecules (free SH) (blue); (b) Butanethiolate SAM showing the predominant C2 (red) component and only a small amount of C3 (blue). Reprinted with permission from Ref. [15]. Copyright (2010) Royal Society of Chemistry.

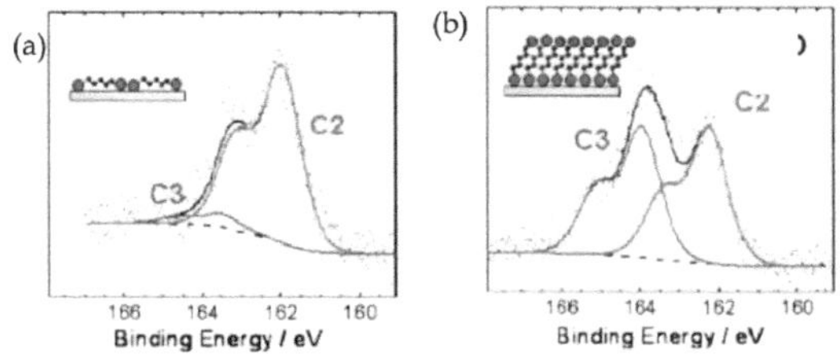

Figure 7.13 XPS spectra (S 2*p* electrons) of alkanedithiol SAMs on Au(111). C2 (red): thiolate, C3 (blue): free SH-terminal group. (a) Butanedithiol SAM: most of the dithiol molecules are in the lying-down configuration. (b) Nonanedithiol SAM: molecules are in standing-up configuration. Reprinted with permission from Ref. [15]. Copyright (2010) Royal Society of Chemistry.

24]. The XPS data for butanethiols (Fig. 7.12a) self-assembling on a surface shows the physisorbed thiol, the free thiol (C3, blue in inset picture), the thiolate, the chemisorbed thiol (C2, red), and the leftover of the atomic sulfur (C1, green). Figure 7.12 shows a more ordered SAM on gold, since C3, which is caused by free physisorbed thiols, is very small and instead the signal from the ordered thiols, C2, is dominant.

Furthermore, the differences between lying-down and standing-up phases can be monitored (Fig. 7.13). Other examples in Fig. 7.13a show dithiols in a lying-down configuration, where just C2 (thiolate) is observed and, therefore, all thiol atoms are chemisorbed on the surface. In Fig. 7.12, the standing-up phase shows the chemisorbed thiol (C2) but also signals from the free terminal SH group in C3.

7.2.3 *How Do Molecules Arrange on a Gold Surface?*

The strong VdW interaction between the spacer units in alkane thiols, described in Section 7.2.1, can give rise to densely packed, crystalline-like two-dimensional assemblies with fully extended alkyl chains when the alkane chain is longer than eight methylene (CH_2) units [1].

In the two-dimensional crystalline case, the molecules on the surface in SAMs can be described by three angles: molecular tilt angle (α), which is the angle between the surface normal direction and the molecular backbone; the angle of rotation of the spacer

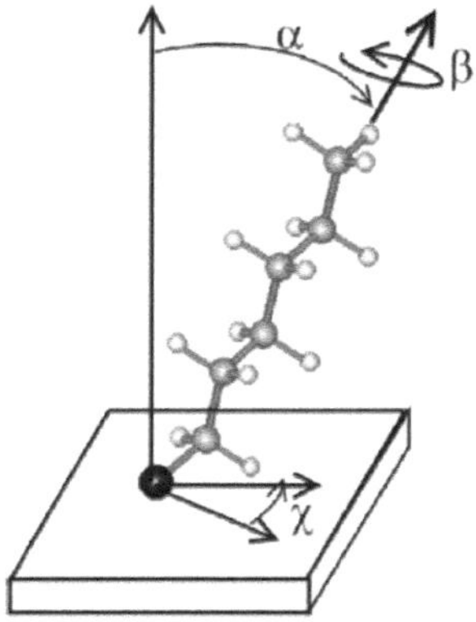

Figure 7.14 Scheme of a hexane thiolate molecule adsorbed on an Au(111) surface in a standing-up configuration. Typical angles are $\alpha = 30°$, $\beta = 55°$, and $\gamma = 14°$. Black circle: sulfur atom; grey: carbon atoms; white: hydrogen atoms. Reprinted with permission from Ref. [14]. Copyright (2005) Royal Society of Chemistry.

unit about the molecular axis (β); and the angle of precession (χ), which gives the tilt direction (derived from the projection of the inclination plane and defined by the substrate normal and the axis of the hydrocarbon chain) [25, 32].

Alkane thiolates on Au(111) in the standing-up, densely packed configuration show the following angles: $\alpha = 30–35°$, $\beta = 55°$, and $\chi = 14°$ (Fig. 7.13). These depend on the substrate and the type of molecule [14, 15, 30]. In comparison, rigid molecules are expected to orient perpendicular to the surface [9].

According to the mechanism of adsorption, thiol heads initially show a "lying-down" phase (not observed for the preparation in liquid phase, where the system evolves directly to the standing-up lattices) for low-coverage and "standing-up" for high-coverage phases (Fig. 7.11) [24]. Some alkane thiols chemisorb at defects directly; some just physisorb. In the second step, the physisorbed molecules undergo S–H cleavage and chemisorb in a lying-down configuration, also called striped phases [33].

Another important detail in SAMs is the formation of structured domains, due to the adsorption on a crystalline gold surface. The thiol heads chemisorb on specific sites of the Au(111) substrate and forms, e.g., a hexagonal lattice. In a high coverage of thiolated molecules on the surface, this is a stable, standing-up lattice with a

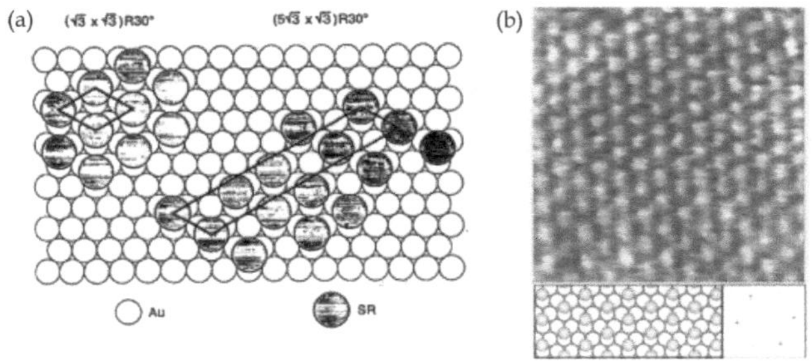

Figure 7.15 (a) Possible surface structures for the $\sqrt{3} \times \sqrt{3}$ $R30^\circ$ and $5\sqrt{3} \times \sqrt{3}$ $R30^\circ$ lattice [21] (b) STM image of the $\sqrt{3} \times \sqrt{3}$ $R30^\circ$ hexanethiolate domain on Au(111). The bottom scheme shows the $\sqrt{3} \times \sqrt{3}$ $R30^\circ$ lattice. Grey circles: alkanethiolate molecules; white circles: gold atoms. Reprinted with permission from Ref. [14]. Copyright (2005) Royal Society of Chemistry.

$\sqrt{3} \times \sqrt{3}$ $R30^\circ$ (R = rotated) arrangement [21, 30, 34, 35]. This means that the structured thiol lattice is rotated 30° with respect to the Au(111) lattice, and the intermolecular spacing in SAM is $\sqrt{3}$ larger than that of the gold atoms (0.288 = R) [32].

Figure 7.15 shows the typical $\sqrt{3} \times \sqrt{3}$ $R30^\circ$ lattice, where the thiols occupy threefold hollow positions on an (unreconstructed) Au(111) [20, 21, 25, 36, 37]. It shows a hexagonal symmetry with 0.5 nm distance between nearest neighbors and a surface coverage of $\theta = 1/3$ [15, 38].

In an STM image (Fig. 7.15) of the $\sqrt{3} \times \sqrt{3}$ $R30^\circ$ lattice, the bright spots represent the position of the thiol molecules and the white dots in the inset indicate the hexagonal lattice.

Besides the stable $\sqrt{3} \times \sqrt{3}$ $R30^\circ$ lattice, the c(4 × 2) superlattice (Fig. 7.15) was detected by IR and diffraction techniques [15]. The superlattice shows the same tilt angle as the $\sqrt{3} \times \sqrt{3}$ $R30^\circ$ but is actually a superlattice of the $\sqrt{3} \times \sqrt{3}$ $R30^\circ$. The unit cell is orthorhombic, and it contains four molecules because it consists of four $\sqrt{3} \times \sqrt{3}$ $R30^\circ$ lattices.

7.2.4 *Adsorption Sites on Gold [Au(111)]*

A simple model of the adsorption of gold includes thiolate binding on three different positions at the gold [Au(111)] surface: atop, bridge, and hollow sites, as depicted in Fig. 7.16.

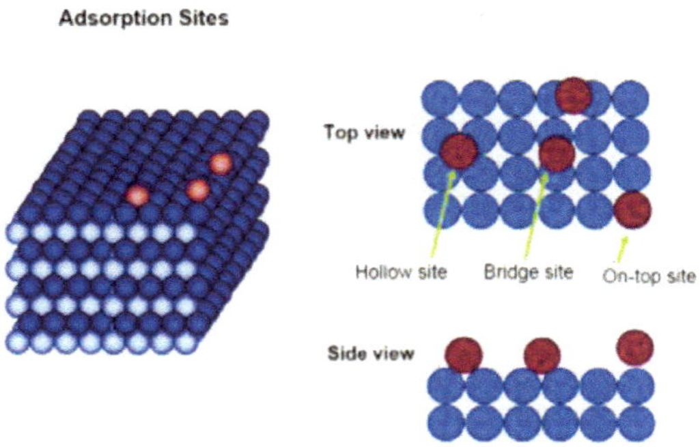

Figure 7.16 Adsorption sites on Au(111) surface. Top and side views show hollow site, bridge site, and on-top site adsorption.

The detailed surface geometry is typically more complicated. Some proposed structures of adsorption sites are disulfide bonding (Fig. 7.17a), a complex that involves an Au adatom binding to the thiolate (Fig. 7.17b), and a polymeric chain structure with monothiolates bridging Au adatoms (Fig. 7.17c).

Jones et al. [39, 40] indicate Au top-site adsorptions of thiolates, which would prove that there are mobile Au–SR complexes on the

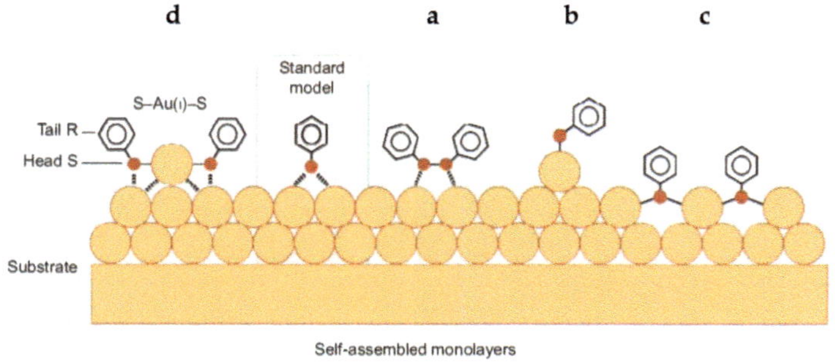

Figure 7.17 Schematics of bonding motifs between thiolates and gold. The standard model is a monothiolate binding at atop, bridge, or hollow site. Other possibilities are: (a) showing disulfide bonding, (b) complexes involving an Au adatom, (c) a polymeric chain structure where monothiolates are bridging Au adatoms, and (d) with the newest results showing a complex of RS–Au(I)–SR with a bridging gold atom in the formal oxidation state +1, found especially in low- and medium-coverage SAM layer. Reprinted with permission from Ref. [24]. Copyright (2012) Nature Publishing Group.

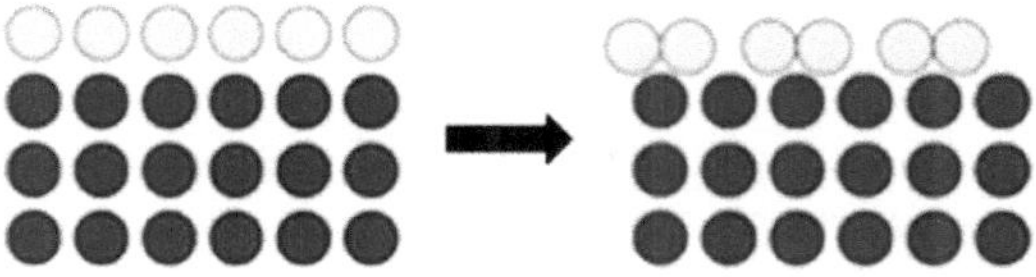

Figure 7.18 Surface reconstruction illustrating how gold atoms on the top atomic layer might move with respect to the crystal lattice.

surface. That means Au adatoms on an Au(111) surface are possible. This process is called surface reconstruction. It is associated with a change in the lateral separation between atoms, as shown in Fig. 7.18.

Further work showed a strong reconstruction of the Au(111) surface, with Au vacancies and/or Au adatoms playing a key role in the binding of thiols to a gold surface [15, 41]. Apparently, a reconstruction of the Au(111) surface occurs as the thiolate SAM forms (chemisorption). The outermost atomic layer of Au(111) adopts the uniaxially compressed hexagonal close packing of the Au–S bond leading to an Au–Au interatomic spacing smaller than that of the underlying bulk [42, 43].

Other surface reconstructions, like the lifting of the "herring-bone" ($22 \times \sqrt{3}$) structure or the appearance of surface vacancy

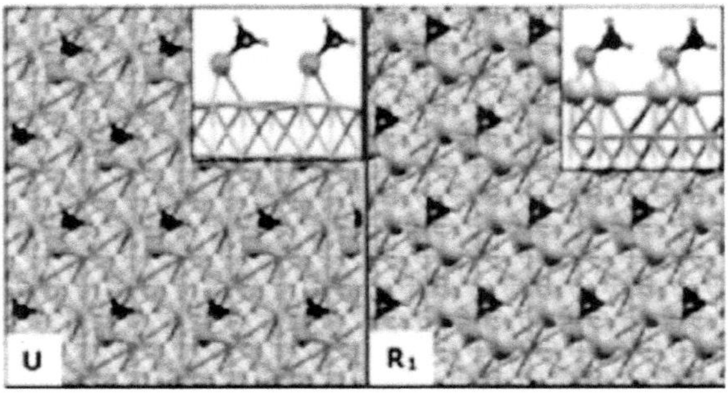

Figure 7.19 Schemes for the $\sqrt{3} \times \sqrt{3}$ $R30°$ methanethiolate lattice on Au(111) as examples for an unreconstructed (U) and a reconstructed (R) Au surface. Shown are the first gold layer (golden), the Au adatoms (lila), the S atoms (green), the C atoms (black), and the H atoms (light blue). Reprinted with permission from Ref. [15]. Copyright (2010) Royal Society of Chemistry.

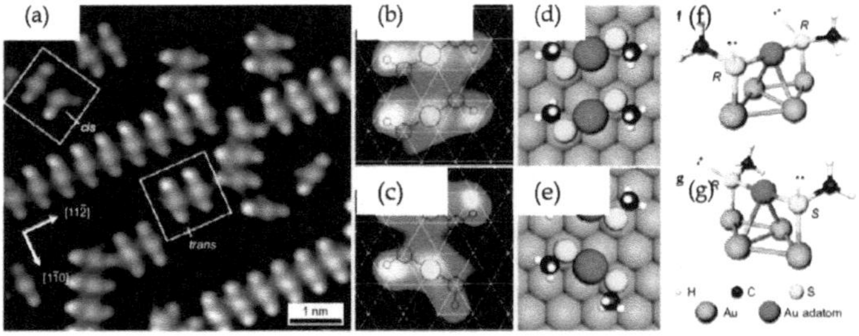

Figure 7.20 (a) STM image of a self-assembled striped pattern, formed by heating a gold crystal pre-dosed with DMSO above 200 K for 10 min. (b) and (d) close-up image of a pair of *trans*-CH_3S-Au-SCH_3 complexes and its atomic model. (c) and (e) Image of the *cis-trans* pair. (f) and (g) show the DFT-optimized structures of the *trans* (f) and the *cis* (g) complexes. Reprinted with permission from Ref. [24]. Copyright (2012) Nature Publishing Group.

islands ("etch pits"), are observed, in line with the mobile adatom theory. Many theoretical models for the $\sqrt{3} \times \sqrt{3}$ $R30°$ and the c(4 × 2) lattice indicate that thiol adsorption promotes a reconstruction of the Au(111) surface with Au adatoms or vacancies.

Recent experiments in STM (see Fig. 7.20) give the evidence of the formation of the RS-Au(1)-SR complexes, mentioned in Fig. 7.16, on Au(111) at low thiolate coverage. The bridging gold atom is in a formal oxidation state of +1 [44]. In these complexes, the central Au atom (red, Fig. 7.20d–g, or large gold adatom, Fig. 7.17d) is coordinated by bonds parallel to the surface. These bonds are between the adatom and the two thiolates, bridging two Au atoms of the first surface layer underneath. Here sulfur adopts gold on-top sites.

The detailed nature of the thiolate–Au(111) interface complex depends on several parameters, such as molecular structure, surface structure, and reaction conditions.

In summary, this section discussed the process of self-assembly, particularly with thiols on a gold surface. The corresponding inter- and intramolecular forces were discussed as well as the process of physisorption and chemisorption, which can lead, at the right experimental conditions, to highly ordered and oriented structures of SAM molecules on the substrate. Furthermore, the adsorption sites on gold with the recent understanding of adatoms and surface reconstructions due to the formation of the thiolate bond are described. In the context of single-molecule electronics, it is important to note that the detailed binding geometry of a thiol to a gold surface can be very complex and needs to be considered when interpreting experimental observations and that multiple geometries are possible.

7.3 Gold and Other Materials

7.3.1 *Why is Gold the Most Prominent Example for Self-Assembly?*

In the early 1980s, Nuzzo and Allara explored SAMs of thiols and dithiols on gold surfaces [45]. The combination of sulfur chemisorbed on an Au(111) surface serves as a model system to understand the process of self-assembly due to several reasons.

- Strong gold–sulfur bond, due to their high affinity [45, 46], resulting in an optimal connection between the metal surface and the head group of the functional molecule.
- SAMs on gold surfaces can be prepared in gas phase and solution.
- Gold surfaces are easy to obtain; thin films or patterns can be produced by lithography.
- Gold films can be directly used for a number of existing spectroscopies and analytical techniques, such as surface plasmon resonance (SPR), quartz crystal microbalance (QCM), and IR (e.g., grazing angle Fourier-transform infrared external reflection spectroscopy).
- Gold is chemically inert, not toxic, and therefore compatible with cells [1].
- Clean and flat surfaces with preferential orientation in (111) can be prepared after thermal annealing [46].

Silver–sulfur has been used widely, but silver is toxic to cells and oxidizes easily in air [47, 48]. Palladium, on the other side, is stable in air and shows catalytic properties and biocompatibility, which makes it also an interesting object [1].

Table 7.1 gives an overview of commonly used metal surfaces and their ligands as well as some metal oxides.

Table 7.1 Metals and metal oxides and their common ligands [1, 29]

Substrate	Ligand	Substrate	Ligand
Au	SH RSSR 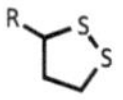 RCSSH $RS_2O_3^-Na^+$ R_3P RSAc RSR 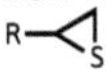 RC≡N RSeH RSeSeR $RO(C{=}S)S^-$ $R_2N(C{=}S)^-$	Si	ROH $RHC{=}CH_2$
Ag	RC≡N RSH RSSR RSeH RCOO-/RCOOH	TiO_2	RCOO-/RCOOH $R_3PO_3^2/RP(O)(OH)_2$ RPO_4^{2-} $RSiX_3$ (S = H, Cl, OCH_2CH_3)
Pd	RSH RSSR R- $N{\equiv}N^+(BF_4^-)$	SiO_x	SiX_3(X = Cl, OMe, OEt) RLi RMgX ROH
Pt	RSH RN≡C	Fe_xO_y	ROH RCOO-/RCOOH
Cu	RSH $RS_2O_3^-Na^+$	Al_2O_3	RCOO-/RCOOH
Zn	RSH		
CdSe	RSH RSeH R_3P $R_3P{=}O$ $R_3PO_3^{2-}/RP(O)(OH)_2$ RCSSH		
CdS	RSeH R_3P $R_3P{=}O$ RSH RSSR		

By choosing different ligands, the process of self-assembly and the strength of the new bonds differ. For example, gold–sulfur (Au–S) (40–50 kcal/mol) is a very strong and stable one. Gold–nitrogen (Au–N) (8 kcal/mol) is already six times weaker, and gold–oxygen (Au–O) (2 kcal/mol) is 25 times weaker [17].

7.3.2 *Silicon Surfaces*

The self-assembly process of alkyltrichlorosilanes on oxidized silicon was originally discovered by Bigelow et al. [49] in 1946 and further developed by Maoz and Sagiv [50]. SAMs can also be spontaneously formed on Si substrates by deposition in vapor phase or in solution [51–53]. Ligands are usually $R(CH_2)nSiX_3$ ($X = Cl$, OCH_3, OC_2H_5, and diazo-compounds) organic molecules, dissolved in alkanes or carbon tetrachlorides [54, 55].

The structure of ligands for silicon oxide surfaces is somewhat similar to what we discussed for sulfur ligands. It consists of a head group with an anchor motif for the substrate (here SiX_3), an alkyl chain as a spacer unit, where the VdW interaction assists the SAM formation, and a functional terminal group [54].

Silicium is spontaneously covered with a native oxide layer, which does not cover the surface evenly. Therefore, a silicon oxide surface is usually generated before the self-assembly. A silicon oxide layer is highly hydrophilic because of its high density of silanol groups (Si–OH). The silanol groups can be used for silanization, or can be activated by halogenation or amination for further functionalization such as alkylation or alkoxylation (see Fig. 7.21).

Figure 7.21 describes the silanization, which occurs in four steps: (1) the silanes get physisorbed on the hydrated surface (physisorption); (2) the silane head group (SiX_3) is hydrolyzed to trihydroxysilane [$Si(OH)_3$] in the adsorbed water layer on the surface (hydrolysis); (3) trihydroxysilane undergoes a condensation reaction and forms covalent bonds to the silicon oxide surface [$Si(OH)_2$] (covalent grafting to the surface); and (4) the self-assembly is driven by VdW interactions of the hydrophobic spacer unit.

The halogenation reactions on surfaces can be carried out using halogenation reagents known from solution-based organic synthesis, e.g., NBS or PCl_5 (see Fig. 7.22a) [56–59]. These

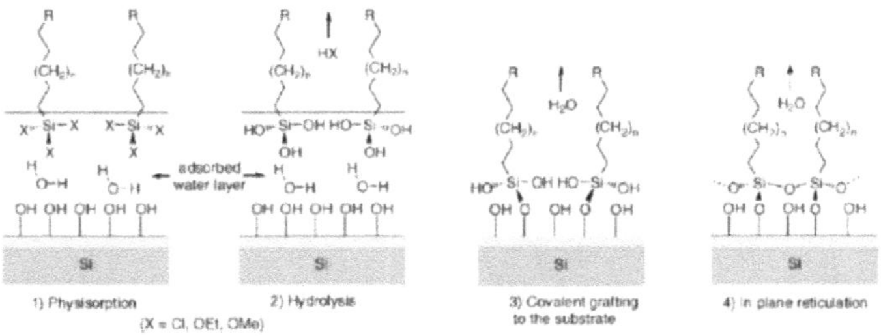

Figure 7.21 Schematic showing the different steps involved in the mechanism of SAM formation on a hydrated silicon surface. Reprinted with permission from Ref. [54]. Copyright (2006) Elsevier.

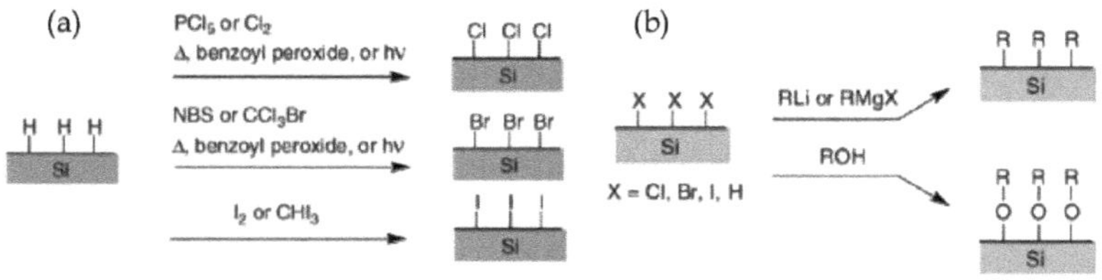

Figure 7.22 (a) Common methods used for silicon halogenation. (b) Formation of monolayers on halogen-terminated Si using organomagnesium, organolithium, or alcohols. Reprinted with permission from Ref. [54]. Copyright (2006) Elsevier.

activated (halogenated) surfaces can be further functionalized with organomagnesium, organolithium, or alcohols (Fig. 7.22) [60, 61].

This type of surface chemistry can be used to introduce different functionalities to the surface by changing the carbon chain and the terminal functional group. A large range of reactive and functional silanes are today available from chemical suppliers.

7.3.3 *Graphene for Self-Assembled Electrodes*

In molecular electronics, gold structures have been the preferred material for electrodes due to their high charge carrier mobility and their noble metal character. But at room temperature, they are often unstable due to the restructuring of the gold surface and low barriers between different gold–sulfur binding sites [62]. To try to circumvent some of these challenges, van der Zant et al. [63]

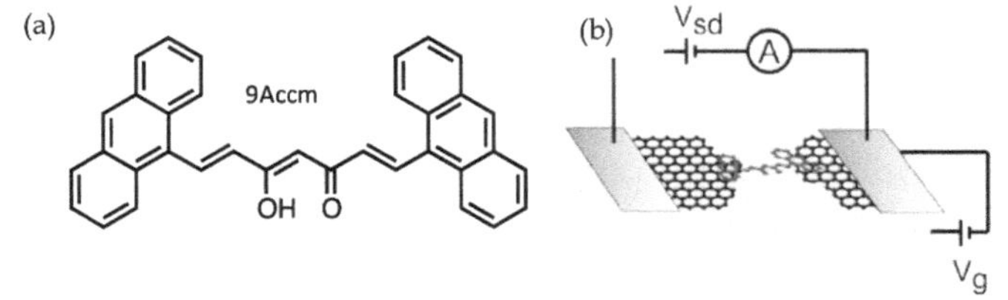

Figure 7.23 (a) Chemical structure of the bridging molecule wire [1,7-(di-9-anthracene)-16-heptadiene-3,5-dione]; (b) schematic picture of a single molecule bridging a graphene nanogap, representing the ideal transistor configuration. Reprinted with permission from Ref. [63]. Copyright (2011) American Chemical Society.

used a few-layer graphene as nanogap electrode material (Fig. 7.23). Graphene can be functionalized via covalent bonds with organic chemistry, but also π–π stacking (induced-dipole–induced-dipole) of aromatic compounds can be used [64].

Another example of self-assembly that tries to circumvent the challenge associated with multiple possible binding geometries for thiols on gold was to use large C_{60} molecules as anchor groups as demonstrated by van der Zant et al. [65]. They showed a way to control the contact between a molecule and a metallic electrode surface by using a conducting molecule, a linear and rigid C_{60}-capped anchor, with well-defined contact geometry (see Fig. 7.24).

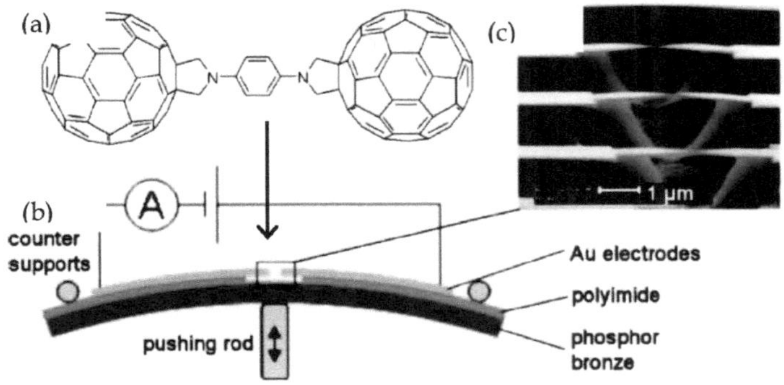

Figure 7.24 (a) C_{60}-anchor molecule; (b) schematic picture of the mechanically controllable break junction; and (c) SEM of the device; three different junctions. Reprinted with permission from Ref. [65]. Copyright (2008) American Chemical Society.

With this approach, a stable chemical anchoring of a molecule was achieved in which the single-molecule junction was occupied via self-assembly C_{60} shows high affinity to noble metals and hybridize strongly with gold [70]. Due to multiple connection sides, C_{60} serves as a strong binder to gold; therefore, a defined chemical bond in the C_{60}-construct serves as the interface between molecule and electrode.

With this approach, they tried to overcome the challenges associated with the use of thiol chemistry for gold surfaces. Amines can be used instead of thiols [66, 67], but the strength of the bond (Au–N) is quite weak [68, 69].

7.4 Summary and Conclusion

This chapter focused on describing the basic processes in self-assembly of molecules on metallic surfaces, since this is a basic tool in single-molecule electronics. Hereby, the knowledge about inter- and intramolecular forces, leading to monolayers and highly ordered structures, was mediated. A deeper look into the detailed adsorption of thiol-anchored molecules on gold surfaces was given, and recent knowledge about adatoms and surface reconstruction due to the formation of the Au–S bond was described.

Futhermore, self-assembly on silicon surfaces with, e.g., alkyl-silanes was described, since silicon is a common surface for microelectronics. Finally graphene was described as a nanogap electrode material. With its help, single molecules can be placed via self-assembly on electrodes.

Problems

1. SAMs of alkyl thiols are formed on gold via a stepwise process. Describe the four steps in the self-assembly process.
2. Describe the three structural motifs a molecule needs to contain to form a functional SAM.
3. Weak intermolecular forces govern self-assembly processes; name and describe these forces.

4. C_{60} terminal groups have been used as an alternative to thiols for single-molecule electronic devices. What are the advantages and disadvantages of using such functional groups?
5. What type of functional groups are typically used for self-assembly on silicon oxide surfaces?

References

1. J. C. Love, L. A. Estroff, J. K. Kriebel, R. G. Nuzzo, and G. M. Whitesides, *Chemical Reviews* **2005**, *105*, 1103–1169.
2. L. M. Smith, *Nature* **2006**, *440*, 283–284.
3. A. S. Reder, N. Spencer, and J. F. Stoddart, *Angewandte Chemie (Int. ed.)* **1994**, *33*, 1286–1290.
4. D. J. Williams, A. J. P. White, J. F. Stoddart, F. M. Raymo, and J. A. Bravo, *European Journal of Organic Chemistry* **1998**, *11*, 2565–2571.
5. J. Harrowfield and J. Lehn, *Chemistry in New Zealand* **2011**, *75*, 170–173.
6. L. R. Harriott, *Proceedings of the IEEE* **2001**, *89*, 366–374.
7. T. Dadosh, Y. Gordin, R. Krahne, I. Khivrich, D. Mahalu, V. Frydman, J. Sperling, A. Yacoby, and I. Bar-Joseph, *Nature* **2005**, *436*, 677–680.
8. http://scienceandscientist.org/biology/Pictures/Top-down_Bottom-up.jpg, "Top-down_Bottom-up," **2012**.
9. A. L. A. Bumm, J. J. Arnold, M. T. Cygan, T. D. Dunbar, T. P. Burgin, L. J. Ii, L. Allara, J. M. Tour, and P. S. Weiss, *Science* **1996**, *271*, 1705–1707.
10. K. Moth-Poulsen, L. Patrone, N. Stuhr-Hansen, J. B. Christensen, J.-P. Bourgoin, and T. Bjørnholm, *Nano Letters* **2005**, *5*, 783–785.
11. Y. Kim, T. Pietsch, A. Erbe, W. Belzig, and E. Scheer, *Nano Letters* **2011**, *11*, 3734–3738.
12. K. V. Vijayamohanan, J. Sharma, M. Aslam, and N. Y. A. K. Chaki, *Proceedings of the National Academy of Sciences of the United States of America* **2001**, *113*, 659–670.
13. N. K. Chaki and K. Vijayamohanan, *Biosensors and Bioelectronics* **2002**, *17*, 1–12.
14. C. Vericat, M. E. Vela, and R. C. Salvarezza, *Physical Chemistry Chemical Physics* **2005**, *7*, 3258–3268.
15. C. Vericat, M. E. Vela, G. Benitez, P. Carro, and R. C. Salvarezza, *Chemical Society Reviews* **2010**, *39*, 1805–1834.

16. P. Chinwangso, A. C. Jamison, and T. R. Lee, *Accounts of Chemical Research* **2011**, *44*, 511–519.
17. E. Pensa, E. Cortés, G. Corthey, P. Carro, C. Vericat, M. H. Fonticelli, G. Benítez, A. A Rubert, and R. C. Salvarezza, *Accounts of Chemical Research* **2012**, *45*, 1183–1192.
18. E. Torres, A. T. Blumenau, and P. U. Biedermann, *ChemPhysChem* **2011**, *12*, 999–1009.
19. M. D. Porter, T. B. Bright, D. L. Allara, and C. E. D. Chidsey, *Journal of the American Chemical Society* **1987**, *109*, 3559–3568.
20. L. H. Dubois and R. G. Nuzzo, *Annual Reviews Physical Chemistry* **1992**, *43*, 437–463.
21. L. H. Dubois, B. R. Zegarski, and R. G. Nuzzo, *The Journal of Chemical Physics* **1993**, *98*, 678.
22. M. Hasan, D. Bethell, and M. Brust, *Journal of the American Chemical Society* **2002**, *124*, 1132–1133.
23. J.-G. Lee, J. Lee, and J. T. Yates, *Journal of the American Chemical Society* **2004**, *126*, 440–441.
24. H. Häkkinen, *Nature Chemistry* **2012**, *4*, 443–455.
25. F. Schreiber, *Progress in Surface Science* **2000**, *65*, 151–257.
26. D. K. Schwartz, *Annual Reviews Physical Chemistry* **2001**, *52*, 107–137.
27. P. Maksymovych, D. Sorescu, and J. Yates, *Physical Review Letters* **2006**, *97*, 1–4.
28. X. Torrelles, C. Vericat, E. Vela, M. H. Fonticelli, A. Mart, and R. C. Salvarezza, *Journal of Physical Chemistry B* **2006**, *110*, 5586–5594.
29. A. Ulman, *Chemical Reviews* **1996**, *96*, 1533–1554.
30. R. G. Nuzzo, B. R. Zegarski, and L. H. Dubois, *Journal of the American Chemical Society* **1987**, *109*, 733–740.
31. L. Kankate, A. Turchanin, and A. Gölzhäuser, *Langmuir* **2009**, *25*, 10435–10438.
32. B. E. Delamarche, B. Michel, H. A. Biebuyck, and C. Gerber, *Advanced Materials* **1996**, *8*, 719–729.
33. R. Staub, M. Toerker, T. Fritz, T. Schmitz-Hübsch, F. Sellam, and K. Leo, *Langmuir* **1998**, *14*, 6693–6698.
34. C. E. D. Chidsey, D. N. Loiacono, T. B. Laboratories, and H. Murray, *Langmuir* **1990**, *1*, 682–691.
35. L. Strong and G. M. Whitesides, *Langmuir* **1988**, *4*, 546–558.
36. M. Walter, J. Akola, O. Lopez-Acevedo, P. D. Jadzinsky, G. Calero, C. J. Ackerson, R. L. Whetten, H. Grönbeck, and H. Häkkinen, *Proceedings of*

the National Academy of Sciences of the United States of America **2008**, *105*, 9157–9162.

37. G. E. Poirier, *Chemical Reviews* **1997**, *97*, 1117–1128.
38. M. R. Buck, J. F. Bondi, and R. E. Schaak, *Nature Chemistry* **2012**, *4*, 37–44.
39. H. Kondoh, M. Iwasaki, T. Shimada, K. Amemiya, T. Yokoyama, T. Ohta, M. Shimomura, and S. Kono, *Physical Review Letters* **2003**, *90*, 1–4.
40. M. G. Roper, M. P. Skegg, C. J. Fisher, J. J. Lee, V. R. Dhanak, D. P. Woodruff, and R. G. Jones, *Chemical Physics Letters* **2004**, *389*, 87–91.
41. G. Liu and G. Yang, *Journal of Physical Chemistry C* **2003**, *107*, 8746–8759.
42. U. Harten, A. Lahee, Peter Toennies J, C. Wöll, *Physical Review Letters* **1985**, *54*, 2619–2622.
43. K. Huang, D. Gibbs, D. Zehner, A. Sandy, and S. Mochrie, *Physical Review Letters* **1990**, *65*, 3313–3316.
44. E. Pensa, E. Cortés, G. Corthey, P. Carro, C. Vericat, M. H. Fonticelli, G. Benítez, A. A. Rubert, and R. C. Salvarezza, *Accounts of Chemical Research* **2012**, *45*, 1183–1192.
45. R. G. Nuzzo and D. L. Allara, *Journal of the American Chemical Society* **1983**, *105*, 4481–4483.
46. C. Vericat, M. E. Vela, G. A. Benitez, J. A. M. Gago, X. Torrelles, and R. C. Salvarezza, *Journal of Physics: Condensed Matter* **2006**, *18*, R867–R900.
47. D. P. Dowling, K. Donnelly, M. L. McConnell, R. Eloy, and M. N. Arnaud, *Thin Solid Films* **2001**, *398–399*, 602–606.
48. V. K. M. Poon and A. Burd, *Burns* **2004**, *30*, 140–147.
49. W. C. Bigelow, D. L. Pickett, and W. A. Zisman, *Journal of Colloid Science* **1946**, *1*, 513–538.
50. R. Maoz and J. Sagiv, *Journal of Colloid Science* **1984**, *I*, 465–496.
51. E. Pavlovic, A. P. Quist, U. Gelius, and S. Oscarsson, *Journal of Colloid and Interface Science* **2002**, *254*, 200–203.
52. J. M. Buriak, *Chemical Reviews* **2002**, *102*, 1271–1308.
53. J. B. Brzoska, I. Ben Azouz, and F. Rondelez, *Langmuir* **1994**, *10*, 4367–4373.
54. D. K. Aswal, S. Lenfant, D. Guerin, J. V. Yakhmi, and D. Vuillaume, *Analytica Chimica Acta* **2006**, *568*, 84–108.
55. M. P. Stewart, F. Maya, D. V. Kosynkin, S. M. Dirk, J. J. Stapleton, C. L. McGuiness, D. L. Allara, J. M. Tour, *Journal of the American Chemical Society* **2004**, *126*, 370–378.

56. A. Bansal, X. Li, I. Lauermann, and N. S. Lewis, *Journal of the American Chemical Society* **1996**, *118*, 7225–7226.
57. A. Bansal, X. Li, S. I. Yi, W. H. Weinberg, and N. S. Lewis, *The Journal of Physical Chemistry B* **2001**, *105*, 10266–10277.
58. B. J. Eves and G. P. Lopinski, *Surface Science* **2005**, *579*, 89–96.
59. X. Zhu, V. Boiadjiev, J. A. Mulder, R. P. Hsung, and R. C. Major, *Langmuir* **2000**, *16*, 6766–6772.
60. J. H. Song and M. J. Sailor, *Journal of the American Chemical Society* **1998**, *7863*, 2376–2381.
61. N. Y. Kim and P. E. Laibinis, *Journal of the American Chemical Society* **1999**, *121*, 7162–7163.
62. F. Prins, T. Hayashi, B. J. A. de Vos van Steenwijk, B. Gao, E. A. Osorio, K. Muraki, and H. S. J. van der Zant, *Applied Physics Letters* **2009**, *94*, 123108.
63. F. Prins, A. Barreiro, J. W. Ruitenberg, J. S. Seldenthuis, N. Aliaga-Alcalde, L. M. K. Vandersypen, and H. S. J. van der Zant, *Nano Letters* **2011**, *11*, 4607–4611.
64. X. Guo, J. P. Small, J. E. Klare, Y. Wang, M. S. Purewal, I. W. Tam, B. H. Hong, R. Caldwell, L. Huang, S. O'brien, et al., *Science* **2006**, *311*, 356–359.
65. C. A. Martin, D. Ding, J. K. Sørensen, T. Bjørnholm, J. M. van Ruitenbeek, and H. S. J. van der Zant, *Journal of the American Chemical Society* **2008**, *130*, 13198–13199.
66. A. M. Moore, A. A. Dameron, B. A. Mantooth, R. K. Smith, D. J. Fuchs, J. W. Ciszek, F. Maya, Y. Yao, J. M. Tour, and P. S. Weiss, *Journal of the American Chemical Society* **2006**, *128*, 1959–1967.
67. C. Li, I. Pobelov, T. Wandlowski, A. Bagrets, A. Arnold, and F. Evers, *Journal of the American Chemical Society* **2008**, *130*, 318–326.
68. C. A. Martin, D. Ding, H. S. J. van der Zant, and J. M. Van Ruitenbeek, *New Journal of Physics* **2008**, *10*, 065008.
69. L. Venkataraman, J. E. Klare, C. Nuckolls, M. S. Hybertsen, and M. L. Steigerwald, *Nature* **2006**, *442*, 904–907.
70. C. Rogero, J. I. Pascual, J. Gomez-Herrero, A. M. Baro, *The Journal of Chemical Physics* **2002**, *116*, 832.

Chapter 8

Molecular Switches

Mogens Brøndsted Nielsen
Department of Chemistry, University of Copenhagen,
Universitetsparken 5 Copenhagen, DK-2100, Denmark
mbn@kiku.dk

Molecules existing in two or more states are particularly attractive as molecular switches for molecular electronics applications. A stimulus may change such molecules between different redox states, between translational isomers, between *cis* and *trans* isomers, between isomers of electrocyclic reactions, or between different tautomers. This chapter describes how a variety of redox- and photoactive molecules can be used for conductance switching. It also describes synthetic protocols for some of the switches.

8.1 Introduction

A molecular switch is a molecule that can be reversibly interconverted between two or more states by external stimuli. Molecules that can be interchanged between high- and low-conducting states play a central role in information storage and logic operations at the molecular level and hence for the development of molecular electronics [1–6]. Thus, switches or transistors are required to

Handbook of Single-Molecule Electronics
Edited by Kasper Moth-Poulsen

ISBN 978-981-4463-38-6 (Hardcover), 978-981-4463-39-3 (eBook)
www.panstanford.com

perform conventional computing operations. Molecular switching may be triggered by a variety of stimuli, such as redox, light, or heat, and may lead to either charging of the molecule, structural changes via isomerization reactions (e.g., *cis-trans* isomerizations, tautomerizations, or pericyclic reactions), or conformational changes. Sometimes the switching ability is not inherently incorporated in the molecule but is a result of its position between electrodes. For example, Danilov et al. [7] reported a nano-electromechanical switch operated by the tunneling of an entire C_{60} molecule. A C_{60} molecule was trapped unsymmetrically in a nanogap between two silver electrodes. Due to the wideness of the gap (ca. 2 nm), the C_{60} molecule interacted strongly with one electrode via direct contact but did not interact with the other electrode. At low temperatures, the device was found to switch between two states with different conductances. A simple mechanical switching where C_{60} rolls from one orientation to another was used to explain the results. The following section of the chapter will cover molecules for which the switching event is an inherent property. It will focus on molecules that have been mobilized as self-assembled monolayers (SAMs) on surfaces or inserted in alkanethiolate monolayers, and usually studied by conducting-probe atomic force microscopy (CP-AFM) or by scanning tunneling microscopy (STM), or on molecules that have been studied as single molecules in two- or three-terminal (source, drain, and gate electrodes) devices. The chosen examples provide an overview of a variety of switching mechanisms. Thiolates or sulfides are most commonly employed as anchoring groups to adhere the molecules to a metal surface (gold or silver), and synthetic protocols will be given for selected molecules.

8.2 Redox-Controlled Switches

Molecules having redox-active groups can exist in different charge states, and switching between these has, in a variety of systems, resulted in conductance switching. Moreover, some systems exhibit hysteretic switching, which means that the system does not respond instantly to the applied stimulus. Such delayed responses are important for the development of data storage.

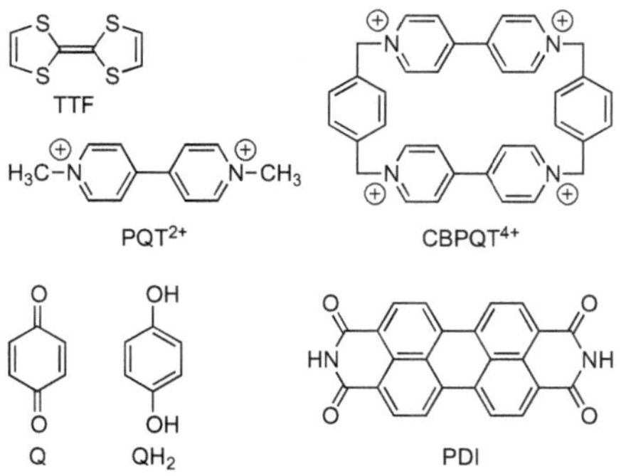

Figure 8.1 Redox-active molecules.

8.2.1 *Redox-Active Molecules*

Figure 8.1 shows a variety of molecules that have found use in molecular electronics due to their redox properties. Tetrathiafulvalene (TTF) is a good electron donor, which can be oxidized reversibly in two one-electron steps (Fig. 8.2). As discussed below, studies on suitable derivatives have shown that the charged species (TTF^{2+}) has a higher conductance (*on*) than the neutral state (*off*). Paraquat (PQT^{2+}) is an electron acceptor, which can be reduced stepwise in two one-electron steps (Fig. 8.2), and studies on a paraquat derivative in a junction have revealed conductance switching upon changing the redox state [8]. The cyclophane cyclobis(paraquat-*p*-phenylene), $CBPQT^{4+}$ (Fig. 8.1), is composed of two paraquat units and is able to encapsulate electron-rich molecules such as TTF in its cavity by the action of donor–acceptor interactions. This ability has been used to generate mechanically interlocked molecules such as catenanes and rotaxanes [9], and an example of a memory device

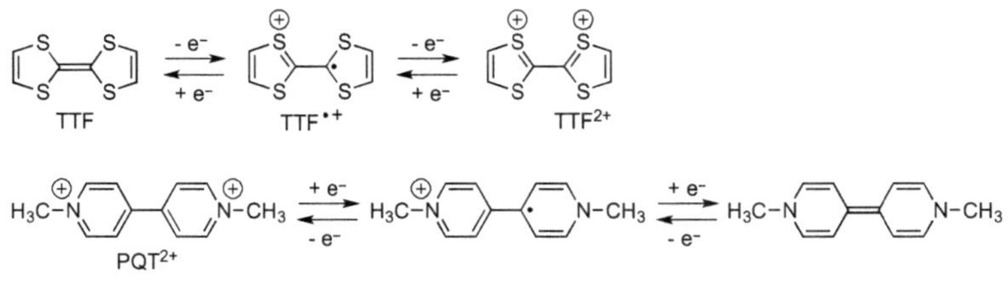

Figure 8.2 Reversible oxidations/reductions of TTF and PQT^{2+}.

based on a rotaxane will be described below. Perylene diimide (PDI, Fig. 8.1) is another electron acceptor, and a derivative functionalized with thiolate-anchoring groups was wired between the source and drain gold electrodes. The current through the molecule was reversibly controlled with a third electrode (gate) over nearly three orders of magnitude [10]. The molecule thus behaved as a single-molecule *n-type* transistor.

8.2.2 *Hydroquinone–Quinone and OPV Switches*

The hydroquinone (HQ)–quinone (Q) redox couple (Fig. 8.1) was used in an electrochemically controlled switch [11]. Thus, a quinone was incorporated as a functional unit in an oligo(phenylenevinylene) (OPV) molecular wire that contains an acetyl-protected thiolate for anchoring onto gold (after cleavage of the acetyl group). The molecule was inserted in an octanethiolate self-assembled monolayer (SAM), as shown in Fig. 8.3. By the application of an electrochemical potential, switching between the quinone and the hydroquinone was achieved. Electrochemical STM was used both to control the oxidation state and to measure the electrical conductivity of the molecule. The fully conjugated hydroquinone state was found to be the high-conducting state, while the quinone state with its disrupted electron delocalization was the low-conducting state. A conductance ratio of greater than 40 was obtained.

An OPV wire can also, by itself, act as a redox-active unit. Thus, a single-electron transistor was constructed from the OPV5 shown in Fig. 8.4 by placing one single molecule, by sublimation, between two gold electrodes in a gated device [12]; the molecule could be taken through eight different charge states (−4, −3, −2, 0, −1, +1, +2, +3) by varying the gate potential. The *tert*-butyl sulfide end-groups provide a weak contact to the electrodes (Coulomb blockade regime).

8.2.3 *Tetrathiafulvalene and Molecular Cruciforms*

An example of functionalization of TTF by thiolate-anchoring groups has been performed according to the route shown in Fig. 8.5

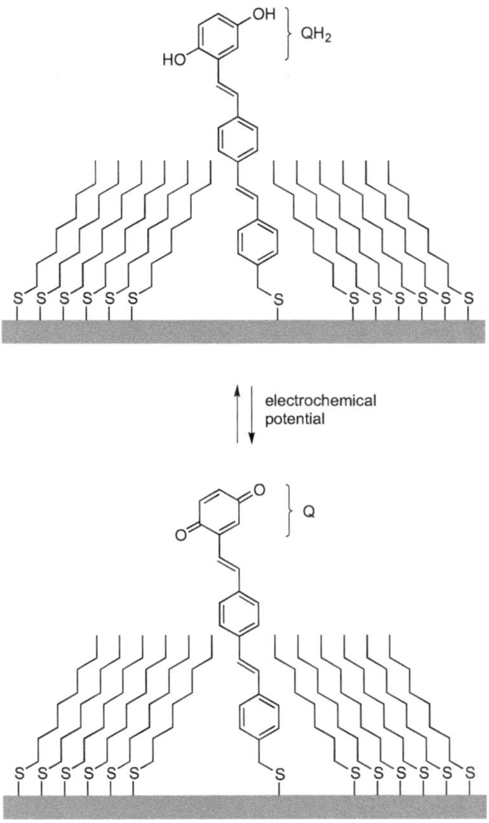

Figure 8.3 Electrochemical switching between QH_2 (*on* state) and Q (*off* state).

[13]. Treating a TTF derivative containing two cyanoethyl-protected thiolate groups [14] with cesium hydroxide caused elimination of acrylonitrile furnishing a TTF-dithiolate, which was then treated with 1-iodo-3-thioacetyl propane. A subsequent reduction in the thioacetyl groups by diisobutylaluminium hydride generated the TTF-dithiol. This molecule was inserted in two-dimensional gold nanoparticle arrays (coated with octanethiols), resulting in interlinked networks of molecular junctions. By chemical oxidation of the TTF unit to the TTF dication using iron(III) chloride, a conductance increase in the networks by up to one order of

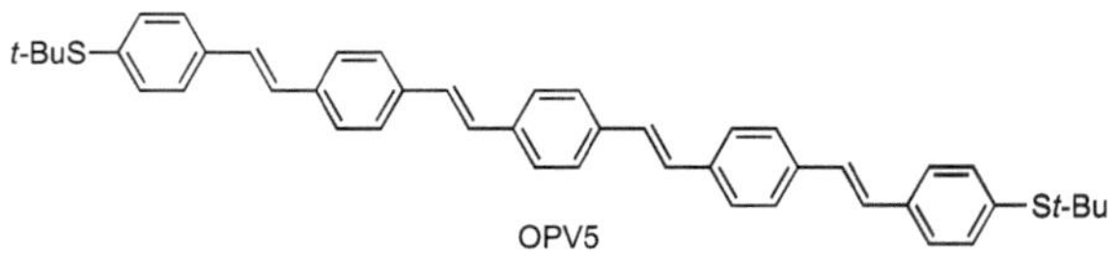

Figure 8.4 Oligo(phenylenevinylene) OPV5.

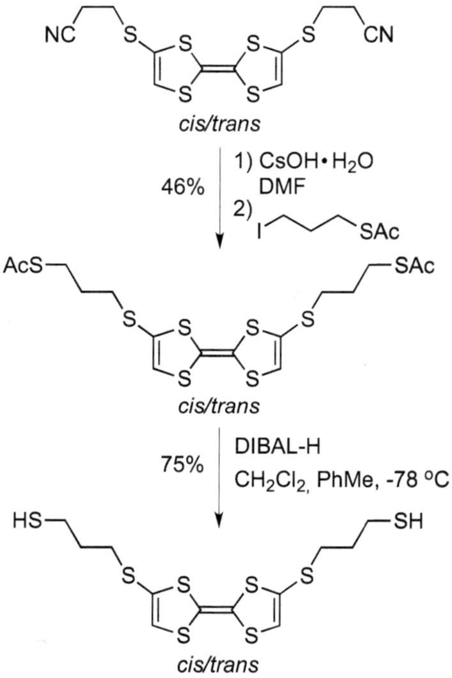

Figure 8.5 Synthesis of TTF-dithiol.

magnitude was observed [13]. Neutral TTF (low-conducting state) could be regenerated by the addition of ferrocene as reducing agent. The low- and high-conducting states could be repeatedly switched by chemical oxidation/reduction, as shown in Fig. 8.6.

Rather than having the redox-active unit along the conduction path, so-called cruciform molecules were designed in which redox-active units are placed orthogonally to the actual wire. The synthesis of a cruciform composed of an extended TTF placed

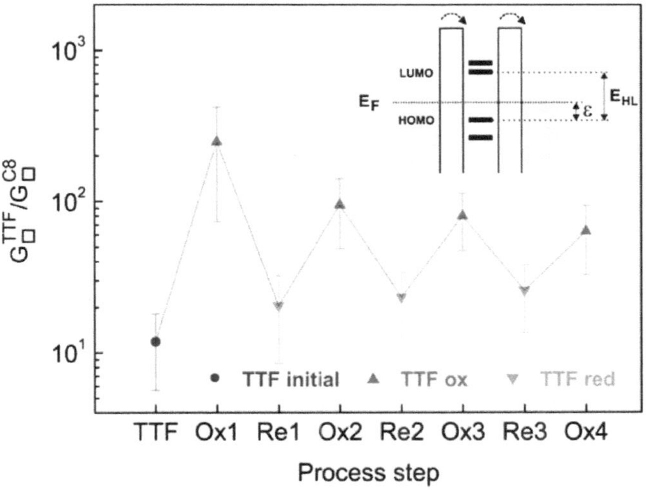

Figure 8.6 Average conductance ratio (between nanoparticles coated with TTF-dithiol/octanethiol and by octanethiol only) for 24 devices during four cycles of oxidation-reduction. The initial neutral TTF state is indicated by a blue circle, while the oxidized TTF dication is shown by red triangles and the regenerated neutral TTF by orange triangles. The inset shows a schematics for charge transport across a molecular junction in a resonant tunneling model. Reprinted (adapted) with permission from Ref. [13]. Copyright (2010) American Chemical Society.

orthogonally to an oligo(phenyleneethynylene) (OPE) wire with SAc anchoring groups is shown in Fig. 8.7 [15]. The central extended TTF moiety of this OPE5-TTF was made (not shown) from Wittig reactions between a terepthalaldehyde precursor and a 1,3-dithiol phosphonium salt containing peripheral methyl ester groups, which were transesterified to propyl esters to enhance the solubility. Stepwise Sonogashira/desilylation reactions provided the OPE5-TTF cruciform. Conducting-probe AFM measurements on SAMs of this molecule showed a ninefold increase in conductance relative to a simple OPE5 wire [16]. The molecule has also been contacted in a three-terminal geometry in which it was reversibly switched between three redox states by a gate electrode, but these results are not yet reported [17].

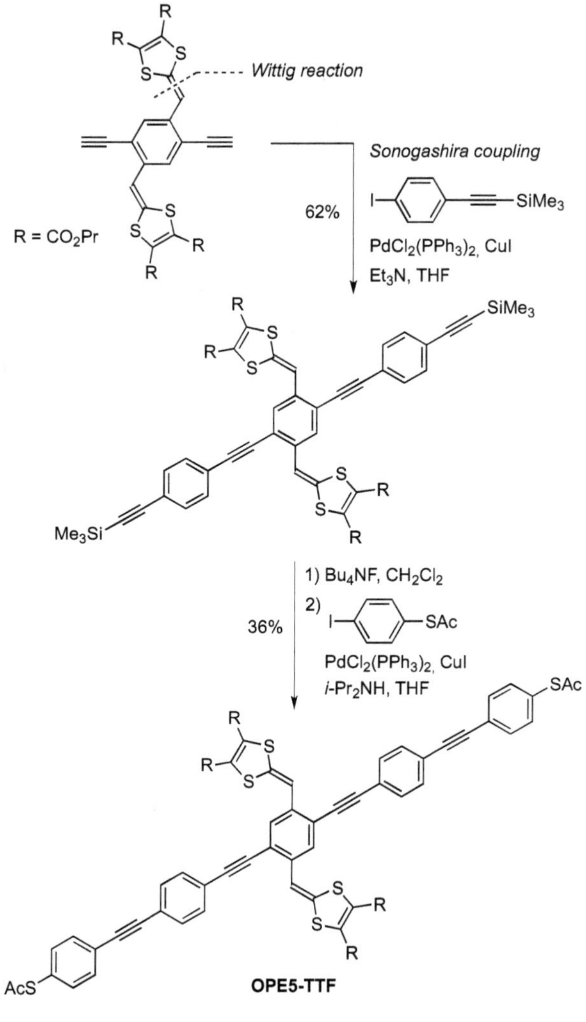

Figure 8.7 Synthesis of OPE5-TTF cruciform molecule.

8.2.4 *Spin-Crossover Metal Complexes*

Electrical manipulation of spin states has been performed in a three-terminal device incorporating a single Mn^{2+} ion coordinated by two terpyridine ligands (Fig. 8.8) [18]. By changing the gate voltage, the terpyridine moiety was reduced (addition of one electron),

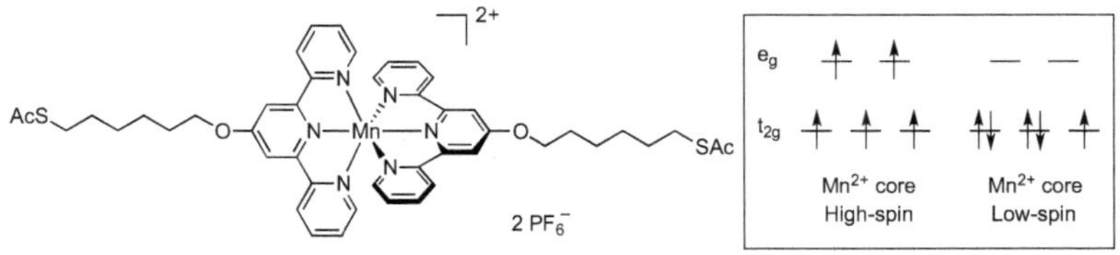

Figure 8.8 High-spin and low-spin states of this manganese complex placed in a three-terminal device were controlled by the gate potential.

which strengthened the ligand-field on the Mn atom. The low-spin configuration of the manganese(II) center was thereby stabilized, and a transition from a high-spin state to a low-spin state was promoted. This transition resulted in a suppression of the sequential tunneling current by spin blockade.

In another example, nanoparticles based on the Fe(II) coordination polymer $[Fe(trz)_3](BF_4)_2$ (trz = triazole) were placed in between two nanometer-spaced electrodes [19]. The particles were ca. 11 nm in diameter and consisted of a spin-crossover Fe(II) core coated with a surfactant shell. It was possible to send current through these particles, and they were found to show bistability in conductance as a function of temperature. A thermal hysteresis loop at room temperature renders the system attractive as a memory element. Moreover, voltage-induced conductance switching was demonstrated with these nanoparticles.

8.2.5 *Bipyridyl-OPE Switch*

Hysteretic switching has been observed in devices incorporating the OPE derivative shown in Fig. 8.9. It contains a central bipyridyl-dinitro unit, which was functionalized by sequential Sonogashira cross-coupling reactions [20]. I–V measurements on isolated molecules [21] and of SAMs [22] revealed a voltage-triggered conductance switching. At a certain bias potential, the molecule switched from a low-conductance state to a high-conductance state. This metastable state was maintained until the bias potential approached 0 V. The exact switching mechanism is, however, not known.

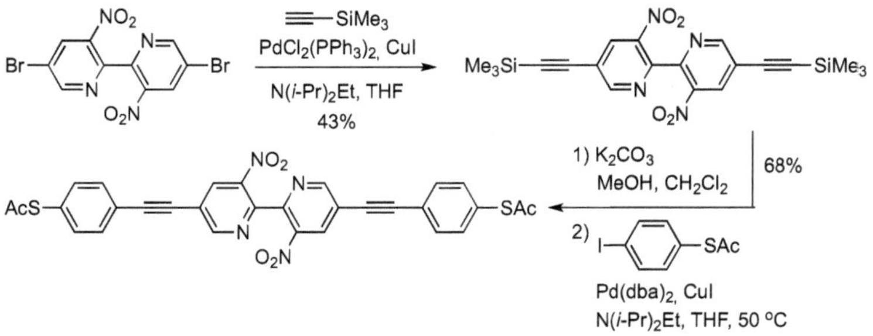

Figure 8.9 Synthesis of bipyridyl-OPE exhibiting hysteretic switching in devices.

8.2.6 *Mechanically Interlocked Molecules*

Another example of hysteresis is provided by switching between translational isomers in interlocked molecules such as catenanes and rotaxanes (Fig. 8.10). For example, the rotaxane shown in Fig. 8.11 was used to store information when integrated in a crossbar architecture [23]. The molecule contains two electron-rich stations: a TTF unit (pyrrolo-annelated TTF) and a dioxynaphthalene (DNP) unit. The cyclic acceptor $CBPQT^{4+}$ is encircling the strongest donor, the TTF unit. The synthesis proceeds via the so-called clipping method (Fig. 8.12). Moreover, the molecule has a hydrophilic and a hydrophobic end, both of which function as stopper groups to prevent dethreading of the cyclic acceptor and allows the molecule to be assembled in a Langmuir–Blodgett film sandwiched between electrodes in crossbar devices. Two different devices have been constructed using this molecule. In one device, the monolayer was

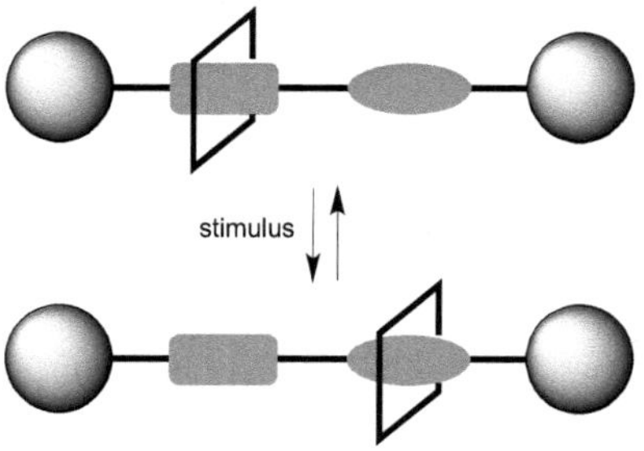

Figure 8.10 Switching between two translational isomers of a rotaxane.

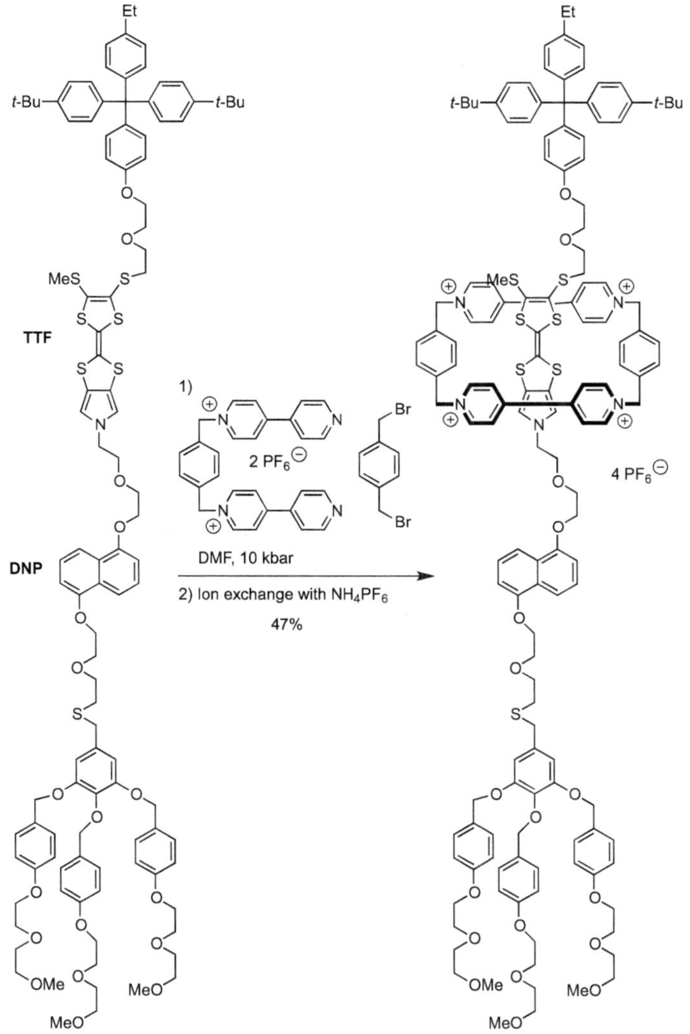

Figure 8.11 Synthesis of rotaxane by the clipping method. The rotaxane contains two donor stations, TTF and DNP, for the cyclic acceptor.

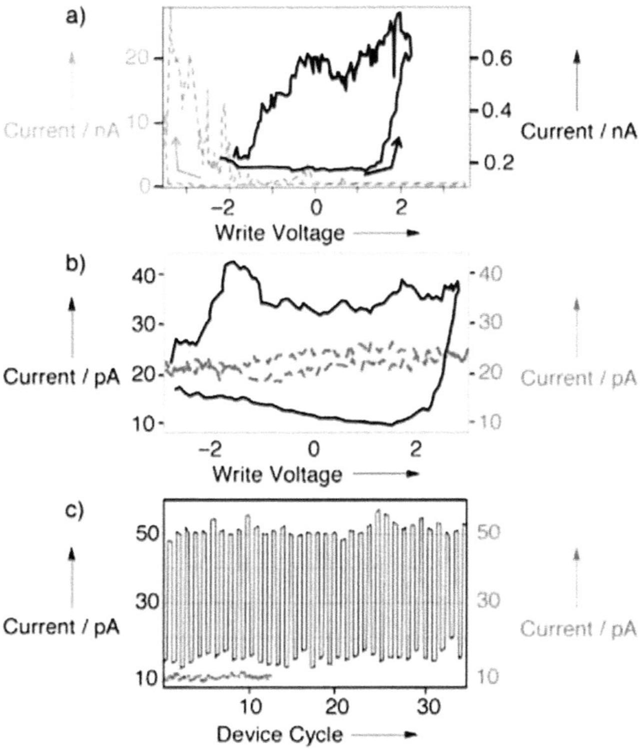

Figure 8.12 Device characteristics (solid black lines) recorded of (a) micrometer- and (b, c) nanometer-scale molecular switch tunnel junctions incorporating the rotaxane monolayer in between two electrodes. The green dashed trace in (a) is the response of the unthreaded dumbbell component; the red dashed traces in (b) and (c) are the responses of eicosanoic acid ($C_{19}H_{39}CO_2H$). The remnant molecular signature hysteresis loop is measured in traces (a) and (b); this is obtained by reading, at low voltage, the response of the junction to a series of voltage pulses mapping out the hysteresis loop. By this loop, the voltages for opening (*ca.* +2 V) and closing (*ca.* −2 V) the switches are defined. The black trace in (c) shows 35 cycles of the nanometer-scale device, corresponding to repeatedly closing and opening the switch, and then recording the current through the junction, at 0.1 V applied bias, after each writing event. Reprinted with permission from Ref. [23]. Copyright (2002) Wiley-VCH Verlag GmbH & Co. KGaA.

sandwiched between a polycrystalline Si bottom electrode and a Ti/Al top electrode (Fig. 8.12) [23]. In the ground state (no voltage applied), $CBPQT^{4+}$ encircles TTF, and it corresponds to the *off* state of the system. By the oxidation of the TTF unit, electrostatic repulsion forces $CBPQT^{4+}$ to move to the DNP station (*on* state). The molecular switch tunnel junction showed two distinguishable signals for the low voltage tunnel current, which depended on the prior applied voltage. The switch can be turned *on* at +2 V and *off* at −2 V, read between 0.1 V and 0.3 V, and recycled several times. The switching event was found to be thermally activated, supporting a mechanism involving molecular motion. When incorporated in another architecture, involving a platinum bottom electrode and a Ti/Al top electrode [24], abrupt switching behavior (with 10^4 larger amplitudes than in the other device) near +1 V and −1.5 V was found to result from electrode switching rather than molecular switching; thus, a device based on the dumbbell itself without the cyclophane exhibited switching.

8.3 Light-Controlled Switches

Light is an attractive external stimulus for triggering molecular conductivity changes, which have stimulated synthesis and studies on a variety of molecular photoswitches. Azobenzenes (ABs) present a class of molecules that undergo light-induced *cis-trans* isomerizations [25], while dithienylethenes (DTEs) [26] and dihydroazulenes (DHAs) [27] are photoswitches undergoing electrocyclic isomerizations to dihydrodithienobenzenes (DHDTBs) and vinylheptafulvenes (VHFs), respectively (Fig. 8.13). To prevent *cis-trans* isomerization of the ethene unit in DTEs, it is often part of a cyclohexene ring. Moreover, DTEs need methyl substituents at the thienyl groups to prevent oxidation of the DHDTB compound. Conversion from DTE to DHDTB requires UV light, while the back-reaction requires visible light. DHA undergoes a light-induced ring-opening to a vinylheptafulvene (VHF). The VHF can exist in s-*cis* and s-*trans* conformations, of which the latter is the most stable. The s-*cis* VHF is thermally returned to DHA, which is the thermodynamically most stable isomer. Most works on DHAs

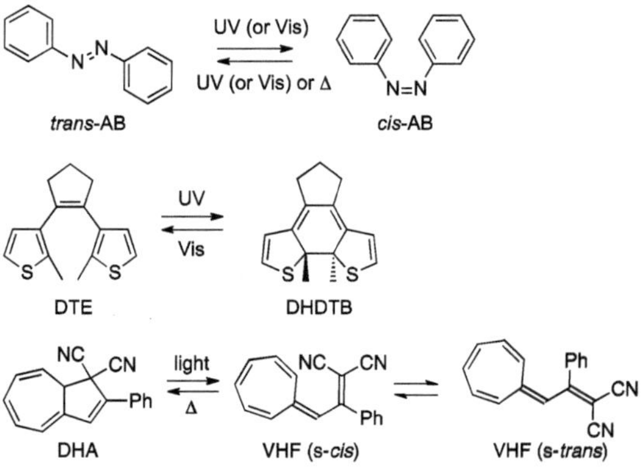

Figure 8.13 Molecular photoswitches.

involve a substituted phenyl substituent at position two and two cyano groups at position 1 (2-phenyl-1,8*a*-dihydroazulene-1,1-dicarbonitrile).

A major obstacle for reversible switching is quenching of the switching behavior via strong coupling of the photoswitch to the electrodes. A possible way of reducing the molecule–electrode interaction is to anchor the molecule via broken conjugation (using a saturated carbon atom between the anchoring group and the molecule), via a meta-substituted thiophenol anchoring group (corresponding to a less efficient cross-conjugated pathway for π-electron delocalization), or via a methylthio anchoring group (exhibiting a weaker electrode binding). The latter two anchoring situations will be covered in the DTE and DHA examples presented below.

8.3.1 *Diazobenzene*

A thiol-anchored azobenzene was prepared according to Fig. 8.14 [28]. The central azo linkage was made in a reaction between 4-nitrosobiphenyl and 4-amino-4'-methylsulfoxybiphenyl. The sulfoxide unit was then subjected to a Pummerer rearrangement upon treatment with trifluoroacetic anhydride, and hydrolysis of this product finally gave the thiol unit. This molecule formed well-ordered SAMs on Au(111), which was studied by conducting AFM

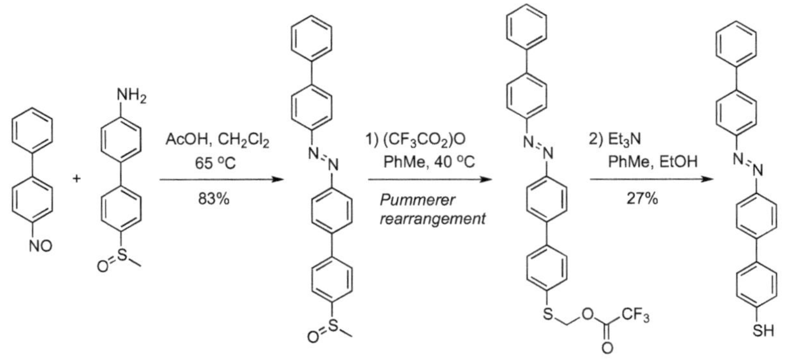

Figure 8.14 Synthesis of AB with thiol anchoring group.

[29]. As shown in Fig. 8.15, a significant increase in current was observed upon UV irradiation by which the *trans*-AB was converted to a *cis*-AB. Analysis of 30 samples provided a current switching ratio of about 30 between the two configurations. The reduced resistance of the *cis* isomer was explained by a reduction in the tunneling barrier length. Thus, the length of the *trans* and *cis* isomers was estimated to 23.6 and 16.0 $\pm$ 1.5 Å, respectively. This length corresponds to the distance from the Au–S linkage to the terminal H of the *trans* isomer, while the distance is from the Au–S linkage to the azo group (or the first carbon atoms above it) for the *cis* isomer. When it comes to electron transport through an entire *cis*-AB, calculations reveal that *cis*-AB is in fact significantly less conducting than *trans*-AB on account of a distorted geometry, which breaks the π-conjugation along the molecule [30, 31].

8.3.2 *Dithienylethenes*

The geometrical difference accompanying conversions between DTE and DHDTB isomers is less significant as that of the *cis*-*trans* isomerization of azobenzenes, which renders wiring between electrodes more advantageous. The synthesis of an important precursor for further functionalization, a dichloro-substituted DTE is shown in Fig. 8.16 [32]. Conversion of this DTE to a boronic acid followed by palladium-catalyzed Suzuki cross-coupling reaction with 1,3-dibromobenzene gave a DTE with two bromophenyl

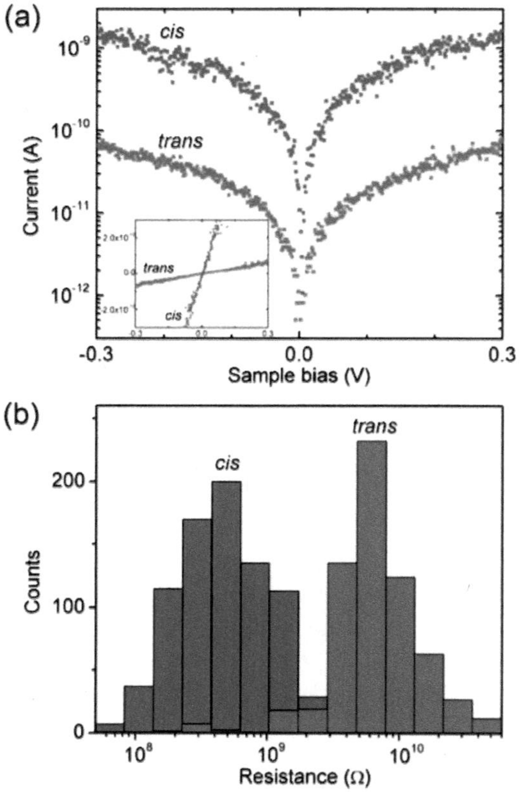

Figure 8.15 I–V traces (a) and resistance histograms (b) of a *cis*-AB (blue) and *trans*-AB (red) (for structure, see Fig. 8.14). Reprinted (adapted) with permission from Ref. [29]. Copyright (2008) American Chemical Society.

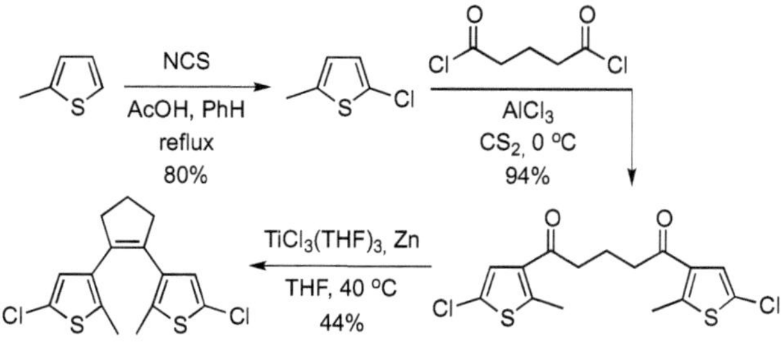

Figure 8.16 Synthesis of DTE with two chloro-substituents.

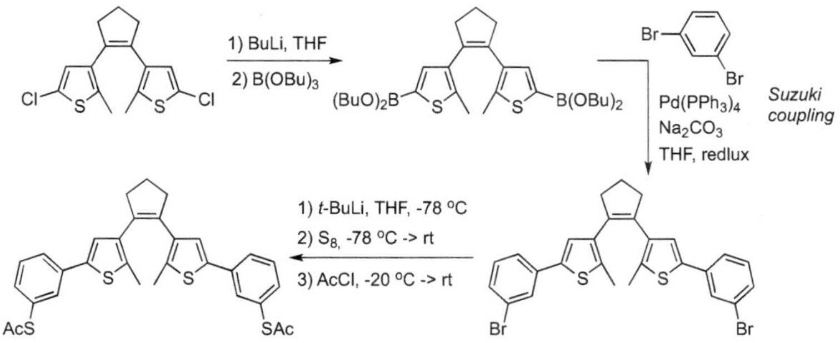

Figure 8.17 Attachment of anchoring groups to a DTE.

functionalities (Fig. 8.17). Lithiation followed by the addition of sulfur and acetyl chloride finally gave a DTE with acetyl-protected thiolate end-groups [33].

A monolayer of this molecule was sandwiched between a gold electrode and a poly(3,4-ethylenedioxythiophene): poly(4-styrenesulphonic acid) (PEDOT:PSS)/Au electrode [34]. By irradiation with UV light (312 nm), the DTE was converted to the closed DHDTB isomer, while irradiation with visible light (532 nm) caused this isomer to return to DTE. The junction showed a clear and reversible conductance switching upon illumination (Fig. 8.18), with DTE being the low-conducting state (*off*) and DHDTB the high-conducting state (*on*); an *on/off* ratio of 16 was demonstrated. The molecule was also bridged between two-dimensional lattices of gold nanoparticles, and again switching between a high-conducting closed form and a low-conducting open form was observed, triggered by UV and visible light, respectively [35]. Reversible *on/off* switching was also observed for an SAM on gold of a related derivative with the SAc anchoring group present on only one aryl group [36]. The cross-conjugated path provided by the *meta* linkage seems to play an important role for the coupling to the electrodes and thereby the reversibility.

8.3.3 *Dihydroazulene*

A particularly attractive property of the DHA-VHF system is that the conversion in either direction occurs quantitatively. Thus, in

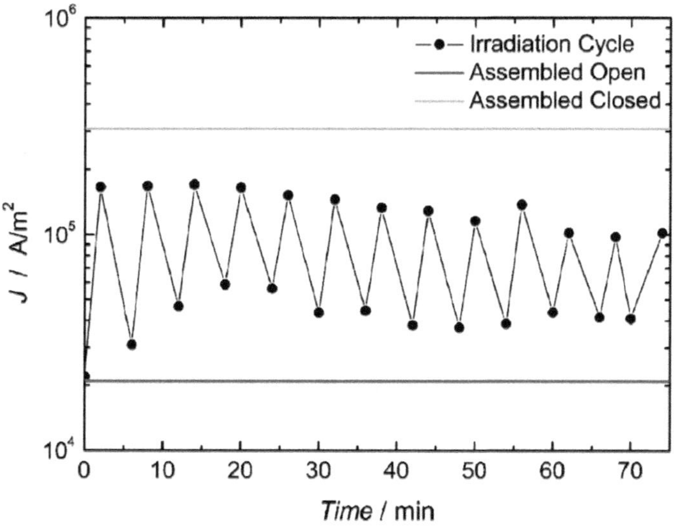

Figure 8.18 Reversible switching of a monolayer (between Au and PEDOT:PSS/Au electrodes) of DTE (open form; low conductance) and DHDTB (closed form; high conductance); conversion from DTE to DHDTB occurs with UV light, while the back-reaction occurs with visible light. Reprinted with permission from Ref. [34]. Copyright (2008) Wiley-VCH Verlag GmbH & Co. KGaA.

solution, DHA is completely converted to VHF upon irradiation as VHF is not photoactive. Upon heating in the dark, VHF can be quantitatively returned to DHA. The system is also easy to access from a synthetic point of view. Thus, DHA is readily prepared in multi-gram scale from a number of high-yielding steps, starting from malononitrile and acetophenone (Fig. 8.19) [37]. Incorporation of a 4-iodo substituent on the phenyl of acetophenone provides a handle for further functionalization of the switch [38]. Thus, an SMe anchoring group was attached by subjecting this compound to a Suzuki cross-coupling reaction (Fig. 8.20) [39].

A single-molecule junction based on this sulfide-anchored DHA was prepared by evaporating the molecule under ultra-high vacuum onto a substrate with a prefabricated nanogap between silver electrodes [39]. The prefabricated nanogap consisted of the source and drain electrodes of silver deposited onto a planar gate electrode made of aluminum metal covered with aluminum oxide prepared

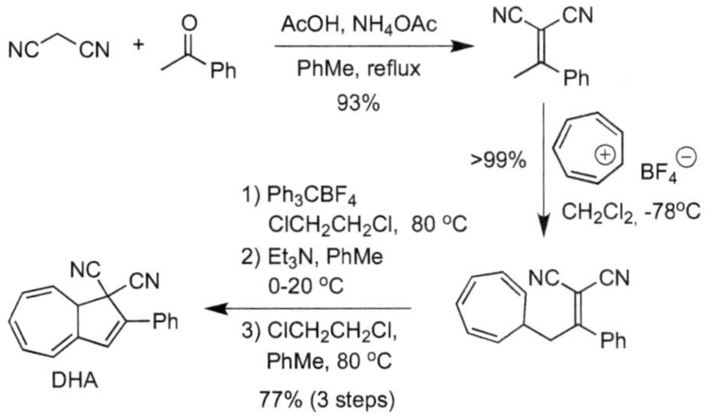

Figure 8.19 Synthesis of DHA.

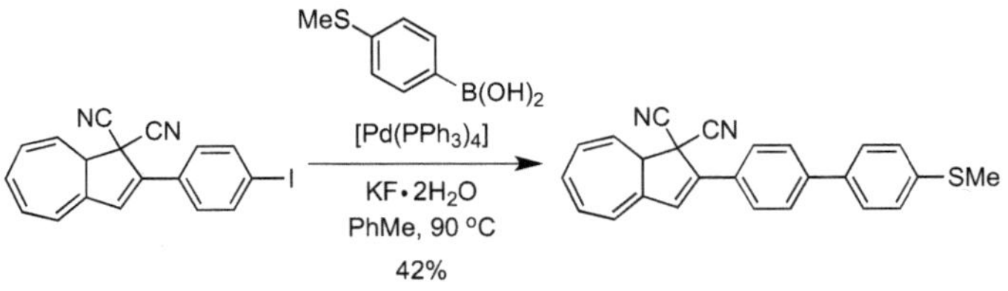

Figure 8.20 Incorporation of sulfide anchoring group.

on a chip of oxidized silicon. Conductance measurements were performed at 4–25 K as a function of the gate voltage, irradiation, bias potential between source and drain, and temperature. The single-molecule device was found to be operating in the Coulomb blockade regime, in which electron transport proceeds via sequential tunneling, i.e., an electron jumps from the one electrode to the molecule, resides there for a time, before jumping to the second electrode. The weak SMe anchoring group is likely responsible for this weak coupling between electrode and molecule. At high bias voltages, switching from the DHA form (high resistivity, *off*) to a more conducting VHF form (low resistivity, *on*) was observed at a gate voltage of around 3 V. At gate voltages not too far from this critical value, the junction was found to be light sensitive. Thus, irradiation of the sample for a few minutes by visual light triggered conversion of DHA to VHF. When the bias voltage was kept low, the sample was found to stay in the VHF form, while for higher bias

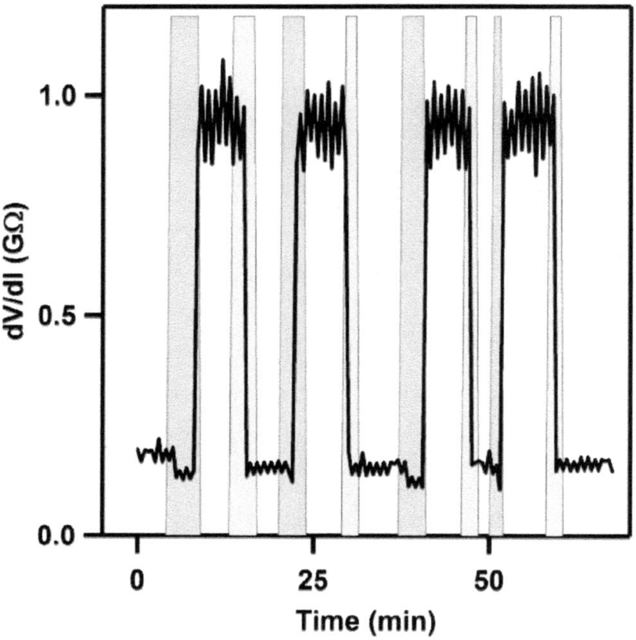

Figure 8.21 A series of *on-off* switchings of SMe-endcapped DHA (see Fig. 8.20) at a gate potential of 2.7 V. The highlighted background designates the time intervals when the sample was subjected to light (yellow) or the temperature was increased from 4 K to 25 K (rose). The differential resistance dV/dI was taken at a bias voltage of 25 mV for both *on* (VHF) and *off* (DHA) states. Reprinted with permission from Ref. [39]. Copyright (2012) Wiley-VCH Verlag GmbH & Co. KGaA.

voltages, the sample would, after some time, reset to DHA. A light–high bias cycle could be repeated many times, and more than 20 *on–off* cycles were obtained. An alternative way to reset the VHF to DHA was to increase the temperature up to 25 K, and several light–heat *on–off* cycles were recorded (Fig. 8.21). Reversible switching between DHA (*off*) and VHF (*on*) was thus successfully controlled by not only light, but also by gate voltage, bias voltage, and temperature. Future studies await to elucidate which state is the most conducting in devices with coherent electron transport, although the exact anchoring positions between the molecule and the electrodes may play a determining role.

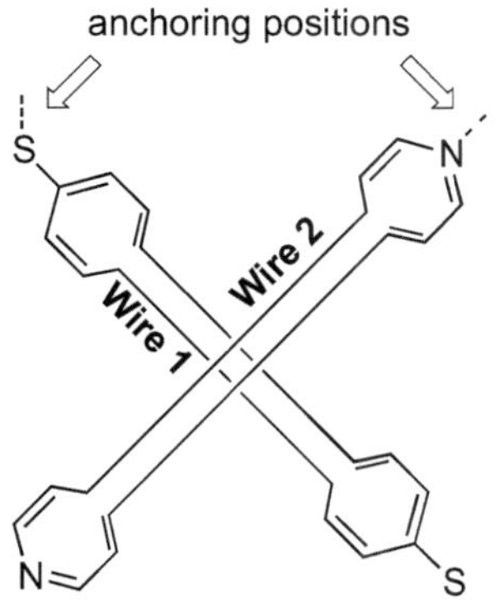

Figure 8.22 Schematic drawing of cruciform motif.

8.4 Coordination-Induced Switches

With the objective to switch between two conducting pathways in the same molecule, cruciform molecules were developed consisting of two different and perpendicularly situated π-conjugated systems (wire 1 and wire 2, Fig. 8.22) endcapped with different anchoring groups (pyridine and thiolate) [40]. The last steps in the synthesis of one such molecule are shown in Fig. 8.23. It consists of a terminally pyridine-functionalized OPV rod and a perpendicular acetyl-protected sulfur-functionalized oligophenylene (OP) rod. The cruciform motif is made by a Suzuki cross-coupling between a diiodo derivative of the central OPV rod and an aryl boronic acid. The trimethylsilylethyl protecting groups were removed by fluoride and the resulting thiolate groups were then acetyl-protected upon treatment with acetyl chloride. When subjecting this molecule to mechanically controllable break-junction experiments (gold electrodes), the conductance histogram showed two conduction peaks in the absence of a deprotecting agent for deacetylation, while only the high-conductance peak was seen when the acetyl groups were cleaved by tetrabutylammonium hydroxide [40]. In the absence of deprotection agent, the pyridine OPV rod of the cruciform is expected to be mobilized between the two electrodes, resulting in the low-conductance peak. Upon this binding, the protected sulfur end-groups should also come close to the electrode and

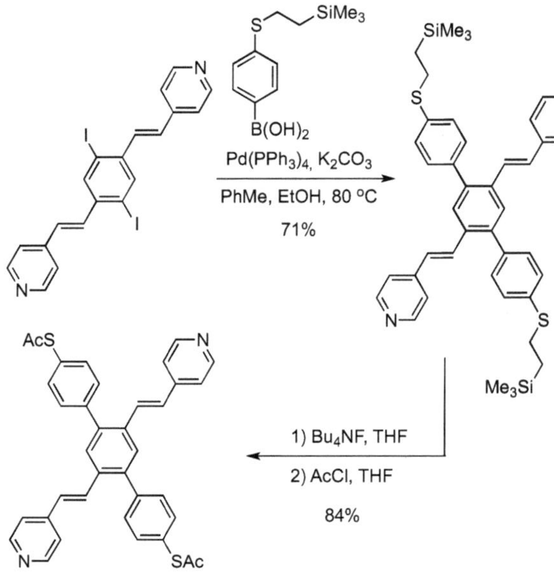

Figure 8.23 Synthesis of cruciform.

the possibility for a spontaneous deacetylation increases. Thus, sometimes a sulfur–gold covalent bond will form at both ends of the OP rod, resulting in the high-conductance peak. When the acetyl groups are cleaved by a deprotection agent, the binding is dominated by the OP rod and hence the high-conductance peak dominates the histogram. The authors suggest that by applying a potential to the liquid cell under appropriate conditions, it might be possible to control binding/unbinding of the pyridine rod and hence alternate between a situation in which both rods can bind and one in which only the sulfur-endcapped rod binds.

8.5 Tautomerization-Induced Switches

Prototropic tautomerization reactions provide another means of changing the electronic properties of the molecule. For example, one such switch has been based on a melamine deposited on a Cu(100) surface (Fig. 8.24) [41]. A tautomerization process is induced by

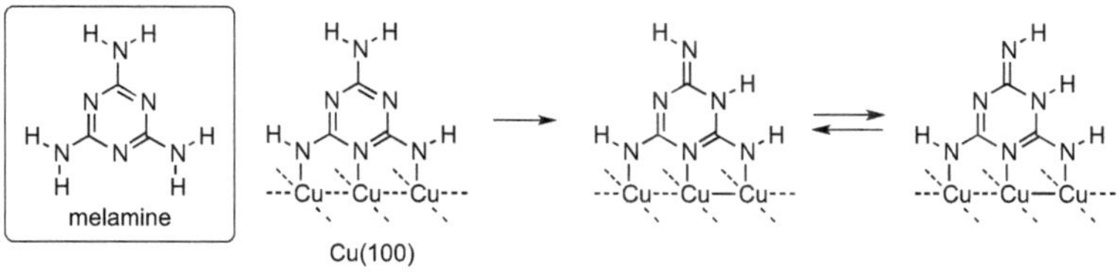

Figure 8.24 Switching of asymmetric melamine tautomer.

applying a high-voltage pulse (+2.3–2.8 V), which results in an asymmetric melamine tautomer. Applying a high-voltage pulse (0.5–1.0 V) again on the modified molecule, the related asymmetric isomer in which only the mutual position of the N–H bonds differs can be obtained. The I–V curves for the molecule before and after it was modified by pulses reveal that the asymmetric tautomer shows significant rectifying effect with a typical ratio of 20–25 at 2 V. Thus, the high-voltage pulse has converted the nonfunctional melamine molecule into a rectifier. Moreover, fluctuation features at higher bias are observed. This behavior was explained as mechanical switching by an out-of-plane rotation of the N–H bond.

8.6 Concluding Remarks

The use of organic molecular switches in computers has obvious potential for dramatically enhancing data storage without needing to increase the size of devices. There is, however, still a long way to use molecular switches in present-day electronics applications, but the many proof-of-concept studies on first prototype molecules have provided fundamental insights into molecular electronics properties, structure-property relationships, and the requirements needed for achieving reversible switching behavior in electrode junctions, although the exact anchoring configuration of molecules is not always well known. The reliable contacting and self-assembly of billions of molecular switches in a device present major challenges. Yet the successful construction of molecular electronic devices incorporating monolayers of rotaxane molecules, as described above, trapped between wires in two-dimensional arrays within crossbar architectures is particularly promising for molecular memory applications. A next important step could be to proceed to crystals with periodic organization of molecular switches.

Problems

1. The anthraquinone wire shown below was developed by Hummelen et al. [42]. Explain how this molecule could possibly function as a redox-controlled switch.

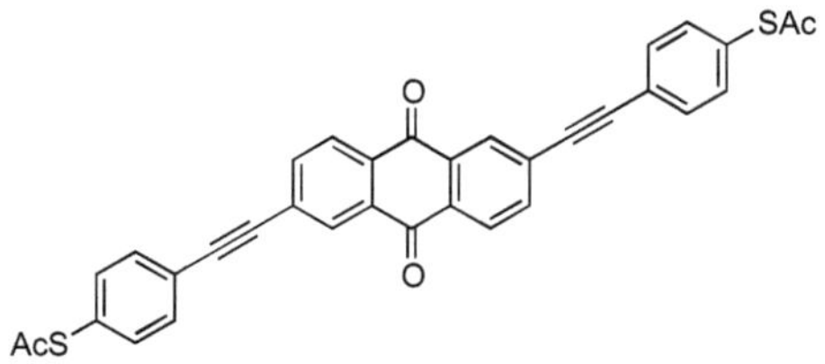

Figure 8.25

2. In the examples provided above, weak coupling between DTE and DHA photoswitches to the electrode was accomplished by using either an SAc anchoring group in a *meta*-configuration on a benzene ring or by an SMe anchoring group. Suggest other ways of anchoring photoswitches to avoid quenching of photoactivity by the electrode. (For a discussion of the strength of molecule–electrode coupling, see Ref. [43]).

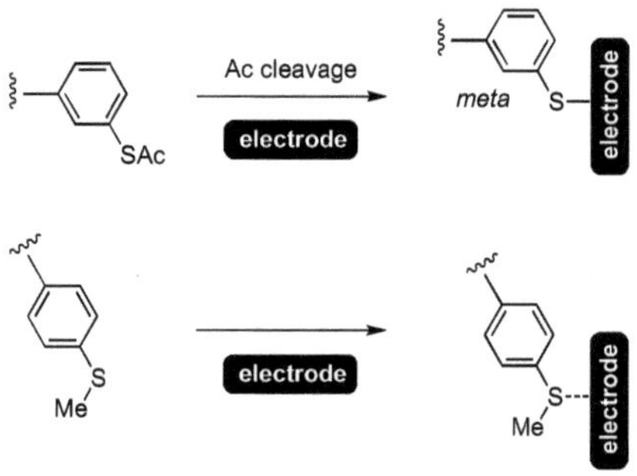

Figure 8.26

3. Liljeroth et al. [44] have observed conductivity switching by a tautomerization reaction of the naphthalocyanine shown below; explain how this tautomerization occurs.

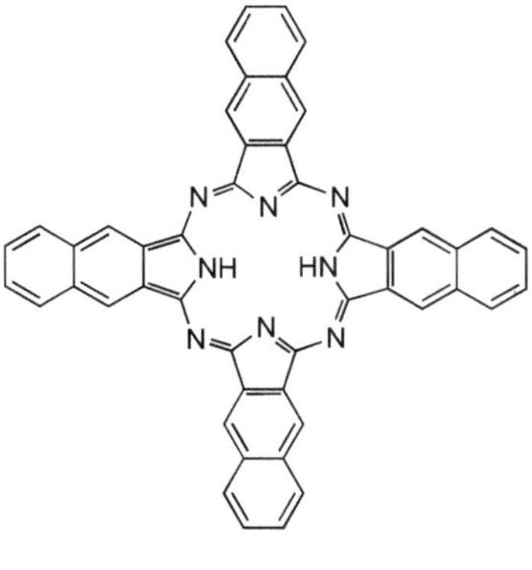

Figure 8.27

References

1. Ward, M. D. (2001). Chemistry and molecular electronics: New molecules as wires, switches, and logic gates, *J. Chem. Ed.*, **78**, pp. 321–328.
2. Carroll, R. L., and Gorman, C. B. (2002). The genesis of molecular electronics, *Angew. Chem. Int. Ed.*, **41**, pp. 4378–4400.
3. Weibel, N., Grunder, S., and Mayor, M. (2007). Functional molecules in electronic circuits, *Org. Biomol. Chem.*, **5**, pp. 2343–2353.
4. Kudernac, T., Katsonis, N., Browne, W. R., and Feringa, B. L. (2009). Nano-electronic switches: Light-induced switching of the conductance of molecular systems, *J. Mater. Chem.*, **19**, pp. 7168–7177.
5. Fuentes, N., Martín-Lasanta, A., de Cienfuegos, L. Á., Ribagorda, M., Parra, A., and Cuerva, J. M. (2011). Organic-based molecular switches for molecular electronics, *Nanoscale*, **3**, pp. 4003–4014.
6. Coskun, A., Spruell, J. M., Barin, G., Dichtel, W. R., Flood, A. H., Botros, Y. Y., and Stoddart, J. F. (2012). High hopes: Can molecular electronics realise its potential? *Chem. Soc. Rev.*, **41**, pp. 4827–4859.
7. Danilov, A. V., Hedegård, P., Golubev, D. S., Bjørnholm, T., and Kubatkin, S. E. (2008). Nano-electromechanical switch operating by tunneling of an entire C_{60} molecule, *Nano Lett.*, **8**, pp. 2393–2398.
8. Gittins, D. I., Bethell, D., Schiffrin, D. J., and Nichols, R. J. (2000). A nanometre-scale electronic switch consisting of a metal cluster and redox-addressable groups, *Nature*, **408**, pp. 67–69.
9. Stoddart, J. F. (2009). The chemistry of the mechanical bond, *Chem. Soc. Rev.*, **38**, pp. 1802–1820.

10. Xu, B., Xiao, X., Yang, X., Zang, L., and Tao, N. (2005). Large gate modulation in the current of a room temperature single molecule transistor, *J. Am. Chem. Soc.*, **127**, pp. 2386–2387.

11. Tsoi, S., Griva, I., Trammell, S. A., Blum, A. S., Schnur, J. M., and Lebedev, N. (2008). Electrochemically controlled conductance switching in a single molecule: Quinone-modified oligo(phenylene vinylene), *ACS Nano*, **2**, pp. 1289–1295.

12. Kubatkin, S., Danilov, A., Hjort, M., Cornil, J., Brédas, J.-L., Stuhr-Hansen, N., Hedegård, P., and Bjørnholm, T. (2003). Single-electron transistor of a single organic molecule with access to several redox states, *Nature*, **425**, pp. 698–701.

13. Liao, J., Agustsson, J. S., Wu, S., Schönenberger, C., Calame, M., Leroux, Y., Mayor, M., Jeannin, O., Ran, Y.-F., Liu, S.-X., and Decurtins, S. (2010). Cyclic conductance switching in networks of redox-active molecular junctions, *Nano Lett.*, **10**, pp. 759–764.

14. Guo, X., Zhang, D., Zhang, H., Fan, Q., Xu, W., Ai, X., Fan, L., and Zhu, D. (2003). Donor-acceptor-donor triads incorporating tetrathiafulvalene and perylene diimide units: Synthesis, electrochemical and spectroscopic studies, *Tetrahedron*, **59**, pp. 4843–4850.

15. Jennum, K., Vestergaard, M., Pedersen, A. H., Fock, J., Jensen, J., Santella, M., Led, J. J., Kilså. K., Bjørnholm, T., and Nielsen, M. B. (2011). Synthesis of oligo(phenyleneethynylene)s with vertically disposed tetrathiafulvalene units, *Synthesis*, pp. 539–548.

16. Wei, W., Li, T., Jennum, K., Santella, M., Bovet, N., Hu, W., Nielsen, M. B., Bjørnholm, T., Solomon, G. C., Laursen, B. W., and Nørgaard, K. (2012). Molecular junctions based on SAMs of cruciform oligo(phenylene ethynylene)s, *Langmuir*, **28**, pp. 4016–4023.

17. Fock, J., Leijnse, M., Jennum, K., Zyazin, A. S., Paaske, J., Hedegård, P., Nielsen, M. B., and van der Zant, H. S. J., Submitted.

18. Osorio, E. A., Moth-Poulsen, K., van der Zant, H. S. J., Paaske, J., Hedegård, P., Flensberg, K., Bendix, J., and Bjørnholm, T. (2010). Electrical manipulation of spin states in a single electrostatically gated transition-metal complex, *Nano Lett.*, **10**, pp. 105–110.

19. Prins, F., Monrabal-Capilla, M., Osorio, E. A., Coronado, E., and van der Zant, H. S. J. (2011). Room-temperature electrical addressing of a bistable spin-crossover molecular system, *Adv. Mater.*, **23**, pp. 1545–1549.

20. Flatt, A. K., Dirk, S. M., Henderson, J. C., Shen, D. E., Su, J., Reed, M. A., and Tour, J. M. (2003). Synthesis and testing of new end-functionalized oligomers for molecular electronics, *Tetrahedron*, **59**, 8555–8570.

21. Lörtscher, E., Ciszek, J. W., Tour, J., and Riel, H. (2006). Reversible and Controllabe Switching of a Single-Molecule Junction, *Small*, **2**, pp. 973–977.

22. Blum, A. S., Kushmerick, J. G., Long, D. P., Patterson, C. H., Yang, J. C., Henderson, J. C., Yao, Y., Tour, J. M., Shashidhar, R., and Ratna, B. R. (2005). Molecularly inherent voltage-controlled conductance switching, *Nature Mat.*, **4**, pp. 167–172.

23. Luo, Y., Collier, C. P. Jeppesen, J. O., Nielsen, K. A., Delonno, E., Ho, G., Perkins, J., Tseng, H.-R., Yamamoto, T., Stoddart, J. F., and Heath, J. R. (2002). Two-dimensional molecular electronics circuits, *ChemPhysChem*, **3**, pp. 519–525.

24. Stewart, D. R., Ohlberg, D. A. A., Beck, P. A., Chen, Y., Williams, R. S., Jeppesen, J. O., Nielsen, K. A., and Stoddart, J. F. (2004). Molecule-independent electrical switching in Pt/organic monolayer/Ti devices, *Nano Lett.*, **4**, pp. 133–136.

25. Bandara, H. M. D., and Burdette, S. C. (2012). Photoisomerization in different classes of azobenzene, *Chem. Soc. Rev.*, **41**, pp. 1809–1825.

26. Tian, H., and Yang, S. (2004). Recent progresses on diarylethene based photochromic switches, *Chem. Soc. Rev.*, **33**, pp. 85–97.

27. Nielsen, M. B., Broman, S. L., Petersen, M. Å., Andersson, A. S., Jensen, T. S., Kilså, K., and Kadziola, A. (2010). New routes to functionalized dihydroazulene photoswitches, *Pure Appl. Chem.*, **82**, pp. 843–852.

28. Pace, G., Ferri, V., Grave, C., Elbing, M., von Hänisch, C., Zharnikov, M., Mayor, M., Rampi, M. A., and Samorì, P. (2007). Cooperative light-induced molecular movements of highly ordered azobenzene self-assembled monolayers, *Proc. Natl. Acad. Sci. U.S.A.*, **104**, pp. 9937–9942.

29. Mativetsky, J. M., Pace, G., Elbing, M., Rampi, M. A., Mayor, M., and Samorì, P. (2008). Azobenzenes as light-controlled molecular electronic switches in nanoscale metal-molecule-metal junctions, *J. Am. Chem. Soc.*, **130**, pp. 9192–9193.

30. Zhang, C., He, Y., Cheng, H.-P., Xue, Y., Ratner, M. A., Zhang, X.-G., and Krstic, P. (2006). Current-voltage characteristics through a single light-sensitive molecule, *Phys. Rev. B*, **73**, 125445.

31. Del Valle, M., Gutiérrez, R., Tejedor, C., and Cuniberti, G. (2007). Tuning the conductance of a molecular switch, *Nat. Nanotechnol.*, **2**, pp. 176–179.

32. Lucas, L. N., van Esch, J., Kellogg, R. M., and Feringa, B. L. (1998). A new class of photochromic 1,2-diarylethenes: Synthesis and switching properties of bis(3-thienyl)cyclopentenes, *Chem. Commun.*, pp. 2313–2314.

33. Kudernac, T., De Jong, J. J., van Esch, J., Feringa, B. L., Dulic, D., van der Molen, S. J., and van Wees, B. J. (2005). Molecular switches get wired: Synthesis of diarylethenes containing one or two sulphurs, *Mol. Cryst. Liq. Cryst.*, **430**, pp. 205–210.

34. Kronemeijer, A. J., Akkerman, H. B., Kudernac, T., van Wees, B. J., Feringa, B. L., Blom, P. W., and de Boer, B. (2008). Reversible conductance switching in molecular devices, *Adv. Mater.*, **20**, pp. 1467–1473.

35. van der Molen, J. S., Liao, J., Kudernac, T., Agustsson, J. S., Bernard, L., Calame, M., van Wees, B. J., Feringa, B. L., and Schönenberger, C. (2009), Light-controlled conductance switching of ordered metal-molecule-metal devices, *Nano Lett.*, **9**, pp. 76–80.

36. Katsonis, N., Kudernac, T., Walko, M., van der Molen, S. J., van Wees, B. J., and Feringa, B. L. (2006). Reversible conductance switching of single diarylethenes on a gold surface, *Adv. Mater.*, **18**, pp. 1397–1400.

37. Broman, S. L., Brand, S. L., Parker, C. R., Petersen, M. Å., Tortzen, C. G., Kadziola, A., Kilså, K., and Nielsen, M. B. (2011). Optimized synthesis and detailed NMR spectroscopic characterization of the 1,8a-dihydroazulene-1,1-dicarbonitrile photoswitch, *Arkivoc*, **ix**, pp. 51–67.

38. Gobbi, L., Seiler, P., Diederich, F., Gramlich, V., Boudon, C., Gissel-brecht, J.-P., and Gross, M. (2001). Photoswitchable tetraethynylethene-dihydroazulene chromophores, *Helv. Chim. Acta*, **84**, pp. 743–777.

39. Broman, S. L., Lara-Avila, S., Thisted, C. L., Bond, A. D., Kubatkin, S., Danilov, A., and Nielsen, M. B. (2012). Dihydroazulene photoswitch operating in sequential tunneling regime: Synthesis and single-molecule junction studies, *Adv. Funct. Mater.*, **22**, pp. 4249–4258.

40. Grunder, S., Huber, R., Wu, S., Schönenberger, C., Calame, M., and Mayor, M. (2010). Oligoaryl cruciform structures as model compounds for coordination-induced single-molecule switches, *Eur. J. Org. Chem.*, pp. 833–845.

41. Pan, S., Fu, Q., Huang, T., Zhao, A., Wang, B., Luo, Y., Yang, J., and Hou, J. (2009). Design and control of electron transport properties of single molecules, *Proc. Natl. Acad. Sci. U.S.A.*, **106**, pp. 15259–15263.

42. van Dijk, E. H., Myles, D. J. T., van der Veen, M. H., and Hummelen, J. C. (2006). Synthesis and properties of an anthraquinone-based redox switch for molecular electronics, *Org. Lett.*, **8**, pp. 2333–2336.

43. Moth-Poulsen, K., and Bjørnholm, T. (2009). Molecular electronics with single molecules in solid-state devices, *Nature Nanotech.*, **4**, pp. 551–556.

44. Liljeroth, P., Repp, J., and Meyer, G. (2007). Current-induced hydrogen tautomerization and conductance switching of naphthalocyanine molecules, *Science*, **317**, pp. 1203–1206.

Chapter 9

Switching Mechanisms in Molecular Switches

Andrey Danilov and Sergey Kubatkin
Department of Microtechnology and Nanoscience, Chalmers University of Technology, 41296 Gothenburg, Sweden
andrey.danilov@chalmers.se

Molecular electronics aims at using tailor-built molecules as active device elements to achieve the desired electronic functionality. Molecular-based electronic switches appear to be the most promising candidates for compact memory arrays with low power consumption. In the past decade, detailed investigations have been performed on a great variety of molecular switches, including mechanically interlocked switches, conformational switches, and redox-active molecules. In this chapter, we will review basic switching mechanisms in molecular switches: thermal fluctuations, current-induced excitations, and quantum tunneling. We will demonstrate how the quantitative information allowing to judge between those different switching mechanisms can be extracted from the data measured on single-molecule devices. We will also discuss how the intrinsic switching properties may be affected when the molecule is bridged to electrodes, and how to distinguish

Handbook of Single-Molecule Electronics
Edited by Kasper Moth-Poulsen

ISBN 978-981-4463-38-6 (Hardcover), 978-981-4463-39-3 (eBook)
www.panstanford.com

whether the switching happens in molecular kernel or at the molecule-to-electrode interface.

9.1 Introduction

Desired or undesired, switching is ubiquitous in single-molecule junctions. If one aims to build a molecular switch, then switching is naturally the target functionality; but unwanted switching in samples meant to be stable junctions is also quite omnipresent. Mastering control over switching is, therefore, one of the main challenges in molecular engineering.

Full understanding of a switching behavior for a molecular switch, same as for any switch in general, reduces to two basic questions:

1. What are the two states (say, On and Off) of a switch?
2. What kind of mechanism triggers transition between two?

Three different design concepts for molecular switches were suggested so far: (1) switches operating by reversible isomerization without pronounced changes in molecular shape like tautomerization [1], opening/closing of conjugation path [2], or intramolecular charge redistribution [3]; (2) switches exploiting rotation or rolling of the whole molecule [4]; and (3) conformational switches based on isomerization between two manifestly different metastable shapes predetermined by molecular design. Two states can differ, e.g., by folding [5, 6] or twisting [7, 8] a part of the molecule around flexible link or by sliding [9] or circumrotation [10] of a molecular component in mechanically interlocked supramolecular structures. Thus, for a dedicated single-molecule switch, the kernel molecule itself typically hints to a proper answer to question 1.

The answer could not be so obvious although if switching comes as an undesirable surprise. Quite often it is not even clear whether the switching should be attributed to the transformation of the molecular kernel itself or to the reconstruction of the molecule-to-electrode interface. As we shall see later, a definite answer to question 2, i.e., the identification of the switching mechanism, could be the way to resolve these uncertainties.

Unfortunately, knowing the answer to question 1 does not imply a definite answer to question 2. Indeed, On-Off transition in the *same* molecular switch can be triggered by temperature, light, or transport current through the junction (consider, e.g., a DHA-VHF switch [11], also discussed in Chapter 8, Molecular Switches).

In this chapter, we will present a general view on molecular switches as bistable systems. We will discuss some universal analysis procedure that allows to identify the switching mechanism in essentially any single-molecule switch from the switching behavior itself, without any presumptions concerning the nature of On and Off states.

Specifically, we will consider the three basic switching mechanisms—thermal fluctuations, current-induced excitations, and quantum tunneling—and how one can deduce the dominating mechanism directly from the switching data.

We will demonstrate that the same analysis procedure allows to find the energy difference between On and Off states and the switching barrier. For switches operating by current-driven mechanism, the electron–phonon coupling strength and the relaxation time for molecular vibrations can be determined, and if the switching is mediated by the excitation of some vibronic mode, the energy of this mode can be found. We will demonstrate how this comprehensive data allow to understand the nature of bistability in a few exemplary systems that were originally meant to be stable junctions.

9.2 Switching Behavior: Stochastic or Deterministic?

We shall start our discussion from the basic question: To what extent is switching behavior in molecular junctions deterministic? If bistability comes as an unwanted hindrance, typically the switching looks completely indeterministic, as depicted in Fig. 9.1 (reproduced from Ref. [12]).

On the other hand, a switch, by design, is a deterministic device where On-Off transition is triggered by some control parameter, e.g., a bias voltage. Figure 9.2 presents a set of current–voltage (I–V) curves for an exemplary C_{60} switch to be discussed later in details.

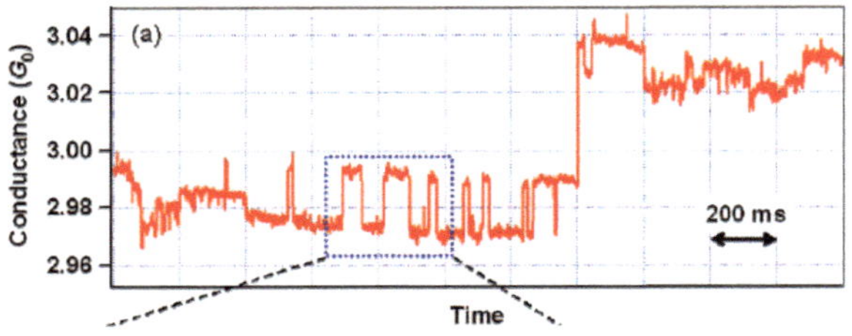

Figure 9.1 A random telegraphic conductance switching at an Au-pentacene–Au nanojunction. Reproduced from Ref. [12].

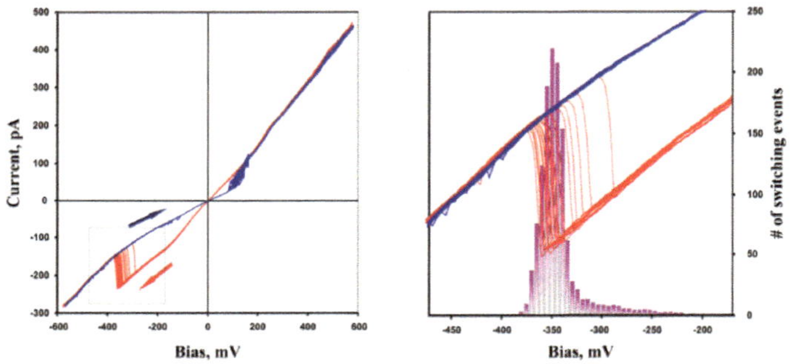

Figure 9.2 On-Off switching in a C_{60}-based switch. Left: a set of I–V curves taken for many trace (blue) and retrace (red) bias sweeps. Right: zoom into On-Off switching area and the histogram of the switching events as a function of the bias voltage. Thirty-two I–V curves are shown, and 1024 traces were used to compose the histogram.

One can see that on a large scale, the switching pattern is absolutely reproducible: At extreme negative biases, the switch is in the Off (low-conductance) state, while at high enough positive biases, the device is always in the On (high-conductance) state. Around zero bias voltage, there is some hysteretic area where the junction is bistable and could be either in the On or the Off state. A single excursion into high (low) bias sets the device, in a deterministic way, into On (Off) state. Nevertheless, the switching is not fully deterministic: There is some distribution of switching voltages from trace to trace presented on the histogram in Fig. 9.2 (right).

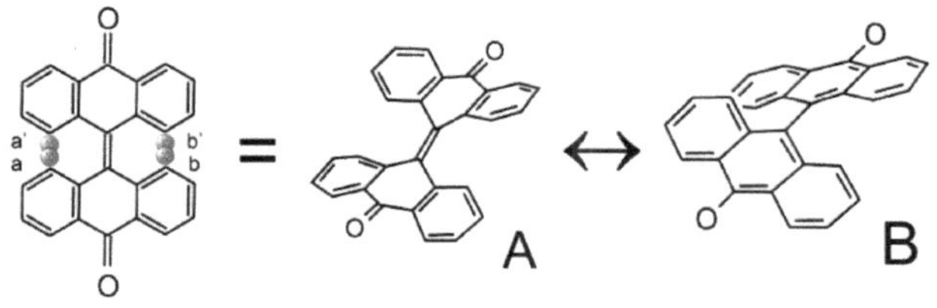

Figure 9.3 The valence structure and the origin of two conformations of bianthrone: in the form **A**, protons **a**′, **b**′ are on top and **a, b** beneath; in the form **B, a, b**′ are on top and **a**′, **b** beneath.

In what follows, we will focus on the detailed analysis of this distribution. It is from this distribution one can deduce a comprehensive information about the switching mechanism and operational parameters for an arbitrary molecular switch.[a]

9.3 Bianthrone Switch

9.3.1 *Experimental Data*

The first example of a single-molecule switch is the bianthrone system (BA), where bistability is evidently encoded in the molecule itself. Indeed, in solution the bianthrone molecule is known to have two well-defined conformations [13], as illustrated in Fig. 9.3.

Bianthrone belongs to a family of overcrowded bistricyclic aromatic enes (BAEs) [14, 15]. A coplanar conformation of the molecule, though favorable for conjugation, is sterically impossible because of overcrowding in the vicinity of the double bond in the middle (in the "fjord" region): Two hydrogen atoms (**a** and **a**′ in Fig. 9.3) cannot share the same space: either **a** or **a**′ should be on top of the other. A similar constraint, of course, applies to the complementary pair **bb**′. Two distinct conformations, **A** and **B**, arise from two ways for the molecule to reduce the strain by out-of-plane

[a]The same analysis can be applied to an apparently stochastic system like the one depicted in Fig. 9.1. In fact, the difference between the junctions presented in Fig. 9.1 and Fig. 9.2 is quantitative, not qualitative: If the switching barrier separating the On and Off states is high enough, then the system can be trapped either in On or Off states (Fig. 9.2); otherwise, it continiously switches between the two states (Fig. 9.1).

deformation: (1) by sacrificing the planarity of the anthrones but conserving the coplanarity around the double bond in the folded conformation **A**, where two protons linked to the same anthrone group are both on top; and (2) by conserving the former and sacrificing the latter in the twisted conformation **B**, where one proton is on top and the other underneath.

This structural constraint, which ensures the existence of two forms **A** and **B**, cannot be overruled by any environmental perturbations (like interaction with substrate and electrodes) as long as the molecule retains its chemical identity. One can, therefore, expect that the bianthrone molecule will reveal a switching behavior also when trapped in nanogap. This is indeed the case. A typical data taken on the bianthrone single-molecule junction[a] are shown in Fig. 9.4 [16].

9.3.2 *General Model for Switching*

A hysteretic behavior like the one in Fig. 9.4 suggests that at extreme negative biases, the ground state is high conductive conformation (**Hi**), while at positive biases, the ground state is low conductive form (**Lo**). When the bias voltage is swept from the negative to the positive direction, the free energy, F_{Hi}, of state **Hi** goes up and at some moment, the state **Hi** becomes metastable. At still higher biases, the barrier that confines the metastable state is suppressed, and eventually the system escapes from the metastable state to a new ground state **Lo**, as depicted in Fig. 9.5.

The free energies in Fig. 9.5 are direct functions of the bias voltage. This is why the switching pattern in Fig. 9.4, as a whole, is reproducible. Nevertheless, for any specific bias sweep, the switching is a stochastic event, which happens as soon as the barrier height U approaches the characteristic energy fluctuations $\langle E \rangle$. The condition $U(V) \sim \langle E \rangle$ (V is the bias voltage) sets the characteristic width of the switching area (i.e., the width of the switching histograms in Fig. 9.4).

[a] In this chapter, we will not go into the details of the fabrication procedure. The reader can instead refer to Chapter 2, Experimental Techniques, of this book and references therein.

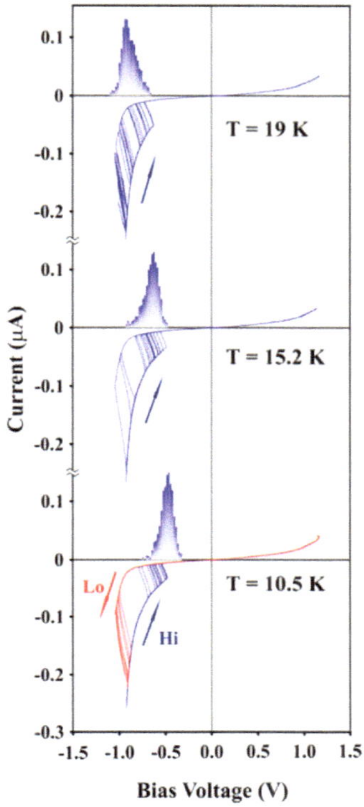

Figure 9.4 A set of I–V curves traced from negative to positive bias (blue) and in the reverse direction (red), taken at 10.5 K. For higher temperatures, 15.2 K and 19 K, only I–V curves taken in the forward direction are shown for clarity (as the areas for forward and reverse switching start to overlap). For each temperature, a few hundred curves were recorded to compose the switching histograms; only 32 representative traces are shown.

To quantify this model, we consider a general system that starts evolving from some metastable state. The barrier that confines the metastable state is time dependent and reduces as the time goes. If the transition is characterized by some escape rate $\Gamma(t)$, then the probability $Q(t)$ that at time $t$the sample is still in the initial state (i.e., that the switching has not happened) is given by a simple rate equation

$$\frac{d}{dt}Q(t) = -\Gamma(t)\,Q(t) \tag{9.1}$$

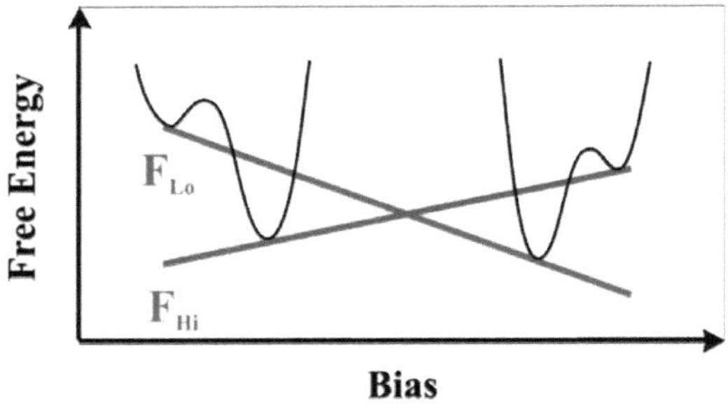

Figure 9.5 Bias dependence of the free energies for states **Hi** and **Lo** consistent with the switching pattern in Fig. 9.4.

Since the bias voltage in the experiment is incremented with a constant sweep rate $V(t) = V_0 + \dot{u}t$, we can convert Eq. 9.1 from the time domain into the voltage domain and rewrite it as

$$\frac{d}{dV}Q(V) = -\frac{1}{\dot{u}}\Gamma(V)\,Q(V) \tag{9.2}$$

This can be easily solved to find the distribution of switching events, which is the probability density $P(V) = \frac{d}{dV}Q(V)$ that the switching will happen at bias V:

$$P(V) = \frac{\Gamma(V)}{\dot{u}}\exp\left[-\int_{V_0}^{V}\frac{\Gamma(V')}{\dot{u}}dV'\right] \tag{9.3}$$

Equation 9.3 gives the distribution of switching events for any given escape rate $\Gamma(V)$. It can be solved for Γ to find the escape rate from the measured distribution of the switching events [17]:

$$\Gamma(V) = \dot{u}\frac{P(V)}{1-\int_{V_0}^{V}P(V')\,dV'} \tag{9.4}$$

In a nutshell, Eq. 9.3 states that for any given (or assumed) escape rate $\Gamma(V)$, one can predict the distribution of switching voltages $P(V)$; while Eq. 9.4 claims that there exists an opposite implication: once the distribution of switching events, $P(V)$, is known from experiment, one can, in a straightforward manner, deduce the switching rate Γ as a function of the bias voltage V.

We will stress that Eq. 9.4 is the direct transform of the experimental data for $P(V)$; no specific assumptions about the nature of **Hi/Lo** states, neither about the switching mechanism, are needed to deduce the switching rate $\Gamma(V)$.

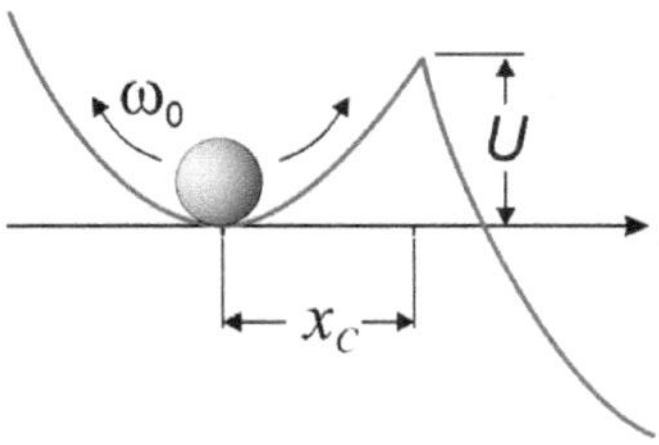

Figure 9.6 A physical model for an escape from the metastable state.

As our final goal is to figure out the physical mechanisms responsible for escape from the metastable state, we will now link the escape rate Γ to the characteristic energy fluctuations $\langle E \rangle$ and to consider all potential contributions to $\langle E \rangle$: thermal fluctuations, current-induced excitations, and quantum ground-state energy.

To this end, we do have to introduce a physical model for a switching event. We assume that, in the most general case, the transition from form **A** to form **B** is parameterized by some reaction coordinate x.[a] The metastable state is confined by the switching barrier U, as sketched in Fig. 9.6.

For simplicity, we will assume that the potential is harmonic out to a coordinate x_c, which corresponds to the barrier top. The reaction coordinate fluctuates around the local energy minima; the system will escape into the ground state if it makes it beyond x_c.

The detailed analysis [18, 19] shows that dispersion of the reaction coordinate is Gaussian, and all stochastic forces from the environment and the transport current contribute to a single parameter $\langle x_0^2 \rangle$, which sets the width of the Gaussian distribution.

The escape rate is, therefore, proportional to the probability

$$\exp\left[-\frac{x_c^2}{\langle x_0^2 \rangle}\right]$$

to find the system at $x = x_c$. Having in mind an attempt rate ω_0, which is the frequency of oscillations around the local minima (see

[a]The physical nature of the reaction coordinate is not relevant for a moment; for the bianthrone molecule, it is linked to the twist angle across the joint between two anthrone forms.

Fig. 9.6), we arrive at

$$\Gamma = \omega_0 \exp\left[-\frac{x_c^2}{\langle x_0^2 \rangle}\right] \tag{9.5}$$

As in the harmonic potential the energy is quadratic in terms of displacement, we can rewrite Eq. 9.5 as

$$\Gamma = \omega_0 \exp\left[-\frac{U}{\langle E \rangle}\right] \tag{9.6}$$

The problem, therefore, reduces to a basic question: How different mechanisms contribute to the average energy of fluctuations $\langle E \rangle$? If the current-induced excitations are insignificant (and this is indeed the case for the bianthrone switch, as we will eventually see from the fit—see the next section) than the system is at the thermal equilibrium and at extreme high temperatures, $\langle E \rangle = k_B T$, while at zero temperature, $\langle E \rangle$ reduces to the energy of the ground-state quantum fluctuations $\frac{1}{2}\hbar\omega_0$. In the most general case,

$$\langle E \rangle = \frac{\hbar\omega_0}{2} \coth\left(\frac{\hbar\omega_0}{2k_B T}\right) \tag{9.7}$$

which is the time-average energy of the quantum oscillator coupled to a heat bath. Note that the characteristic frequencies for molecular vibrations are in the THz range, and $f = 1$ THz corresponds to $hf/k_B = 50$ K, which means that for cryogenic temperatures, the ground-state oscillations are on the same page or even exceed the thermal fluctuations, and one should use a complete formula (9.7).

Having the theoretical model completed, we can now turn to the analysis of experimental data.

9.3.3 *Data Analysis*

Following the discussed procedure, we first convert an original data for switching histograms into plots for escape rates Γ; the result is presented in Fig. 9.7.[a]

---

[a] For calculations, it is convenient to convert binned histograms in Fig. 9.4 into equivalent representation

$$P(V) = \sum_i \exp\left[-(V - V_i)^2/\delta V^2\right],$$

where δV is the bin size, V_i is the switching voltage for trace number i, and i enumerates all I–V traces.

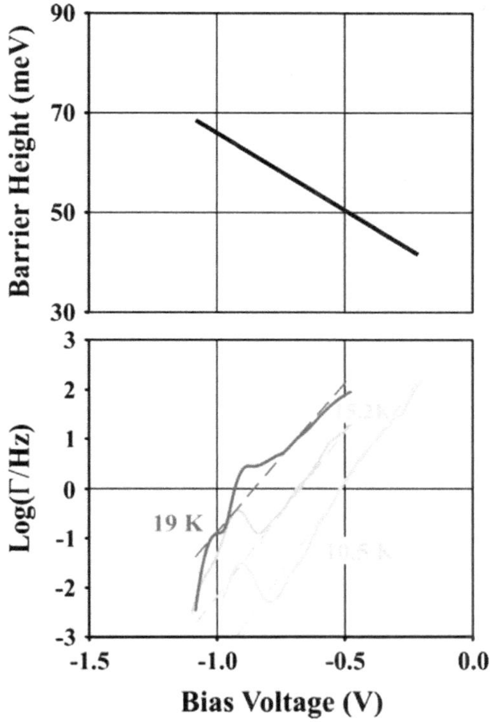

Figure 9.7 The bottom panel presents switching rates extracted from the switching histograms in Fig. 9.4; the fits based on the formulas (9.9) are depicted with dashed lines. The top panel shows the reconstructed barrier height $U(V)$.

Each histogram in Fig. 9.4 defines the switching rate within the switching area at corresponding temperature. As there are no switching events at lower or higher biases, the switching rates at these bias voltages are, of course, not defined.[a]

One can see that the switching rates increase roughly as the exponent of the bias voltage (note that the bottom panel is the logarithmic plot). Knowing from Eq. 9.6 that $U \sim \log \Gamma$, we conclude that the barrier height is roughly a linear function of the bias voltage:

$$U(V) = U_0 + \alpha V \tag{9.8}$$

[a]An excessive noise on Γ-plots in Fig. 9.7 at bias voltages around -1 V is also due to the limited number of switching events in this area.

Collecting Eqs. 9.6–9.8 together, we have for the tunneling rates

$$\Gamma \approx \frac{\omega_0}{2\pi} \exp\left[-\frac{U_0 + \alpha V}{\frac{1}{2}\hbar\omega_0 \coth\left(\frac{1}{2}\hbar\omega_0/k_B T\right)}\right] \tag{9.9}$$

The fit based on Eq. 9.9 is depicted in Fig. 9.7 with dashed lines. The top panel in Fig. 9.7 presents the reconstructed barrier height U, versus the bias voltage, as defined by Eq. 9.5. The fit returns the values for three physical parameters: $\hbar\omega_0 = 3.6$ meV, $U_0 = 40$ meV, and $\alpha = 0.03e$ (e is an elementary charge).

First of all, we see that $\hbar\omega_0/k_B \approx 42$ K, which means that even at $T = 19$ K (the highest temperature in Fig. 9.4), $\frac{1}{2}\hbar\omega_0 \approx k_B T$. The quantum ground-state fluctuations are as important as the thermal fluctuations. At lower temperatures, the bianthrone escapes from the metastable state predominantly by quantum tunneling. This surprising result has important implications for potential device applications, which we will discuss in more details in the next chapter.

The parameter $\alpha = 0.03e$ implies that in states On and Off, the bianthrone molecule is charged slightly differently—the difference being ~3% of an electron charge. When the molecule switches, some charge goes to/from the electrode and bias-induced electrostatic forces contribute to the total switching energy. In the first approximation, this effect is linear in terms of the bias voltage, as suggested by Eq. 9.8.

Surprisingly, the switching barrier is only 49–70 meV, which is a dramatic reduction as compared to the switching barrier for bianthrone in solution [13, 20].

To address this apparent contradiction, Lara-Avila et al. [16] carried out detailed quantum-chemical calculations for bianthrone molecule in solid-state environment. It was found that pronounced barrier suppression arises if one takes into account the interaction between the charge on the molecule and the image charges on the substrate. It appears that the transition state is less rigid than any of the conformations. This allows the atomic charges in the anion to approach closer their images, so that the electrostatic interaction compensates the distortion energy, thus lowering the transition barrier for the anion at a metal surface down to ca. 0.1 eV, in good agreement with experimental findings.

9.4 C_{60} Junction

9.4.1 *Experimental Data*

Our next example will be C_{60} trapped in silver nanogap. This case is especially interesting for us because the fullerene junction was not expected to be switchy by design. There are, of course, no feasible structural conformations for C_{60} cage. As we will see, it is by the identification of the switching mechanism and relevant switching parameters that one can conclude the mere nature of the On and Off states in this system.

As in the case of bianthrone switch, we will not discuss the fabrication procedure; an interested reader should refer to the original work [4]. Here we will go straight to the experimental results presented in Fig. 9.8.

Qualitatively, the switching pattern for the C_{60} switch is similar to that of the bianthrone switch: Obviously, the sample has two states with different conductance. At 4 K, switching between the two states is histeretic. For bias scans taken in the negative to positive direction, the sample starts from low-current state (**Lo**) and, soon after zero bias, switches to high-conducting state (**Hi**). For return scans, the system stays in the high-current state down to -400 mV. At 14 K, the hysteretic area shrinks, and at 20 K, there is a narrow bias window where the conductance switches between two states at very high rate.

Following the same procedure as described in the previous section, the comprehensive statistics based on more than 10^5 switching events was collected, and the resulting switching histograms for 10 different temperatures are presented in Fig. 9.9 (for **Lo**$\rightarrow$**Hi** switching events, i.e., for bias sweeps taken in the negative to positive direction; the reverse **Hi**$\rightarrow$**Lo** switching will be discussed later). The same transformation based on Eq. 9.4 was then applied to get the plots for the switching rates Γ, which are shown in Fig. 9.9 (bottom). As explained earlier, at any given temperature, the switching histogram is localized around specific bias voltage, so that the corresponding plot for Γ is defined within the same vicinity.

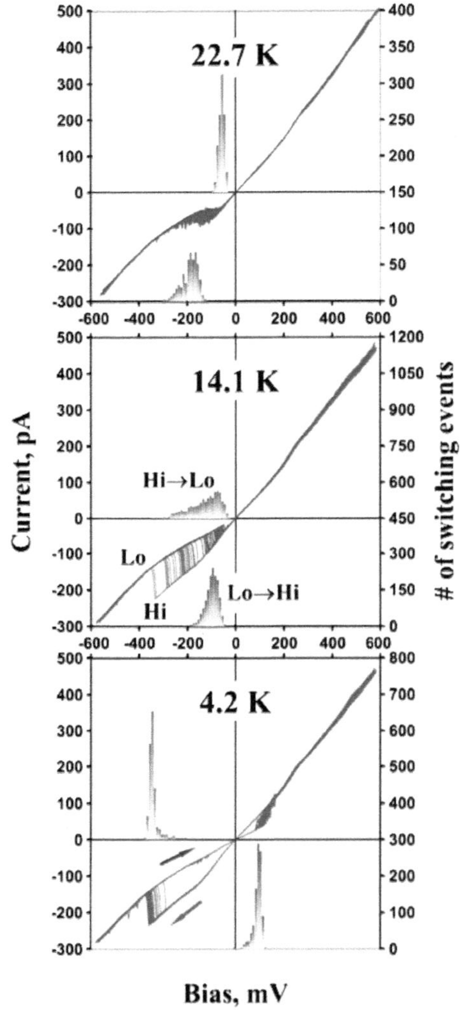

Figure 9.8 Current–voltage characteristics of C_{60} single-molecule junction at three different temperatures. For every temperature, the 32 representative voltage ramps are shown; a few thousand ramps were taken to collect the switching voltage distribution presented on histograms.

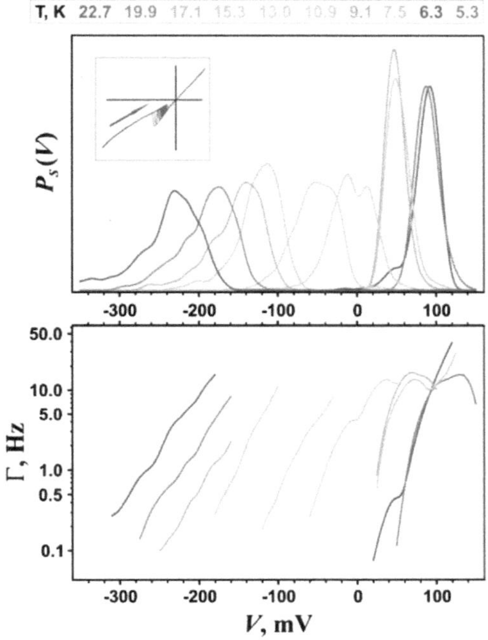

Figure 9.9 Top: observed probability that the C_{60} junction will switch from state **Lo** to state **Hi** at a certain voltage, when the voltage is swept from negative to positive voltages (a few exemplary I–V curves are show on inset). Bottom: escape rates generated using Eq. 9.4.

9.4.2 *Switching Rates: Forward Switching*

Compared to the bianthrone case, we will now add one extra ingredient to the model: the heating associated with a current running from one electrode to the other through the molecule. This is, in part, motivated by the fact that for $T = 10.9$ K, the switching events take place around voltage $V = 0$ (see Fig. 9.9), and the observed probability has a pronounced dip at zero bias, i.e., when little or no current is moving through the system. This suggests that the current is actually helping the system getting over the barrier the more the voltage and the associated current is different from zero.

When an electron is added to C_{60}, the equilibrium position of the reaction coordinate is shifted by some amount, and the coordinate experiences a force.[a] When the extra electron leaves the molecule, the original equilibrium position is restored, but an amount of energy is transferred to the C_{60} in the process. A model containing similar physics was used in the analysis of the Eigler switch [18], [19], where it was shown that in the presence of a transport current, the time-average oscillator energy reads:

$$\langle E \rangle = \frac{\hbar\omega_0}{2} \coth\left(\frac{\hbar\omega_0}{2k_B T}\right) + \alpha\left(|eV| - \hbar\omega_0\right)\Theta\left(|eV| - \hbar\omega_0\right) \quad (9.10)$$

where $\alpha = \left(G/e^2\right)\left(\tilde{E}/\eta\right)$; G is the sample conductance, $\tilde{E}$ is the average energy supplied by one tunneling electron, and η is the relaxation rate of the mechanical oscillations. The current-induced correction to the average energy is, therefore, the energy supplied by passing electrons during the relaxation time of vibrations associated with the reaction coordinate; the Θ-function in Eq. 9.10 accounts for the fact that electrons cannot excite the reaction coordinate if the bias-supplied energy $|eV| < \hbar\omega_0$.

To summarize, in the presence of transport current, the escape rate is given by Eq. 9.6 with an average energy $\langle E \rangle$ in the generalized form (Eq. 9.10) (for detailed calculations, see the Supporting Information in Ref. [4]).

The model now includes two fitting parameters for $\langle E \rangle$: ω_0 and α, which is manageable, but the barrier height $U(V)$ is some unknown function of the bias voltage. One can, in principle, try to guess this function (introducing a few more fitting parameters for $U(V)$) and to make a fit, based on a *single* formula for $U(V)$, matching *all nine* plots for Γ in Fig. 9.9. However, there is a more straightforward way to recover the switching barrier: instead of fitting many curves for Γ, we can try to compose a single plot for the barrier height as a function of the bias voltage. To this end, we first solve Eq. 9.6 for U:

$$U(V) = -\langle E(V) \rangle \ln \frac{\Gamma(V)}{\omega_0} \quad (9.11)$$

We see that, for a given choice of two fitting parameters ω_0 and α, each $\Gamma(V)$ plot in Fig. 9.9 generates a segment for a $U(V)$ plot. Each segment is defined for a limited bias range $\{V\}_{T=T_i}$, where the

---

[a] Note that at this stage, we do not need to specify the physical nature of the reaction coordinate: It could be the C_{60} displacement, rotation, or even the distorion of the C_{60} cage.

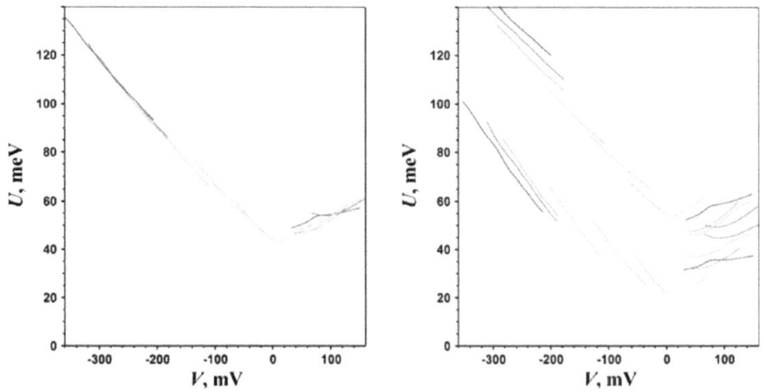

Figure 9.10 Left: the barrier height for **Lo→Hi** transition in C_{60} junction. The parameter $\hbar\omega_0$ is 2.5 meV, $\alpha = 0.02$. Right: similar plots generated for two other choices of parameters ω_0 and α. Only one specific pair of fitting parameters makes all segments to collapse into a single common line.

corresponding plot $\Gamma(V)$ for the temperature $T = T_i$ is defined. We expect that, with a proper choice of ω_0 and α, all segments should collapse into a smooth common curve for $U(V)$. This indeed happens for one (and just one) particular choice of (ω_0, α) pair, as shown in Fig. 9.10.

9.4.3 *Switching Rates: Reverse Switching*

In a similar way, one can analyze histograms for switching in the opposite, **Hi→Lo**, direction. The switching histograms for the reverse transition are shown in Fig. 9.11 (top). One can clearly see that at the three highest temperatures, the switching histograms have a clear onset at approximately 35 mV. A behavior like this suggests that the backward switching is mediated by the excitation of the $\hbar\Omega = 35$ meV mode, so that the tunneling electrons start to pump energy into reaction coordinate as soon as the bias exceeds $\hbar\Omega/e$.

We, therefore, will modify the formula for an average energy $\langle E\rangle$ accordingly:

$$\langle E\rangle = \frac{\hbar\omega_0'}{2}\coth\left(\frac{\hbar\omega_0'}{2k_B T}\right) + \alpha'\left(|eV| - \hbar\Omega\right)\Theta\left(|eV| - \hbar\Omega\right), \quad (9.12)$$

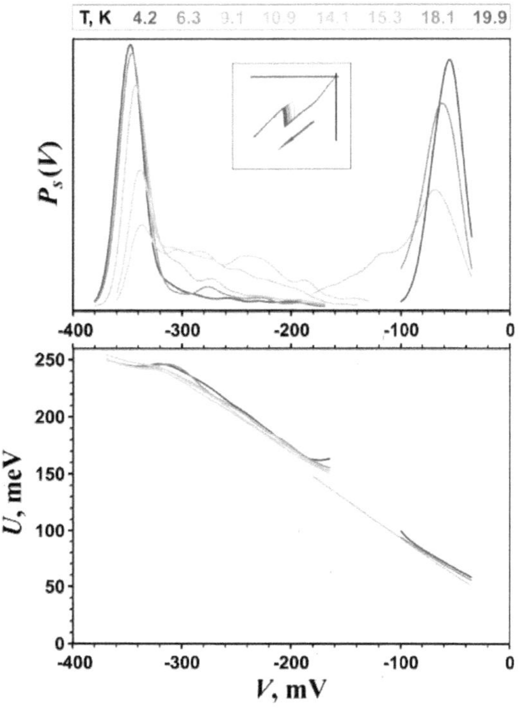

Figure 9.11 Top: switching histograms for the reverse **Hi→Lo** transition, when the voltage is swept from negative to positive voltages (a few exemplary I–V curves are show on inset). Bottom: switching barrier for the reverse transition reconstructed using the refined model (9.12) with an effective temperature determined by current generated heating through coupling to the vibration mode with $\hbar\Omega = 35$ meV. The parameter $\hbar\omega_0'$ is 2.5 meV, $\alpha' = 0.02$.

where the renormalized coupling constant α' implicitly includes the intermode ($\Omega \to \omega_0'$) coupling. The switching barrier for the reverse transition reconstructed using the refined model (9.12) is shown in Fig. 9.11.

9.4.4 *Potential Landscape and On-Off Hysteresis*

Having a reconstructed potential landscape along the reaction coordinate for both forward (Fig. 9.10) and reverse (Fig. 9.11) transitions, we can now combine these results and reconstruct the

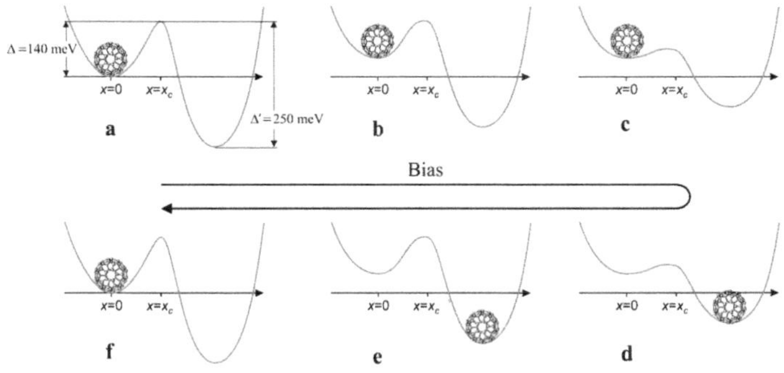

Figure 9.12 A complete hysteretic cycle (a–f) for C_{60} switch. (a) System is in metastable state. As the bias is ramped toward positive values, the metastable state is getting shallow (b) and eventually C_{60} escapes into the ground state (c→d). During the reverse bias ramp, the metastable state is restored (e) and, as soon as the bias exceeds $\hbar\Omega$, C_{60} is reset back to the metastable state (f). Within the hysteretic region states, (b) and (e) coexist as two bistable options [(b) = **Lo** and (e) = **Hi**].

sequence of events for a complete hysteretic cycle, sketched in Fig. 9.12.

Initially, at extreme negative bias, C_{60} is in the metastable state. As the bias is swept in the negative to positive direction, the metastable trap gets more and more shallow and, eventually, the system falls into the ground state. For the opposite bias ramp, the metastable trap is restored, but an average energy is not enough to overcome the (much higher) barrier for the reverse transition until the energy of tunneling electrons exceeds $\hbar\Omega$. Excitation of the mediating mode $\hbar\Omega$ opens an extra channel for pumping energy into the reaction coordinate and promotes the backward switching.

For **Lo**→**Hi** switching, we assumed that the 2.8 meV vibration mode associated with the reaction coordinate is coupled directly to the tunneling electrons. This does not need to be the most general scenario. Some vibrational modes with higher energies may, in fact, be stronger coupled to the moving electrons. This energy may then, through the intermode coupling, "triggle down" to the reaction coordinate and ultimately help the system to pass over the energy barrier. It is known that the charged C_{60} molecule is Jahn–Teller distorted. The Jahn–Teller active modes are the ones with the highest electron–phonon coupling. The energy of the lowest Jahn–

Teller active mode Hg(1) for C_{60} in vacuum is around 34 meV [21], [23]; it seems to be natural to associate the mediating coordinate with this breathing mode. Note that the $\hbar\Omega$ mode is active in the ground state (**Hi**, Fig. 9.12e) and not active in the metastable state (**Lo**, Fig. 9.12b). We shall discuss the reason for this asymmetry in the next section.

9.4.5 *What are the On and Off states?*

So far, we have made no detailed assumptions about the nature of the reaction coordinate. All statements (Eqs. 9.6, 9.10, and 9.12) are irrelevant to what kind of physical motion corresponds to this coordinate. However, as we shall see, the operating parameters ω_0, Ω, and U, extracted from the fit, leave very little room for interpretation.

As it was discussed, at low temperatures $T < \frac{1}{2}\hbar\omega_0/k_B = 14.5$ K the dominant switching mechanism is tunneling,[a] and one can use a simplified formula

$$\Gamma = \omega_0 \exp\left[-\frac{2U}{\hbar\omega_0}\right] = \omega_0 \exp\left[-\frac{x_c^2}{\langle x_0^2\rangle}\right] \tag{9.13}$$

From this, we can estimate the dispersion of the ground state as

$$\langle x_0^2\rangle = -x_c^2/\ln\left(\Gamma/\omega_0\right) \tag{9.14}$$

From Eq. 9.14 we can estimate the mass of the tunneling object: as the switching actually happens when Γ is in the range $T_{\text{sweep}}^{-1} \sim 1 - 10$ Hz (where $T_{\text{sweep}} \sim 1$ s is a full bias sweep time), we can estimate the logarithm in the former formula as ~25. The maximum possible value for $x_c \sim 0.3$ Å can be estimated from the fact that the sample conductance in Fig 9.8 only changes by a factor of 2, and we arrive at $\langle x_0^2\rangle \sim 0.06$ Å and the corresponding tunneling mass $M = \hbar/\omega_0 \langle x_0^2\rangle > 400 m_p$ (m_p is the proton mass). This result rules out all switching mechanisms involving the rearrangement of

[a] At 4 K and 100 mV bias, e.g., the different contributions to the total oscillator energy are according to Eq. 9.11: 1.4 meV from the quantum mechanical ground-state oscillations, ~0.001 meV from the thermal fluctuations, and ~0.7 meV from the current-induced heating.

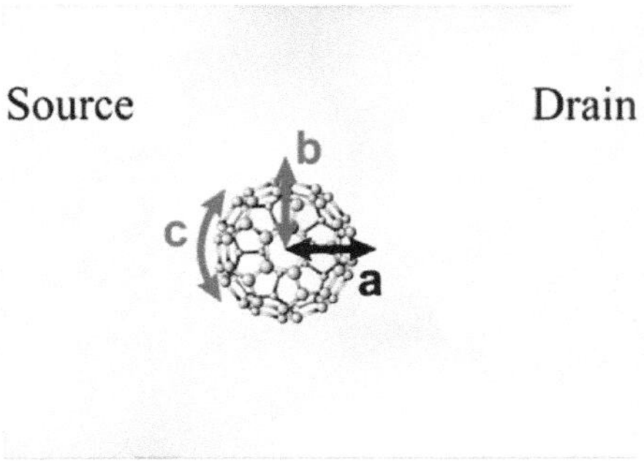

Figure 9.13 Possible mechanical transformations for C_{60} in nanogap: shuttling between source and drain electrodes (a), sliding along one of the electrodes (b), and rolling (c).

metal atoms in the electrodes and links the On-Off switching to a mechanical motion of the C_{60} cage itself.

One can, in principle, consider many possibilities for C_{60} motion in nanogap: shuttling over the gap between source and drain electrodes, sliding along one of the electrodes etc., as sketched in Fig. 9.13.

However, for any transformation involving the displacement of C_{60} mass center, the corresponding energy barrier is at least 0.8 eV [22] (for lateral displacement as in Fig. 9.13b) or even higher: 2 eV for desorption [23] (needed for shuttling as in Fig. 9.13a), which is well above the observed barrier height of 40–140 meV. On the other hand, the barrier for C_{60} rotations, which was calculated to be 90 meV for C_{60} on Au(110) [24] and experimentally found to be 58 meV in a C_{60} film [25], is in the right range. In fact, rotation is the only known feasible C_{60} motion at such low energies. We can, therefore, compellingly identify the reaction coordinate with C_{60} orientation.

Knowing that the difference between On and Off states is C_{60} orientation we can now understand why the 35 mV breathing mode is active in the ground state and inactive in the metastable state (as it was discussed earlier, it is this asymmetry that makes the device bistable).

The Jahn–Teller distortion of C_{60} anion goes along the hexagon-hexagon axis. When C_{60} faces the metal with hexagon facet the Jahn–

Teller distortion goes in the direction normal to the surface and coupling to the rolling vibrations is suppressed by symmetry; when C_{60} faces the metal with pentagon this symmetry constrain does not apply.

9.4.6 *Switching by Tunneling: Approaching the Classical Limit*

As we have already discussed, due to the very high (THz) frequencies of intramolecular vibrations, at cryogenic temperatures, the energy of ground-state fluctuations can exceed that of the thermal fluctuations. Therefore, at low enough temperatures, escape from metastable state happens predominantly by quantum tunneling. This result, in a way, runs against the plain intuition: As the C_{60} mass exceeds the electron mass more than one million times, one would naively expect that such a heavy object could not tunnel. It is instructive to take a closer look at the corresponding numbers.

In a fact, the tunneling probability itself, i.e., the probability to penetrate the barrier in a single attempt, is extremely small: for C_{60}, it is in the range of $\exp[-2U/\hbar\omega_0] = \exp[-25] \sim 10^{-11}$, which means that C_{60} has to make 100 billion attempts before it can actually tunnel. On the natural time scale, the metastable state is actually more stable than the solar system. It is a very high attempt rate $\sim 10^{12}$, which makes the system switchy.

Equation 9.13 can be rewritten as $\Gamma = \omega_0 \exp[-2\tilde{n}]$, where $\tilde{n}$ is the number of quantized states in the metastable trap. For C_{60} switch, $\tilde{n} \sim 13$ (see Fig. 9.14).

What if, for some other molecular switch, $\tilde{n}$ is even smaller, say $\tilde{n} \approx 6$? It is easy to see then that the switching rate will be in nanosecond range, and this is for how long one can store information when such a switch is deployed as a memory element. We see that the quantum tunneling sets the fundamental limits for operating parameters for a classical memory element: To store information in a bistable system, one either needs high enough switching barrier or a massive toggle; which implies trading the switching time for switching energy or vice versa. Essentially, operating parameters for the C_{60} switch are already approaching the physical limits.

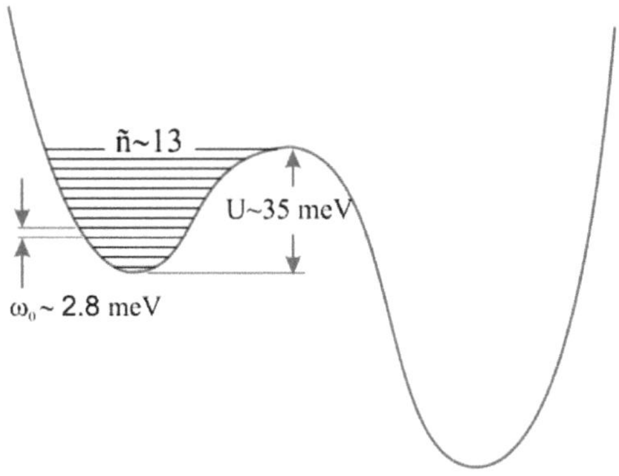

Figure 9.14 Escape from the metastable state in the C_{60} switch happens when the switching barrier is about 35 meV, which corresponds to ~13 quantized energy levels in the metastable trap.

9.5 Switching Behavior of OPV3 Junction

9.5.1 *Experimental Data*

The last example we will consider is the switching behavior of single-molecule OPV3 junction. An OPV (oligophenylenevinylene) molecule, presented in Fig. 9.15, is a conjugated polymer, where $\pi - \pi$ coupling makes a HOMO[a] delocalized across the whole molecule.

In simple words, an OPV is a molecular wire, which, if strongly coupled to electrodes, makes a high conductive junction [26]. To facilitate strong coupling, the π-conjugated moiety was terminated with thiol groups (see Fig. 9.15), which are known to form covalent bonds to a gold electrode. By design, a rigid π-kernel linked to

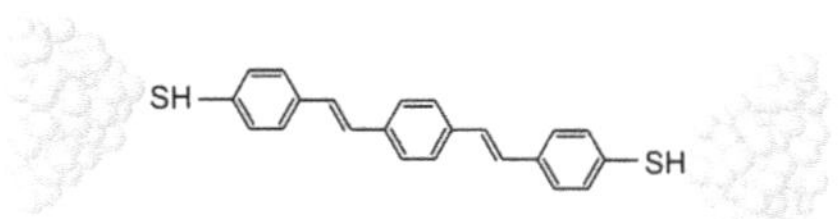

Figure 9.15 Schematic representation of OPV3 molecule placed in the nanogap between gold electrodes. The terminal SH-groups provide strong coupling to the electrodes.

---

[a]HOMO = highest-occupied molecular orbital

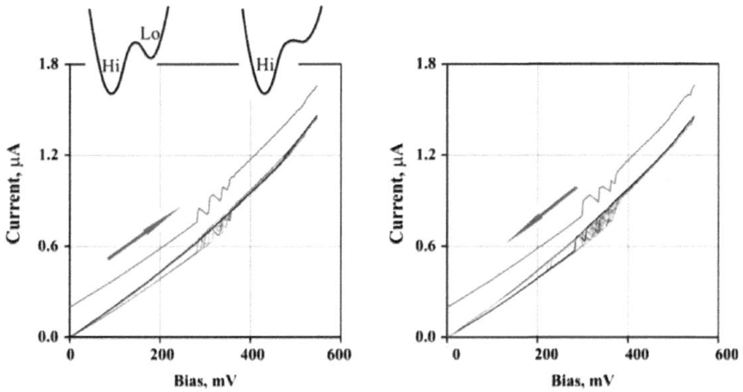

Figure 9.16 I–V characteristic of an OPV3 device showing switching in both sweeping directions. Two representative traces are shown offset: one for sweeping voltage in the positive direction (left) and one for sweeping in the negative direction (right). Inset on right-top shows the potential landscape at low and high bias voltages.

the electrodes with firm and directional covalent bonds is arguably the best one can do to implement a stable molecular junction. Nevertheless, in experiment, the junction was found to be switchy, as shown in Fig. 9.16. In this chapter, we will demonstrate how the analysis of the switching statistics allows to identify reconstruction of the molecule-to-electrode interface as a source of bistability in this system.

9.5.2 *Data Analysis*

Two features emerge from the data set in Fig. 9.16: At low biases, the system could be in one of two (**Lo** or **Hi**) conducting states, while at high biases, only state **Hi** is feasible; there is a bias window around ~300 mV where the system switches between the two states many times. A switching pattern like this suggests that at low biases, the transition barrier is so high that the system is permanently trapped in one of two states; as the bias voltage increases, the barrier is suppressed, and at around 300 mV, the switching becomes possible in both (**Hi**→**Lo** and **Lo**→**Hi**) directions. Eventually, at high enough biases, the barrier is completely suppressed so that the metastable trap disappears and only state **Hi** is feasible (see inset in Fig. 9.16).

On the way back (for bias scans taken in high-to-low direction, as in Fig. 9.16, right), the system always starts evolution in high-conducting state, because this is the only state available at high biases and is eventually frozen either in state **Hi** or **Lo** at zero bias.

From the analysis perspective, this situation is different from the switching pattern in Fig. 9.2, where the switching happens only once for any bias scan. An assumption that the switching happens only once was a prerequisite for our analysis (9.1–9.4). Indeed, Eq. 9.1 accounts for switching in one direction only. Now we shall generalize our model to a situation where switching goes in both directions.

The generalization is rather straightforward: Eq. 9.1 should be replaced with a rate equation accounting for switching rates Γ_{Hi-Lo} and $\Gamma_{Lo\rightarrow Hi}$ in both directions, so that the probability P_{Lo} to be in the low-conducting state evolves as

$$\frac{dP_{Lo}}{dt} = -\Gamma_{Lo\rightarrow Hi}\left(V\left(t\right)\right)P_{Lo} + \Gamma_{Hi\rightarrow Lo}\left(V\left(t\right)\right)P_{Hi} \tag{9.15}$$

with a similar equation for the low state, but because $P_{Lo} + P_{Hi} = 1$, we need only solve a single differential equation. Same as for Eq. 9.1, we can convert Eq. 9.15 from the time domain into the voltage domain:

$$\dot{u}\frac{dP_{Lo}\left(V\right)}{dV} = -\left(\Gamma_{Hi}\left(V\right) + \Gamma_{Lo}\left(V\right)\right)P_{Lo}\left(V\right) + \Gamma_{Hi\rightarrow Lo}\left(V\right) \tag{9.16}$$

This equation has a formal solution (for the bias ramps starting at $V = 0$)

$$P_{Lo}\left(V\right) = p\left(V\right)\left\{\int_0^V \frac{dV'}{\dot{u}}\left[p\left(V'\right)\right]^{-1}\Gamma_{Hi\rightarrow Lo}\left(V'\right) + P_{Lo}\left(0\right)\right\} \tag{9.17a}$$

$$p\left(V\right) = \exp\left(-\int_0^V \frac{dV'}{\dot{u}}\Gamma\left(V'\right)\right) \tag{9.17b}$$

Assuming some specific mechanisms for **Hi→Lo** and **Lo→Hi** transitions, one can, in principle, find how P_{Lo} evolves as the bias voltage is ramped, and, eventually, predict the distribution of switching events. However, as one can see in Fig. 9.16, for bias voltage above ~400 mV, a switching to the low-conducting state is immediately followed by the switching back to the high-conducting state, indicating that the time spent in the low-conducting state is too short for the measurement system to resolve it. Obviously, if the switching happens in both **Hi→Lo** and **Lo→Hi** directions, then, due to a finite bandwidth of the measurement setup, not all switching events can be resolved, and one cannot extract complete switching histograms from experimental data.[a]

To make comparison with the model, we can instead compare measured current–voltage characteristics averaged over many traces: the current is simply extracted as

$$\langle I\,(V)\rangle = G_{Hi}P_{Hi}\,(V) + G_{Lo}P_{Lo}\,(V) \tag{9.18}$$

where G_{Hi} and G_{Lo} are the conductances of the high and low states, respectively.

As the system is bistable at low biases, we will treat I–V curves starting in **Lo** (**Hi**) conducting state separately: First we solve Eq. 9.17a,b with the initial condition $P_{Lo}\,(V_i) = 1$ (the system starts evolution in state **Lo**); the resulting current $I_{Lo}\,(V)$ is the predicted current, averaged over all I–V curves starting in the low-conducting state. In a similar way, solving Eq. 9.17a,b for the initial state $P_{Lo}\,(V_i) = 0$, we arrive at the current $I_{Hi}\,(V)$, which is an average over all I–V traces starting in the high-conducting state.

The crucial ingredients still missing are the formulas for switching rates Γ_{Hi-Lo} and $\Gamma_{Lo\to Hi}$, which are not known a priori. In simple cases, such as when some specific switching mechanism dominates in Γ_{Hi-Lo} and one (possibly different) in $\Gamma_{Lo\to Hi}$, a proper fit can be found simply by the trial-and-error approach. In particular, for the OPV3 switch, it was found that the **Lo→Hi** transactions are thermally activated:

$$\Gamma_{Lo\to Hi} = \omega_0 \exp\left[\left(U_0 - \alpha V\right)/k_{\mathrm{B}}T\right] \tag{9.19}$$

and the opposite **Hi→Lo** transition is current driven:

$$\Gamma_{Hi\to Lo} = \frac{\gamma G}{e}\left(|V| - \frac{\hbar\Omega}{e}\right)\Theta\left(|V| - \frac{\hbar\Omega}{e}\right) \tag{9.20}$$

where G is the conductance, Ω is the frequency of the vibronic mode, which mediates current-induced heating, and γ is an effective

[a] If, for a given bias sweep direction, the switching goes in one direction only, then there is only one switching event for any bias ramp, like in Fig. 9.2. In this case, all switching events are, of course, registered and the switching histogram is well defined.

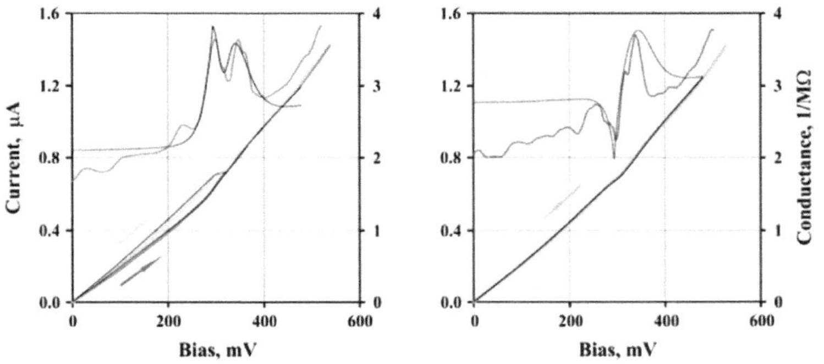

Figure 9.17 Fitting procedure for the OPV3 switching device. First, the average current (red curve and purple, the derivative on the left panel) for the up-sweep (starting in the low-conducting state) is used to determine the four fitting parameters by four extrema points in the derivative. Then, these numbers are used to calculate the average current in the different sweep directions (all black curves). The calculated average currents are compared to the experimental ones: up-sweep starting in **Lo** (left panel, red), starting in **Hi** (top panel, cyan), and down-sweep (right panel, green). In addition, the derivative of the down-sweep is shown (right panel, red) and compared to the extracted model curve (right panel, blue).

electron–vibron coupling. The fit based on the model (9.17a,b) with switching rates (9.19, 9.20) is presented in Fig. 9.17.

To make a fit, the average current in the positive sweep direction starting in the state **Lo** was first calculated by the numerical integration of Eq. 9.17 with switching rates given by Eq. 9.19 and Eq. 9.20. This is the red curve in Fig. 9.17. The derivative of this curve (purple curve) was used to fix the effectively four independent parameters:

$$\frac{\dot{u}}{\omega_0}\exp\left[-\frac{U_0}{k_B T}\right] = 5.7 \times 10^{-10}\text{mV} \tag{9.21a}$$

$$\hbar\Omega = 300\,\text{meV} \tag{9.21b}$$

$$\alpha = -0.021e \tag{9.21c}$$

$$\gamma = 1.1 \times 10^{-12} \tag{9.21d}$$

As for bianthrone and C_{60} switches, we find that the potential of the well shifts with the bias voltage corresponding to a small fraction of an electron charge, which again is a reasonable value of the effective charge on the molecule. The extracted value of γ implies that one out of $\sim 10^{12}$ tunneling electrons triggers a switching event. Finally, given $T = 4.2$ K and $\dot{u} = 53$ mV/s, we estimate the switching barrier to be 10–20 meV.

9.5.3 *The On and Off States of the OPV3 Junction*

Having all the relevant parameters for the OPV3 switch quantified, we can now discuss the nature of the On and Off states in this junction. While there is nothing special about the parameters α, γ, and U_0, the energy of vibronic mode $\hbar\Omega = 300$ meV is remarkably high. First of all, such a high value indicates that this is the energy of some local mode: all collective modes (when the whole molecule is twisted or bended) have energies well below 100 meV. All local modes, associated with stretching or bending of some specific chemical bonds, have well-known and tabulated energies; consulting the table one finds that 300 meV is the energy of the **S-H** stretch mode.[a]

This result is direct evidence that switching in OPV3 junction is linked to a proton in the **SH**-group bound to the gold electrode. Once the proton is attached to **S**, it can detach when the successive tunneling electrons emit an **H-S** vibron. On the other hand, once the proton is detached, it is no more subjected to current-induced excitations, which means that the reattachment of the proton can happen only via thermal excitation. This scenario naturally explains why current activation (9.20) is possible in only one direction.

We can plausibly identify the metastable (detached **H**) state as a proton taken up by a nearby Lewis base (a lone pair of aluminum oxide), as shown in Fig. 9.18. Indeed, the activation energy for the [Lewis acid] → [Lewis base] reaction is on the same page as the switching barrier of 10–20 meV.

In fact, this is very low energy, as compared to most protonation–deprotonation reactions. It is due to the low activation barrier that

[a]Two closest alternatives are C-H stretch mode at 365 meV, which is too high, and CH2 twist at 161 meV, which is obviously too low.

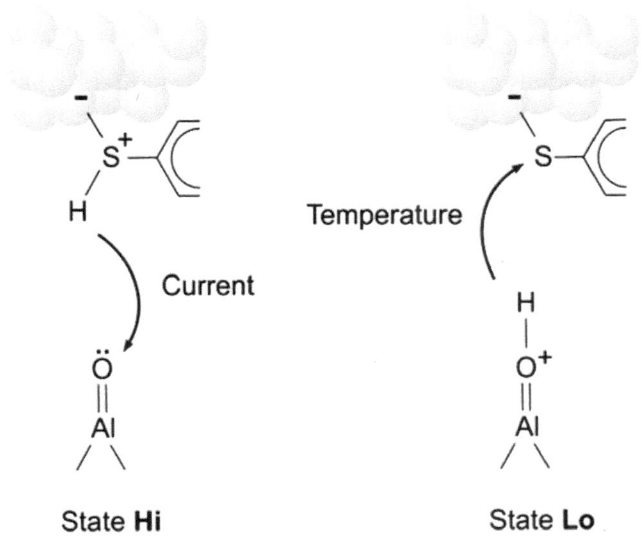

Figure 9.18 Switching mechanism in the OPV3 junction. Left: The S-H bond formed in state **Hi** can be excited by the electron-transfer reaction as described in the text, resulting in switching **Hi**→**Lo**. An eliminated proton is taken up by a nearby Lewis base shown as a lone pair of aluminum oxide. Right: This state can relax thermally to the ground state **Hi**, where the hydrogen binds to sulfur instead of the Lewis acid.

the metastable state (**Lo**) can relax to the ground state (**Hi**) via thermal activation.

9.6 Summary

If the areas for forward and backward transitions in a molecular switch do not overlap (like for the bianthrone switch and C_{60} at low temperatures), then there is only one switching event for each bias ramp. In this case, all switching events can be registered and one can compose a complete histogram for the distribution of the switching voltages. From this distribution, one can, in a straightforward way, extract the switching rates Γ as a function of the bias voltage, temperature, etc. Knowing the functional dependence $\Gamma(V, T)$, one can deduce the relative contribution of three main switching mechanisms: thermal fluctuations, quantum mechanical tunneling, and current-induced excitations.

Complications arise if the areas for On–Off and Off–On transitions overlap, as in the OPV3 junction, so that there is a bias window where the system switches back and forth between two states. In this situation, not all switching events can be unambiguously identified due to the finite bandwidth of the measurement system; therefore, one cannot compose complete switching histograms. The best strategy in this situation is to model an average current (averaged over many I–V traces) assuming some particular switching mechanisms, i.e., some specific functional dependence for Γ. Practically, this strategy works well if only one specific mechanism dominates for each switching direction.

Apart from the switching mechanisms, the model fitted to experimental data returns a full set of operating parameters for a molecular switch: the energy difference between On and Off states and the switching barrier. If the current-induced excitations contribute to switching, then the electron–phonon coupling can be extracted, and if the coupling is mediated by the excitation of some vibronic mode, then the energy of this mode can be found. The cumulative information extracted from the analysis is so comprehensive that very little or no uncertainty is left for plausible interpretation.

In conclusion, we would like to stress that the whole analysis presented in this chapter is applicable to only the data measured on single-molecule devices. For an array of molecules (say, coupled in parallel), the dispersion of per-element parameters (like switching barrier variation from molecule to molecule) also contributes to the width and the shape of the switching histograms. As this dispersion is not known a priori, the whole analysis procedure becomes undefined. Not surprisingly, it is much easier to infer the underlying physics when interrogating a single molecule.

Problems

1. The bianthrone molecule has two conformations, shown in the following figure.

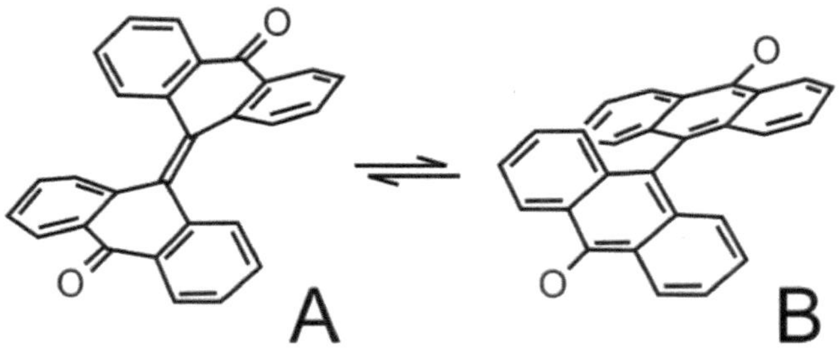

Figure 9.19

At room temperature, form B accounts for approximately 1% of all molecules in solution. When the temperature suddenly drops from 25°C to 20°C, a new equilibrium concentration is reached in about 2 h.

Use these numbers to estimate:

a. The energy difference between A and B conformations.
b. The transition barrier height (= the energy of transition state).

Show that your results are consistent with the following experimental data [16]:

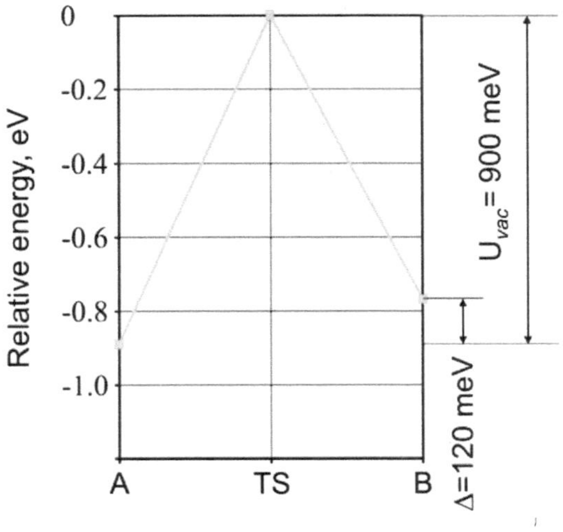

Figure 9.20

Hint: assume that all molecular vibrations are in THz range.

2. Assume the simplest situation:
   a. Molecular switch starts evolution in metastable state (1) at the bias voltage $V = 0$ and eventually switches to ground state (0) as the bias goes negative.
   b. For all $V < 0$, $\Gamma_{0\to1} = 0$ (there is no reverse switching).

Consider two different switching mechanisms discussed in the chapter:

A. Thermally activated transitions (the switching barrier being a linear function of the bias voltage V):

$$\Gamma \sim \omega_0 \exp\left[-\frac{(U_0 - qV)}{k_B T}\right]$$

B. The transitions are induced due to excitations of the molecular vibrations by the incident current:

$$\Gamma \sim (e|V| - \hbar\Omega) \cdot \Theta(e|V| - \hbar\Omega)$$

(Transition being triggered by the excitation of vibronic mode with the energy $\hbar\Omega$).

For each scenario (A and B), work out the expected distribution of switching events. Discuss the qualitative difference between two switching bias histograms.

Consider experimental data from [Danilov et al., *Nano Lett.*, 2006, **6**, 2184] (the bias was swept in the right-to-left direction):

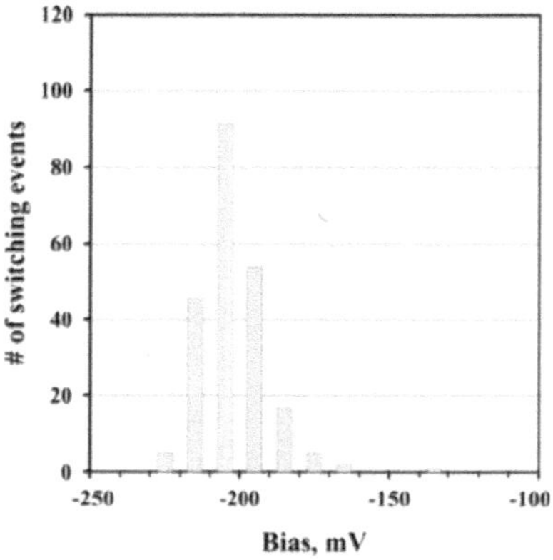

Figure 9.21

Which model (A or B) better fits the experiment?

3. Consider the excitation of C_{60} bouncing vibrations by tunneling electrons. When C_{60} is placed on a substrate, it sits in a van der Waals potential well. The parameters of this trap were estimated in Ref. [27] and are presented in a figure:

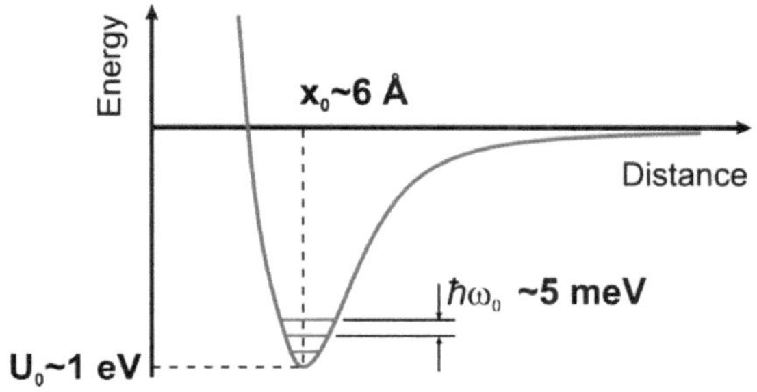

Figure 9.22

When the tunneling electron loads C_{60} with a charge e, an image charge force between the electron and its image shifts an equilibrium position by some distance δ:

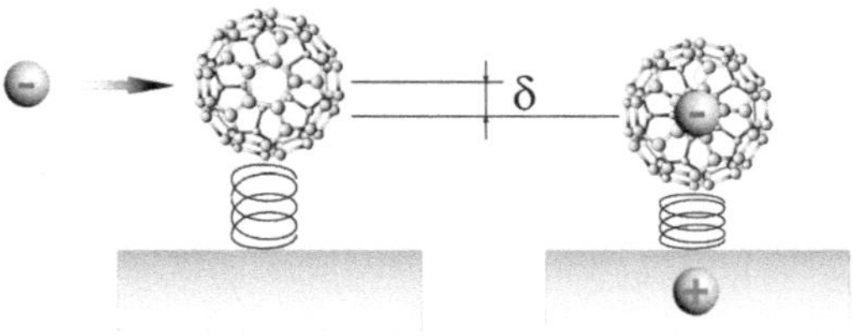

Figure 9.23

If δ is on the same page or higher than the ground-state dispersion of C_{60} coordinate $\bar{x}_0$, then tunneling is suppressed due to a finite overlap of nuclear wave functions before and after tunneling (the Franck–Condon principle).

For parameters as presented in the figure, estimate δ and the relative probabilities for the following events:

a. Electron tunnels without excitation of C_{60} vibration
b. Electron tunnels and excites C_{60} into $n = 1$ vibrational state
c. Electron tunnels and excites C_{60} into $n = 2$ vibrational state

4. Use the results from the previous problem to estimate how much energy C_{60} gets when a single transport electron passes through (assume that the tunneling events are so rare that after each excitation, C_{60} has enough time to relax to the ground state before the next electron comes).

References

1. Pan, S., Fu, Q., Huang, T., Zhao, A., Wang, B., Luo, Y., Yang, J., and Hou, J., Design and control of electron transport properties of single molecules, *Proc. Nat. Acad. Sci. USA* 2009, 106, 15259.
2. He, J., Chen, F., Liddell, P.A., Andreasson, J., Straight, S.D., Gust, D., Moore, T.A., Moore, A., L., Li, J., Sankey, O.F., and Lindsay S.M., Switching of a photochromic molecule on gold electrodes: Single-molecule measurements, *Nanotechnology* 2005, 16, 695.
3. Fleming, C., Long, D.-L., McMillan, N., Johnston, J., Bovet, N., Dhanak, V., Gadegaard, N., Kögerler, P., Cronin, L., and Kadodwala, M., Reversible electron-transfer reactions within a nanoscale metal oxide cage mediated by metallic substrates, *Nat. Nanotechnol.* 2008, 3, 289.
4. Danilov, A.V., Hedegård, P., Golubev, D.S., Bjørnholm, T., and Kubatkin, S.E., Nanoelectromechanical switch operating by tunneling of an entire C_{60} molecule, *Nano Lett.* 2008, 8, 2393.
5. Choi, B.-Y., Kahng, S.-J., Kim, S., Kim, H., Kim, H.W., Song, Y.J., Ihm, J., and Kuk, Y., Conformational molecular switch of the azobenzene molecule: A scanning tunneling microscopy study, *Phys. Rev. Lett.* 2006, 96, 156106.
6. Kumar, A.S., Ye, T., Takami, T., Yu, B.-C., Flatt, A.K., Tour, J.M., and Weiss, P.S., Reversible photo-switching of single azobenzene molecules in controlled nanoscale environments, *Nano Lett.* 2008, 8, 1644.
7. Donhauser, Z.J., Mantooth, B.A., Kelly, K.F., Bumm, L.A., Monnell, J.D., Stapleton, J.J., Price D.W., Jr., Rawlett, A.M., Allara, D.L., Tour, J.M., and Weiss, P.S., Conductance switching in single molecules through conformational changes, *Science* 2001, 292, 2303.
8. Loppacher, Ch., Guggisberg, M., Pfeiffer, O., Meyer, E., Bammerlin, M., Lüthi, R., Schlittler, R., Gimzewski, J.K., Tang, H., and Joachim, C., Direct determination of the energy required to operate a single molecule switch, *Phys. Rev. Lett.* 2003, 90, 066107.

9. Luo, Y., Collier, C.P., Jeppesen, J.O., Nielsen, K.A., Delonno, E., Ho, G., Perkins, J., Tseng, H.-R., Yamamoto, T., Stoddart, J.F., and Heath, J.R., Two-dimensional molecular electronics circuits, *ChemPhysChem* 2002, 3, 519.
10. Collier, C.P., Mattersteig, G., Wong, E.W., Luo, Y., Beverly, K., Sampaio, J., Raymo, F.M., Stoddart, J.F., and Heath, J.R., A [2]catenane-based solid state electronically reconfigurable switch, *Science* 2000, 289, 1172.
11. Lara-Avila, S., Danilov, A.V., Kubatkin, S.E., Broman, S.L., Parker, C.R., and Nielsen, M.B., Light-triggered conductance switching in single-molecule dihydroazulene/vinylheptafulvene junctions, *J. Phys. Chem. C* 2011, 115, 18372–18377.
12. Kihira, Y., Shimada, T., Matsuo, Y., Nakamura, E., and Hasegawa, T., Random telegraphic conductance fluctuation at Au-pentacene-Au nanojunctions, *Nano Lett.* 2009, 9, 1442.
13. Korenstein, R., Muszkat, K.A., and Sharafy-Ozeri, S., Photochromism and thermochromism through partial torsion about an essential double bond. Structure of the B colored isomers of bianthrones, *J. Am. Chem. Soc.* 1973, 95, 6177.
14. Biedermann, P.U., Stezowski, J.J., and Agranat, I. Conformational space and dynamic stereochemistry of overcrowded homomerous bistricyclic aromatic enes: A theoretical study, *Eur. J. Org. Chem.* 2001, 15.
15. Biedermann, P.U., Stezowski, J.J., and Agranat, I. *Overcrowded Polyciclic Aromatic Enes*, pp. 245–322 in *"Advances in Theoretically Interesting Molecules,"* Vol. 4, JAI Press (1998).
16. Lara-Avila, S., Danilov, A., Geskin, V., Bouzakraoui, S., Kubatkin, S., Cornil, J., and Bjørnhol, T.J., Bianthrone in a single-molecule junction: Conductance switching with a bistable molecule facilitated by image charge effects, *Phys. Chem. C* 2010, 114, 20686.
17. Fulton, T.A., and Dunkleberger, L.N., Lifetime of the zero-voltage state in Josephson tunnel junctions, *Phys. Rev. B* 1974, 9, 4760.
18. Brandbyge, M., and Hedegård, P., Theory of the Eigler switch, *Phys. Rev. Lett.* 1994, 72, 2919.
19. Brandbyge, M., Hedegård, P., Heinz, T.F., Misewich, J.A., and Newns, D.M., Electronically driven adsorbate excitation mechanism in femtosecond-pulse laser desorption, *Phys. Rev. B* 1995, 52, 6042.
20. Kikuchi, O., and Kawakami, Y., An MO study of conformational behavior of bianthrone and its ions, *J. Mol. Str. Theochem.* 1986, 137, 365.

21. Manini, N., and Tossatti, E., Exact zero-point energy shift in the en⊗(nE), t⊗(nH) many-modes dynamic Jahn–Teller systems at strong coupling, *Phys. Rev. B* 1998, 58, 782.

22. Perez-Jimenez, A.J., Palacious, J.J., Louis, E., SanFabian, E., and Verges, J.A., Analysis of scanning tunneling spectroscopy experiments from first principles: The test case of C_{60} adsorbed on Au(111), *ChemPhysChem* 2003, 4, 388.

23. Tomita, S., Andersen, J.U., Bonderup, E., Hvelplund, P., Liu, B., Nielsen, S.B., Pedersen, U.V., Rangama, J., Hansen K., and Echt O. Dynamic Jahn–Teller effects in isolated C_{60}- studied by near-infrared spectroscopy in a storage ring, *J. Phys. Rev. Lett.* 2005, 94, 053002.

24. Chavy, C., Joachim, C., and Altibelli, A. Interpretation of STM images: C_{60} on the gold (110) surface, *Chem. Phys. Lett.* 1993, 214, 569.

25. Johnson, R.D., Yannoni, C.S., Dorn, Y.C., Salem, J.R., and Bethune, D.S., C_{60} rotation in the solid state: Dynamics of a faceted spherical top, *Science* 1992, 255, 1235.

26. Danilov, A., Kubatkin, S., Kafanov, S., Hedegård, P., Stuhr-Hansen, N., Moth-Poulsen, K., and Bjørnholm, T., Electronic transport in single molecule junctions: Control of the molecule-electrode coupling through intramolecular tunneling barriers, *Nano Lett.* 2008, 8, 1.

27. Park, H., Park, J., Lim, A.K.L., Anderson, E.H., Alivisatos, A.P., and McEuen, P.L., Nanomechanical oscillations in a single-C_{60} transistor, *Nature* 2000, 407, 57.

Chapter 10

Thermoelectricity in Molecular Junctions

Shubhaditya Majumdar,[a] Won Ho Jeong,[b] Pramod S. Reddy,[b] and Jonathan A. Malen[a]

[a] *Department of Mechanical Engineering, Carnegie Mellon University, 5000 Forbes Ave Pittsburgh, PA 15213, USA*
[b] *Department of Mechanical Engineering, University of Michigan, 2350 Hayward Street Ann Arbor, MI 48109, USA*
shubhadm@andrew.cmu.edu, artjwh@umich.edu, pramodr@umich.edu, jonmalen@cmu.edu

10.1 Introduction

New devices built from hybrid materials composed of organic and inorganic components are being researched as replacements to conventional inorganic semiconductors [1]. These devices exploit the transport properties of hybrid materials that emerge due to the juxtaposition of organic molecules with discrete states and inorganic semiconductors with continuous band structures. An exciting prospect for hybrid materials is their use in thermoelectric energy conversion creating an electric current by applying a temperature gradient in a material, thus converting heat directly

Handbook of Single-Molecule Electronics
Edited by Kasper Moth-Poulsen

ISBN 978-981-4463-38-6 (Hardcover), 978-981-4463-39-3 (eBook)
www.panstanford.com

into electricity. In this chapter, we will discuss the details of this phenomenon in molecular junctions and provide models to elucidate its behavior.

Studying the internal structures of hybrid devices is complicated due to the presence of nanoscale interfaces between the organic and the inorganic components. However, it has proven insightful to examine the interaction between one or a few molecules with macroscopic inorganic contacts and to understand the electronic and thermal transport at the interfaces that dominate these hybrid materials.

Although electrical conductance is still the most studied transport property in molecular junctions, other transport properties such as thermopower have garnered interest over recent years. These studies address many open scientific questions on the nature of charge transport at the organic–inorganic interface and inspire new directions for research in hybrid thermoelectric materials [2].

Thermoelectric devices directly convert thermal energy into electricity without any moving parts. This lack of mechanical complexity gives them an advantage in terms of reliability over devices exploiting conventional thermodynamic power cycles. They hold great promise in scavenging waste heat into useful power though their ubiquitous use is limited by the costs and efficiencies of current devices. To put this into perspective, the global energy supply was 143,851 TWh in 2008 and 147,899 TWh in 2010 out of which only 20,181 TWh and 28,005 TWh corresponded to global electricity production respectively [3, 4]. The total energy supply for electricity production at all power plants was 4,398,768 ktoe (kilo ton of oil equivalent) or 51,158 TWh in 2008 [3]. So these power plants operated at an efficiency of 39% with the rest 61% of the energy being lost as waste heat which has the potential to be in part scavenged by thermoelectric energy conversion.

The efficiency of a thermoelectric device is related to a combination of material properties and can be expressed as a function of the thermoelectric figure of merit ZT:

$$ZT = \frac{S^2 \sigma T}{\kappa} \tag{10.1}$$

where S is the thermopower (a.k.a the Seebeck coefficient), σ is the electrical conductivity, κ is the thermal conductivity and T is the absolute temperature. The thermoelectric effect was historically detected across a junction between two dissimilar conductors. This was discovered by Thomas J. Seebeck in 1822 when he succeeded in creating an electromotive force through the application of a temperature gradient [5]. A complementary effect was discovered by Leonard Peltier in 1834, wherein an electric current was able to create a temperature difference across a metallic junction.

In bulk materials, S and σ are primarily dependent on electronic structure, and κ has contributions from both electrons and phonons. However, S and σ are typically opposing functions of carrier concentration [6–8]. Thus, optimizing the carrier concentration is key to maximizing $S^2\sigma$ and ZT. Successful attempts in reducing the phonon contribution to κ (without jeopardizing $S^2\sigma$) have enabled recent incremental gains in the value of ZT [9–12].

It is well known that σ and S depend on carrier concentration $n(E_F)$ and $d[\ln(n(E))]/dE|_{E=E_F}$, respectively [6], where E is energy and E_F is the Fermi energy or the chemical potential μ at $T = 0$ K. In many 3D materials, the electronic band structure is approximately parabolic with respect to E, which leads to an electronic density of states $D(E) \approx \sqrt{E}$ [13]. Thus, $n(E) = \int_0^{E_F} D(E)\,dE \propto E^{3/2}$ and $d[\ln(n(E))]/dE \approx E^{-1}$. Hence, the parabolic band structure results in opposing behaviors of $n(E)$ and $d[\ln(n(E))]/dE$. As a material is doped, E_F, $n(E_F)$ and σ increase but S and $d[\ln(n(E))]/dE|_{E=E_F}$ decrease; e.g., metals have high $n(E_F)$ and thus have very high σ but very low S. This behavior is not seen if the chemical potential μ of the material lies within the band gap. Additionally, looking at the form of ZT, it is seen that a small ratio of κ to σ is also desired. In metals, where κ has dominant contributions from electrons, the ratio $\kappa/\sigma T$ is well approximated by the Lorenz number ($L = (\pi k_B/e)^2/3$). This ratio is smaller in semiconductors where phonons contribute more significantly to κ.

In fully 3D materials, the competition between these effects is optimized in different materials at specific operating temperatures. These conventional thermoelectric materials include bismuth telluride (Bi_2Te_3)-based materials operating around 450 K and below, lead telluride (PbTe)-based materials operating around 1000 K, and silicon germanium (SiGe)-based materials operating around 1300 K.

The latter two materials are mainly used for energy generation while Bi_2Te_3 is also used for thermoelectric refrigeration.

An interesting question to ask is if we are not constrained by the parabolic band structure, what is the optimum electronic structure to maximize ZT? Mahan and Sofo [14] find that this is given by a delta-shaped transport distribution function, which can be practically realized when charge transport takes place through a single energy level. Such a mechanism of charge transfer gives the highest value of ZT by optimizing the values of S, σ, and κ. Single-level transport is hard to achieve in bulk materials, as they possess continuous energy bands but can be potentially realized through organic–inorganic heterojunctions as the molecules have discrete electron energy levels. The thermal conductance of the junction is poor compared to metal–metal interfaces due to a large mismatch between the discrete vibrational states of the molecules and the substrates [15]. To quantify the performance of these junctions, ZT can be expressed in a different form:

$$ZT = \frac{S^2 G_e T}{G_{th}} \qquad (10.2)$$

where G_e is the electronic conductance and G_{th} is the thermal conductance. To understand the origin of thermoelectricity in molecular junctions, it is important to first study the mechanism of the coupling between the molecular and substrate energy states. When molecules are isolated, they have discrete electronic energy states. Among these, the highest occupied molecular orbital (HOMO) and the lowest unoccupied molecular orbital (LUMO) are key in electronic transport through the molecule.

In addition to electronic orbitals, both the contacting electrodes and the molecules possess vibrational energy levels In the case of the macroscopic electrodes (a.k.a contacts), there are a large number of modes whose corresponding eigenfrequencies form a continuous range. The vibrational energy of each of these modes is quantized and these quanta that travel through the crystal are referred to as phonons. The frequency range of phonons is proportional to $\sqrt{k/m}$, where k is the bond stiffness and m is the atomic mass. As for the molecules, due to the lack of periodicity, discrete vibrational modes exist at energies also related to $\sqrt{k/m}$.

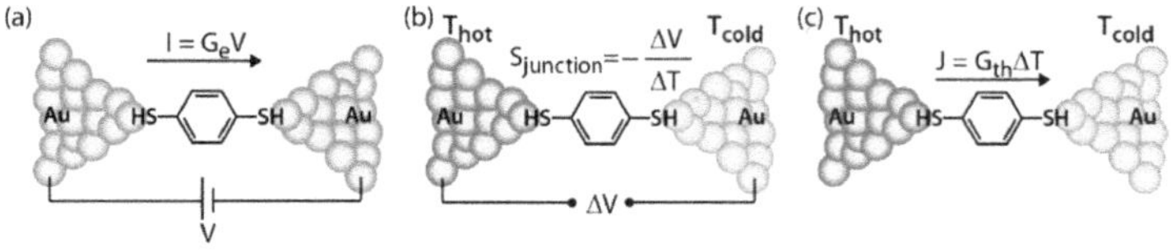

Figure 10.1 Schemes for probing thermoelectric property of molecular junction. (a) Electronic conductance G_e can be measured by looking at current–voltage characteristics. (b) Thermopower S can be measured by analyzing the open-circuit thermoelectric voltage ΔV on the application of a temperature difference ΔT. (c) Thermal conductance G_{th} can be measured applying a heat flux J across the junction and then measuring the temperature difference ΔT across it (Adapted with permission from Lee et al. [16]).

When the molecules come in contact with the substrates, the continuous band of energy levels in the substrates interacts with the discrete molecular states for both the electronic and vibrational cases. The position of the Fermi level with respect to the HOMO and LUMO greatly influences the electronic and thermoelectric properties of the junction. Likewise, the overlap of vibrational states between the contacts and the molecules influences the junction thermal conductance due to phonons/atomic vibrations.

The different configurations in which these transport properties of a molecular junction can be studied are shown in Fig. 10.1. A voltage bias can be applied to study electron transport (Fig. 10.1a); an open-circuit voltage can be generated due to the junction thermopower S by applying a temperature difference (Fig. 10.1b); and lastly, a heat current can be set up through the application of a temperature differential (Fig. 10.1c).

To comprehensively explain the theory of thermoelectric power generation by molecular junctions, we will first provide analytical forms of the various parameters that affect ZT. The following sections will be divided into a brief introduction to the Landauer theory for electronic and thermal conductance, and a derivation of the Seebeck coefficient across a molecular junction. This will be followed by a description of computational and experimental endeavors to predict and measure S and ZT in real molecular systems.

10.2 Electronic Conductance

From the Landauer theory the current $I_{L\to R}$ associated with electron flow between two electrodes left (L) and right (R) is given by:

$$I_{L\to R} = \int_{-\infty}^{\infty} (f_L(E) - f_R(E))v(E)eD(E)\,dE \tag{10.3}$$

where e is the electronic charge, v is the electron velocity, $D(E)$ is the density of electron energy states and, $f_L(E)$ and $f_R(E)$ are the occupation of electrons in the left and right electrodes respectively, given by the Fermi–Dirac statistics: $f_{L/R}(E) = 1\Big/\left(1 + \exp\left(\frac{E-\mu_{L/R}}{k_B T_{L/R}}\right)\right)$. For a 1D, free electron gas, the density of electronic states [13] is given by $D(E) = 2/hv(E)$. Thus, substituting this back into Eq. 10.3 (and dropping the notation E), we get:

$$I_{L\to R} = \frac{2e}{h}\int_{-\infty}^{\infty} (f_L - f_R)\,dE \tag{10.4}$$

To obtain a simplified analytical expression for $(f_L - f_R)$, we analyze this configuration at a zero temperature difference between the contacts and a small bias voltage V:

$$f_L - f_R = eV\left(-\frac{\partial f}{\partial E}\bigg|_{E=\mu}\right) = eV\delta(E-\mu) \tag{10.5}$$

Using Eq. 10.5, we obtain a simple expression for the electronic current at zero temperature and small bias voltage:

$$I_{L\to R} = \frac{2e^2V}{h}\int_{-\infty}^{\infty} \delta(E-\mu)\,dE \tag{10.6}$$

$$I_{L\to R} = \frac{2e^2V}{h} \tag{10.7}$$

This is a statement of Ohm's law where V and I are related by the electronic conductance:

$$G_e^o = \frac{2e^2}{h} \tag{10.8}$$

where G_e^o is known as the quantum of electronic conductance and represents the largest value of conductance possible across a junction consisting of a 1D chain of atoms in the absence of scattering mechanisms (ballistic transport). Surprisingly, it is not a function of any material properties.

10.3 Thermal Conductance

In a solid-state crystalline material, electrons and phonons carry the heat. Thus, the contribution to the total heat current J and thermal conductance by each carrier G is given by:

$$J = J_e + J_p \tag{10.9}$$

$$G_{th} = G_{th,e} + G_{th,p} \tag{10.10}$$

where the subscript e denotes electron and p denotes phonon contributions.

The 1D heat current between the contacts due to phonons can be derived from the Landauer theory as was done for the electronic current in Eq. 10.3:

$$J_{L\to R,p} = \sum_i \int_0^{\infty} \frac{dk}{2\pi} \hbar\omega_i(k) v_{gi}(k)(n_L - n_R) \tag{10.11}$$

where i is the polarization, k is the phonon wave vector, v_g is the group velocity, and n_L and $n_R$are the Bose–Einstein distribution functions for the phonon population in the left and right electrodes respectively: $n_{L/R}(\omega) = 1/\left(\exp\left(\frac{\hbar\omega}{k_B T_{L/R}}\right) - 1\right)$. On comparing Eq. 10.3 and Eq. 10.11, we see that their general forms are very similar, except here we quantify the amount of heat energy (given by $\hbar\omega$) being transported by phonons and earlier we looked at the amount of charge (given by e) transported by electrons. Another notable difference is that for phonons, we have written $J_{L\to R,p}$ as an integral over wave-vectors k instead of energy E. But we now have both frequency and wave vector in the integrand of Eq. 10.11. To simplify this situation, we convert all the terms to frequency using the dispersion relation $d\omega_i(k)/dk = v_{g_i}(k)$. On simplifying Eq. 10.11 we get:

$$J_{L\to R,p} = \sum\nolimits_i \int_{\omega_i(0)}^{\infty} \frac{d\omega}{2\pi} \hbar\omega_i(n_L - n_R) \tag{10.12}$$

We note that the assumption of Debye dispersion (a common approximation used for the dispersion relations, which gives a quadratic density of states for 3D materials) is not necessary to derive Eq. 10.12. As in the case of electronic conductance, we draw a similar analogy to simplify the expression for the difference in phonon population between both substrates. Here, instead of a small energy difference eV, we have a small temperature difference ΔT that drives the heat current:

$$n_L - n_R = \frac{\partial n}{\partial T}\Delta T = \frac{x_i e^{x_i}}{(e^{x_i}-1)^2}\frac{\Delta T}{T} \quad (10.13)$$

where $x_i = \hbar\omega/k_B T$. Thus the complete expression for the heat current is as follows:

$$J_{L\to R,p} = \frac{k_B^2 T}{h}\Delta T \sum_i \int_{x_i(0)}^{\infty} \frac{x_i^2 e^{x_i}}{(e^{x_i}-1)^2} dx_i \quad (10.14)$$

The Sommerfeld expansion [17] can be used to determine the behavior at low T, which gives the quantum of thermal conductance due to phonons as:

$$G^0_{Th,p} = \frac{J^0_{L/R,p}}{\Delta T_{L/R}} = \frac{k_B^2 \pi^2 T}{3h} \quad (10.15)$$

To derive Eq. 10.15, we consider only one polarization where $x_i(0) = 0$, corresponding to an acoustic phonon branch.

Similarly, the thermal energy current due to electrons can be written as:

$$J_{L\to R,e} = \frac{2}{h}\int_{-\infty}^{\infty} (E-\mu)(f_L - f_R)\, dE \quad (10.16)$$

If we write $(f_L - f_R)$ in terms of ΔT and substitute back into Eq. 10.16, we get:

$$f_L - f_R = \frac{\partial f}{\partial T}\Delta T = \frac{x e^{x}}{(e^{x}+1)^2}\frac{\Delta T}{T} \quad (10.17)$$

where $x_i = (E-\mu)/k_B T$

$$J_{L/R,e} = \frac{2k_B^2 T}{h}\Delta T \int_{-\infty}^{\infty} \frac{x^2 e^{x}}{(e^{x}+1)^2} dx \quad (10.18)$$

Just like in the case of energy transport due to phonons, this expression can be evaluated at low T using the Sommerfeld expansion [17] to provide us with the quantum of thermal conductance due to electrons:

$$G^0_{Th,e} = \frac{J^0_{L/R,e}}{\Delta T} = \frac{2k_B^2 \pi^2 T}{3h} \quad (10.19)$$

It can be noted that $G^0_{Th,e} = 2G^0_{Th,p}$, which is due to the electron spin degeneracy of two. A summary of the above-derived expressions is given in Table 10.1.

Table 10.1 Summary of transport property expressions

		Population	Density of states (1D)	Quantum of conductance $G^0_{e/th}$
Charge	Electrons	$f = \frac{1}{e^{\frac{E-\mu}{k_B T}}+1}$	$\frac{2}{h\nu(E)}$	$\frac{2e^2}{h}$
Heat	Phonons	$n = \frac{1}{e^{\frac{E}{k_B T}}-1}$	$\frac{1}{2\pi v_g}$	$\frac{k_B^2 \pi^2 T}{3h}$
	Electrons	$f = \frac{1}{e^{\frac{E-\mu}{k_B T}}+1}$	$\frac{2}{h\nu(E)}$	$\frac{2k_B^2 \pi^2 T}{3h}$

10.4 Molecular Junctions

The energy and charge transport properties of molecules are quite different from solid crystalline contacts, and their unique coupling can be exploited to synthesize tunable transport characteristics. For molecular junctions, the properties are intimately related to the relative position of the molecular energy states with respect to the contacts' continuous energy levels. When a molecule is connected to two contacts, its discrete energy levels spread due to its interaction with the contacts. All electron energy levels below the chemical potential μ of the contacts are occupied, and all levels above it are empty, as shown in Fig. 10.2a. Upon applying an external bias voltage V across the junction, the chemical potentials in the contacts shift leading to an energy difference of eV (Fig. 10.2b). Each contact tries to maintain an average number of electrons at a particular energy given by $2D(E)\, f_{L/R}(E)$. Over a small range in energies eV, this creates a difference in f_L and f_R leading to a net current.

The existence of a molecule with discrete energy levels leads to a nonideal junction between the continuous contacts. Thus, to model this nonideality, when writing the Landauer energy formalism for a molecular junction, an energy-dependent transmission function $\tau_e(E)$ is added. The value of $\tau_e(E)$ ranges between 0 and 1, and it is effectively an energy-dependent transmission probability of electrons incident from the contacts. Thus, Eq. 10.4 can be modified to get the electrical current in a molecular junction:

$$I_{L\to R} = \frac{2e}{h}\int_{-\infty}^{\infty} \tau_e(E)\,(f_L - f_R)\, dE \qquad (10.20)$$

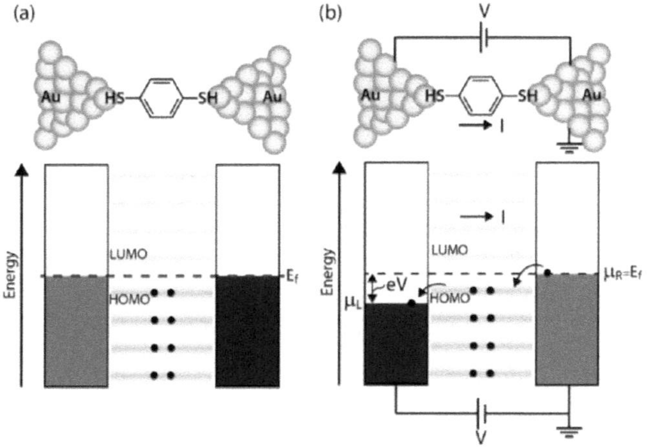

Figure 10.2 Electronic conductance across a molecular junction. (a) When there is no bias voltage, for identical contacts, the chemical potential is the same for both (and equal to the Fermi energy E_f at zero temperature). In the molecule the electrons occupy all available energy states till the HOMO and all states from the LUMO are empty. The discrete electronic energy levels of the molecule spread after interacting with the continuous levels in the contacts. (b) In the presence of a bias voltage V, the chemical potentials of the contacts now differ by eV. It also leads to a net current across the junction (Adapted with permission from Lee et al. [16]).

This can be simplified to get:

$$I_{L\rightarrow R} = \frac{2e^2 V}{h}\tau_e(E = \mu) \tag{10.21}$$

$$G_e = \frac{2e^2}{h}\tau_e(E = \mu) \tag{10.22}$$

For low bias conditions, the quantity $f_L - f_R$ is only nonzero at $E = \mu$, and as a result, Eq. 10.22 shows that the junction conductance is maximized if the transmission function had peaks at μ. A similar approach is applied to the analysis of thermal transport across the junction. If we repeat the analysis of Eq. 10.21 and Eq. 10.22 for the heat current due to electrons, we get:

$$J_{L\rightarrow R,e} = \frac{2}{h}\int_{\infty}^{\infty} (E-\mu)(f_L - f_R)\,\tau(E)\,dE \tag{10.23}$$

$$J_{L\rightarrow R,e} = \frac{2k_B^2 T}{h}\Delta T_{L/R}\int_{-\infty}^{\infty}\frac{x^2 e^x}{(e^x+1)^2}\tau(xk_B T + \mu)dx \tag{10.24}$$

The relation for the heat current due to phonons from Eq. 10.14 is now:

$$J_{L\rightarrow R,p} = \sum_i \int_0^{\infty} \frac{d\omega}{2\pi} \hbar\omega_i \tau_p(\omega_i)(n_L - n_R) \tag{10.25}$$

where the subscript p indicates that only transport due to phonons is considered. As always in the case of phonons, we are summing over all polarizations and integrating for all frequencies where $\tau_p(\omega)$ denotes the phonon transmission function at a particular frequency. There is no concept of the Fermi energy here and transmission through all frequency modes must be considered.

10.4.1 *Thermoelectricity*

For a molecular junction, when the energy derivative of the transmission function is nonzero in the region of the Fermi energy, there will be a net diffusion of electrons between the contacts for a temperature gradient across the junction (or when the substrates are maintained at different temperatures). This can be visualized by inspecting the Landauer formalism for the electronic current (driven by temperature differentials) as shown in Fig. 10.3.

The individual terms of the integrand of Eq. 10.20 have been plotted in Fig. 10.3. Figure 10.3a shows the variation of energy with the Fermi–Dirac distribution at two temperatures T_L and T_R corresponding to the temperatures of the two contacts. It can be seen in Fig. 10.3b that f_L–f_R is symmetric about the Fermi energy. But when it is multiplied by a transmission function (in the case shown in Fig. 10.3a,c, Lorentzian transmission function is used which is a simple approximation for a molecular junction), the integrand can be made to be asymmetric about the Fermi energy, as shown in Fig. 10.3d. This leads to a nonzero value of the electric current across the junction. Without the asymmetry of the transmission function about the Fermi energy, the integral is zero, and the current due to a temperature difference between the contacts cannot be established.

The accumulation of charge on the electrodes leads to an open-circuit thermoelectric voltage across the junction. Thermopower or the Seebeck coefficient S relates this open-circuit voltage and

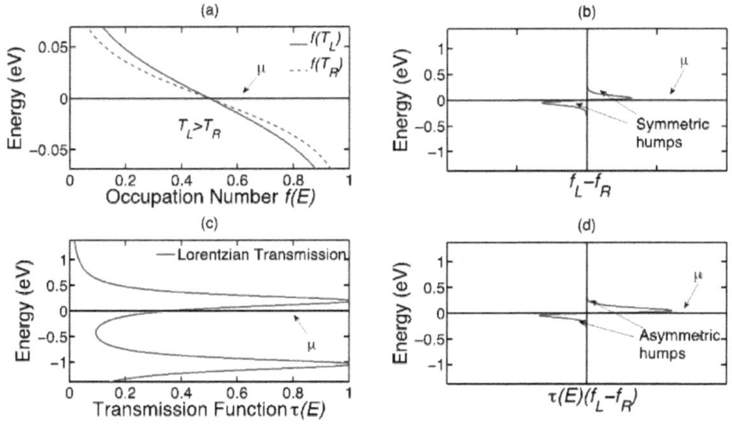

Figure 10.3 Origin of thermoelectric current/voltage. (a) Fermi–Dirac distribution at two different temperatures T_L and T_R. (b) Difference of Fermi–Dirac distributions f_L and f_R. (c) A Lorentzian transmission function (toy model). (d) Multiplication of $\tau_e(E)$ and $(f_L - f_R)$ giving an indication of whether nonzero thermoelectric current exists; area under the curve is important as seen in Eq. 10.3 and Eq. 10.20.

temperature difference as follows:

$$S = -\frac{\Delta V}{\Delta T} \tag{10.26}$$

where ΔV is the induced thermoelectric voltage differential and ΔT is the applied temperature difference. A relation between S and $\tau(E)$ is needed to understand what kind of junction would have high thermopower. This relationship was developed by Butcher [17] and is derived by setting $I_{L\to R} = 0$ (since S is defined at open circuit) in Eq. 10.20. Then the occupation difference between the contacts is expanded with respect to T and μ with reference to contact L as:

$$I_{L\to R} = 0 = \frac{2e}{h}\int_{-\infty}^{\infty} (f_1 - f_2)\,\tau_e(E)\,dE \tag{10.27}$$

$$0 = \frac{2e}{h}\int_{\infty}^{\infty}\left[\left(\left.\frac{\partial f}{\partial \mu}\right|_{\mu=\mu_L}\right)\Delta\mu + \left(\left.\frac{\partial f}{\partial T}\right|_{T=T_L}\right)\Delta T\right]\tau_e(E)dE \tag{10.28}$$

To simplify this expression, we modify and approximate a few terms as:

$$\frac{\partial f}{\partial T} = -\frac{E-\mu}{T}\frac{\partial f}{\partial E} \tag{10.29}$$

$$\frac{f}{\partial\mu} = -\frac{\partial f}{\partial E} \approx \delta(E-\mu) \tag{10.30}$$

Substituting these back into Eq. 10.28, we get:

$$\frac{\Delta\mu}{\Delta T} = \frac{1}{\tau_e(E) \mid E=\mu_L} \int_{-\infty}^{\infty} \tau_e(E) \frac{(E-\mu_L)}{T_L}\frac{\partial f_L}{\partial E} dE \tag{10.31}$$

The $\tau_e(E)$ in the integrand can be expanded about μ_L as follows to get:

$$\tau_e(E) \approx \tau_e(E)\,|_{E=\mu_L} + (E-\mu_L)\left.\frac{\partial\tau_e(E)}{\partial E}\right|_{E=\mu_L} \tag{10.32}$$

$$\frac{\Delta\mu}{\Delta T} = \int_{-\infty}^{\infty} \frac{(E-\mu_L)}{T_L}\frac{\partial f_L}{\partial E} dE + \frac{1}{\tau_e(E)\,|_{E=\mu_L}} \left(\left.\frac{\partial\tau_e(E)}{\partial E}\right|_{E=\mu_L}\right) \int_{-\infty}^{\infty} \frac{(E-\mu_L)^2}{T_L}\frac{\partial f_L}{\partial E} dE \tag{10.33}$$

The first integral is symmetric about μ_L and thus equals zero and the second can be evaluated through the use of the Sommerfeld expansion [17] to obtain $\pi^2 k_B^2 T_L/3$. The change in the chemical potential results in the voltage $\Delta\mu_{L/R} = -|e|\Delta V$, and we get:

$$S = -\frac{\Delta V}{\Delta T} = -\frac{\pi^2 k_B^2 T_L}{3|e|}\left.\left(\frac{1}{\tau_e(E)}\frac{\partial\tau_e(E)}{\partial E}\right)\right|_{E=\mu} \tag{10.34}$$

It can be seen that S is related to the derivative of the transmission function at the chemical potential. Whereas it was seen earlier that the electronic conductance is related to the absolute value of the transmission function at the chemical potential. This result parallels the energy dependence of σ and S in semiconductors though $\tau_e(E)$, rather than $n(E)$, links their behavior. Figure 10.4 shows how S would vary if a Lorentzian transmission is considered with peaks at the molecular energy levels. A more detailed description of transmission functions is discussed in Section 10.5

10.5 Transmission Functions and *ZT*

10.5.1 *Electronic Transmission Functions*

The electronic transmission function is of great importance to the thermoelectric properties in a molecular junction. It exists in the

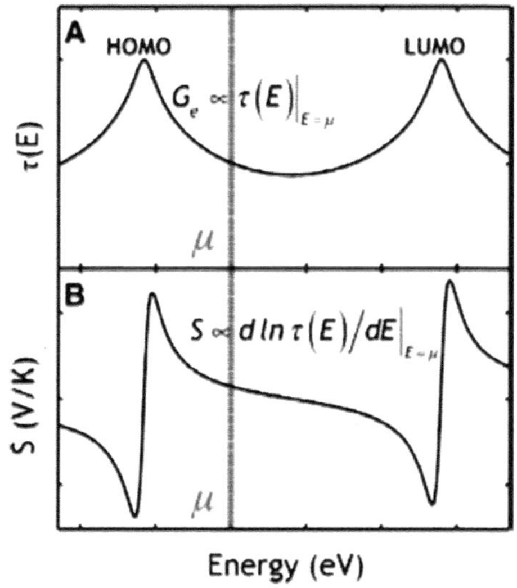

Figure 10.4 Plot of a Lorentzian transmission function (toy system) with peaks at the HOMO and LUMO of a molecule and the corresponding thermopower S as derived from Eq. 10.34 (Adapted with permission from Malen et al. [18]).

analytical expressions for all terms characterizing ZT. From a qualitative standpoint, it can be understood that it depends on the alignment of the discrete molecular orbitals with the delocalized, continuous states in the contacts.

Mahan and Sofo [14] suggested that a Dirac-delta-shaped density of states near the chemical potential for a bulk material would result in the highest ZT value for a conventional thermoelectric device. Such an argument was also made for the case of molecular junction devices [19]. However, subsequent research [20, 21] has shown that this would result in zero power output for the device. A transmission function resulting in a high power output for the thermoelectric device is more desirable. In this context, instead of the Carnot efficiency limit (as in the case of a reversible device with Dirac-delta density of states/transmission functions) the Curzon–Ahlborn efficiency limit [22] of maximum power output is more relevant and given by:

$$\eta_{CA} = \frac{\eta_C}{2} + \frac{\eta_C^2}{8} + O\left(\eta_C^3\right) + \ldots \tag{10.35}$$

where η_C is the Carnot efficiency. Nakpathokun et al. [20] suggested that this might be achieved by approximating the transmission function as a Lorentzian function:

$$\tau_{m,e}(E) = \frac{(\Gamma_m/2)^2}{(E - E_m)^2 + (\Gamma_m/2)^2} \tag{10.36}$$

where m represents the index of the mth orbital, Γ_m is the full width at half maximum of the transmission function, and E_m is the energy about which the Lorentzian is centered. Physically, E_m is related to the HOMO or LUMO energy, and Γ_m is related to their strength of coupling with the contacts. It was shown that the power output would become zero for a very sharp Lorentzian (or a Dirac-delta function). The parameter Γ_m and energy separation between E_m and E_f could be chosen to achieve maximum power output. Kittel et al. [13] and Datta [23] also proposed a similar approximation but, they assumed the Lorentzian to have asymmetric coupling with contacts 1 and 2 such that $\Gamma_{m,1} \neq \Gamma_{m,2}$ and the transmission function is now:

$$\tau_{m,e}(E) = \frac{4\Gamma_{m,1}\Gamma_{m.2}}{4(E - E_m)^2 + (\Gamma_{m,1} + \Gamma_{m,2})^2} \tag{10.37}$$

If all the orbitals are treated as independent tunneling pathways, the total transmission function for the molecule and the corresponding thermopower S can be written as:

$$\tau_e(E) = \sum_{m=1}^{M} \tau_{m,e}(E) \tag{10.38}$$

$$S = \sum_{m=1}^{M} \frac{\pi^2 k_B^2 T}{3|e|} \frac{2(\mu - E_m)\,\tau_{m,e}(\mu)}{\Gamma_{m,1}\Gamma_{m,2}} \tag{10.39}$$

Nevertheless, the Curzon–Ahlborn limit could not be achieved by assuming a single Lorentzian transmission function. This is because Lorentzians have a long low energy tail that allowed the conductance of electrons through the junction opposite in direction to the thermally excited electrons [20].

To overcome this complication, Karlstrom et al. [21] suggested taking advantage of a two-level quantum interference system where both energy levels would be located on the same side of the chemical potential. These two levels were coupled to the contacts

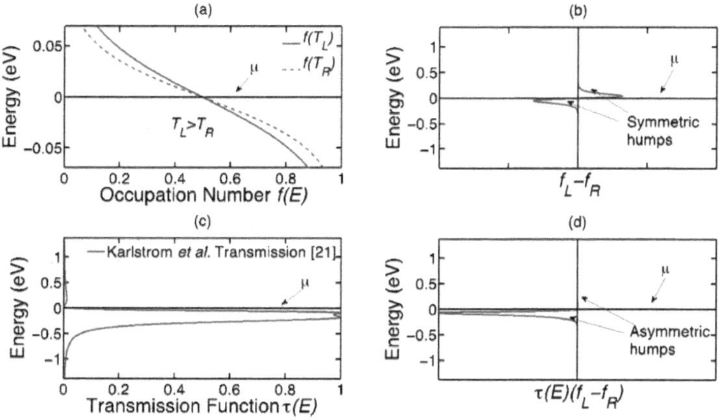

Figure 10.5 Origin of thermoelectric current/voltage. (a) Energy of electrons as a function of Fermi–Dirac distribution. (b) Difference of Fermi–Dirac distributions f_L and f_R. (c) Transmission function as proposed by Karlstrom et al. [21]. (d) Multiplication of $\tau_e(E)$ and $(f_L - f_R)$ giving an indication of whether nonzero thermoelectric current exists.

with different parities, and their coupling strengths differing by a factor a^2. Approximating each tunneling pathway as a Lorentzian, the complete transmission function could be written as:

$$\tau_e(E) = \Gamma^2 \left| \frac{1}{(E-\mu) + E_1 + i\Gamma} - \frac{a^2}{(E-\mu) + E_2 + ia^2\Gamma} \right|^2 \tag{10.40}$$

where E_1 and E_2 are the two energy levels. It can be shown that for $E_2 \sim a^2 E_1$, the transmission function is zero at the chemical potential and is large for a finite range of energies above it. Also, since the function has a finite width, a net power output can be achieved and it was shown that efficiencies close to the Curzon–Ahlborn limit could be realized. A similar analysis as presented in Fig. 10.3, is shown in Fig. 10.5 using $\tau_e(E)$ as given by Eq. 10.40.

10.5.2 *Phonon Transmission Functions*

Thermal boundary conductance across an interface has been analytically studied for quite some time [24, 25]. In this regard, phonon transmission functions for interfaces between perfect crystals have been the most popular area of study with molecular junctions only recently explored. For crystals, the Diffuse Mismatch Model (DMM) and Acoustic Mismatch Model (AMM) are widely used in calculating transmission functions to describe different regimes of phonon propagation. The DMM is used in the regime where phonons scatter diffusely at an interface, i.e., transmitted phonons have no memory of the properties (polarization, wave vector) of the incident phonons. It is applicable when the temperature of the system is quite high. The AMM parallels Snell's law of refraction of light. It assumes the phonons are waves, and their transmission and reflection at the interface depend on the relative mismatch of the acoustic impedance of the two media. This is applicable at low temperatures when the dominant phonon wavelengths are long and frequencies are low.

The DMM can be used for both elastic and inelastic scattering situations. Elastic scattering means that the energy of the phonon is conserved throughout the scattering event, i.e., a phonon approaching the interface with frequency ω will leave it with the same frequency after scattering. For inelastic scattering, this rule does not apply—any phonon can scatter into any other phonon state/mode. When using the DMM, the rule of detailed balance needs to be invoked, which states that for a small temperature difference at the interface, the phonon flux at the interface is equal from both sides.

Let us say the interface is created by left (L) and right (R) substrates. From the Landauer theory (Eq. 10.11 and Eq. 10.25) and by invoking the principle of detailed balance considering the total phonon flux at the interface, we get an expression for the transmission function $\tau_{p,inel}$ in the case of the inelastic scattering as:

$$\tau_{p,inel} = \frac{\sum_i \int_{k,R} \hbar\omega_{i,R} k_{i,R}^2 v_{i,R} n_0 dk_{i,R}}{\sum_i \int_{k,R} \hbar\omega_{i,R} k_{i,R}^2 v_{i,R} n_0 dk_{i,R} + \sum_i \int_{k,L} \hbar\omega_{i,L} k_{i,L}^2 v_{i,L} n_0 dk_{i,L}} \tag{10.41}$$

where i is the polarization, k is the phonon wave vector, v is the phonon group velocity, ω is the phonon frequency and n_0 is the Bose–Einstein distribution. In the elastic scattering limit, this analysis must be applied on a per frequency basis since energy is conserved. Thus, Eq. 10.41 can be simplified to get the transmission function $\tau_{p,el}$:

$$\tau_{p,el}(\omega) = \frac{\sum_i [k_{i,R}(\omega)]^2}{\sum_i [k_{i,R}(\omega)]^2 + \sum_i [k_{i,L}(\omega)]^2} \tag{10.42}$$

The inelastic transmission coefficient is different from the elastic in two ways. First, it is not a function of frequency due to the integrals within the summations. Second, the equilibrium phonon distributions do not cancel out, and thus $\tau_{p,inel}$ is a function of temperature.

A modification to the DMM by Duda et al. [26] was applied to self-assembled monolayer (SAM) interfaces with crystalline substrates. In this setup, the heat flux for the SAM side was modified by assuming it to be similar to bulk polyethylene having Dirac-delta density of states at the vibrational frequencies. The modified transmission function τ_{SAM} was:

$$\tau_{SAM} = \frac{\frac{1}{A}\frac{1}{L}\sum_j \hbar\omega_j n_0 v_{PE}}{\frac{1}{8\pi^2}\sum_i \int_k \hbar\omega_i k_i^2 |v_i| n_0 dk_i + \frac{1}{A}\frac{1}{L}\sum_j \hbar\omega_j n_0 v_{PE}} \tag{10.43}$$

where A is the area per ligand in the SAM, L is the length of a molecule, v_{PE} is the speed of sound in polyethylene, and j is the index summing over the discrete frequencies of the molecule.

The DMM however, has some unavoidable deficiencies. There is an inaccurate account of the phonon flux approaching and leaving the interface, an over-simplified description of the scattering mechanism of phonons/vibrations at the interface, and an absence of the account of the interaction/bonding strength of the atoms across the interface. These turn out to be quite complicated to overcome analytically, and thus to provide a more detailed quantification of thermal conductance of molecular junctions, various computational studies have been performed.

10.5.3 *Computational Studies of Thermal Transport in Molecular Junctions*

Numerical and computational studies of thermal transport in molecules have been conducted for almost half a century now. Fermi et al. [27] indicated the possibility of infinite thermal conductivity in a 1D chain, and this has been subsequently quite extensively researched. Length independence of chain thermal conductance (or a diverging thermal conductivity with length) has been shown by Casher and Lebowitz [28], who used a nonequilibrium statistical mechanics approach on a model system of harmonically linked particles, and also by Lepri et al. [29, 30], who showed that the thermal conductivity of 1D lattices would diverge as $N^{0.4}$, where N is the number of a lattice points.

For a realistic system, Segal et al. [31], using a generalized Langevin approach, showed length independence of thermal conductance of molecular junctions composed of simple alkane chains when the number of carbon atoms exceeded 10. However, their study also predicted an increasing thermal conductance for chains having five carbon atoms or less. Their results were consistent over a wide temperature range as well. Henry and Chen [32] investigated the thermal conductance of a junction having a single polyethylene chain and found thermal conductance to be three orders of magnitude higher than that of bulk polyethylene junctions, due to reduction in scattering events in 1D systems as opposed to 3D. These results were also partially verified by Shen et al. [33] through experiments on aligned polyethylene fibers of diameters 50–500 nm. Sasikumar and Keblinski [34] used nonequilibrium molecular dynamics to understand the effects of chain conformation on thermal conductance. They found a strong dependence of conductance on the number gauche conformations; an increase in their number reduced the conductance. They also studied the effect of kinking and found straight chains to be more favorable for thermal transport.

10.5.4 *ZT and Thermoelectric Efficiency*

Using the prior analytical analysis of the transport characteristics of molecular junctions, it is possible to calculate the thermoelectric efficiency η_{TE}. From Eq. 10.2, we know that $ZT = S^2 G_e T / G_{th}$. Using Eq. 10.39 for the expression for S and the assumption that $G_e T / G_{th,e} \approx 3e^2 / k_B^2 \pi^2$ (for metals, this comes from the Lorenz number, which is a ratio of κ and σ), the complete expression of $Z_i T$ for the ith molecular orbital (assuming a Lorentzian transmission) in the weak coupling limit between molecule and contacts such that $(\mu - E) \gg \Gamma^2$ is given by:

$$Z_i T = \frac{S^2 G_e T}{G_{th,e} + G_{th,p}} = \frac{S^2 G_e T}{G_{th,e}(1+R)} \approx \frac{4\pi^2 k_B^2 T^2}{3\,(\mu - E_i)^2 (1+R)} \tag{10.44}$$

where R is $G_{th,p}/G_{th,e}$. The expression shows that as $\mu - E_i$ decreases, $Z_i T$ rapidly increases. This challenges us to identify a molecule-contact system where μ and E_i of the molecular orbitals are well aligned.

To calculate η_{TE}, let the temperature difference across the molecular junction be $T_H - T_C$ (similar to the analysis of a heat engine). Then the maximum possible efficiency η_{TE} is given by:

$$\eta_{TE} = \frac{T_H - T_C}{T_H} \frac{\sqrt{1+ZT_m} - 1}{\sqrt{1+ZT_m} + T_C/T_H} \tag{10.45}$$

where $T_m = (T_H + T_C)/2$ is the average temperature of the junction. We know the Carnot efficiency is $\eta_c = 1 - T_C/T_H$, and thus it is seen that the thermoelectric efficiency is limited by the Carnot efficiency. This is quite intuitive since thermoelectric devices are essentially heat engines. If $ZT \to \infty$, then $\eta_{TE} \to \eta_C$.

In recent years, there has been significant interest in theoretical studies of ZT. Murphy et al. [35] predict that very high ZT values can be achieved for weakly coupled molecular orbitals if their energies were of the order $k_B T$ from μ of the contacts. In this case, ZT is limited by phonon contributions to the thermal conductance with the highest possible $ZT \sim (G_{th}^0/G_{th,p})$. They show that for weak coupling, the Lorenz number goes to zero violating the Weidemann–Franz law, while S remains nonzero. This result is similar to the Mahan and Sofo [14] model for bulk thermoelectrics. Studies carried out by Finch et al. [36] showed that high ZT could also be achieved for junctions where the transmission function exhibited Fano resonances [37] in the region of μ, Fano resonances occur when the molecule has side groups attached to it such as an oxygen atom. Ke et al. [38] demonstrated a similar effect for real molecules using density functional theory calculations where high S values were achieved for large molecules with orbitals well aligned with μ. From their findings, the following behaviors can be observed: S increases monotonically with increasing length of molecule (due to increase in the electron tunneling resistance across the molecule), and S decreases with increasing electronegativity of the end groups (attributed to an electron-withdrawing effect, also discussed in Section 10.6.2, which shifts μ away from the HOMO).

10.6 Experimental Techniques for Probing Transport Properties of Molecular Junctions

In the past decade, a wide variety of experimental techniques [39–44] have been developed to study the charge transport properties of single and multiple molecular junctions. Some of the widely used methods are electromigrated break junction (EBJ) technique [39], mechanically controllable break junction (MCBJ) technique [40], scanning tunneling microscope break junction (STM-BJ) technique [41–45], and contact probe atomic force microscope (CP-AFM) technique [46, 47]. Among these techniques, the STM-BJ technique, which was invented by Xu and Tao [42], provides one of the most convenient approaches for performing charge transport measurements in a large number of two-terminal single-molecule junctions in a short time enabling a statistical interpretation of results. Further, with some modification, the STM-BJ technique also enables studies on the thermoelectric properties [2, 48–51] of single-molecule junctions. In addition to this technique, the CP-AFM technique can also be used for probing the charge and the thermoelectric properties of multiple molecular junctions [52, 53]. Here, we will focus on these two experimental techniques.

10.6.1 *Formation of Metal–Molecule–Metal Junction*

The first step in both STM-BJ and CP-AFM techniques is the creation of a monolayer of chemically bound organic molecules on a metal

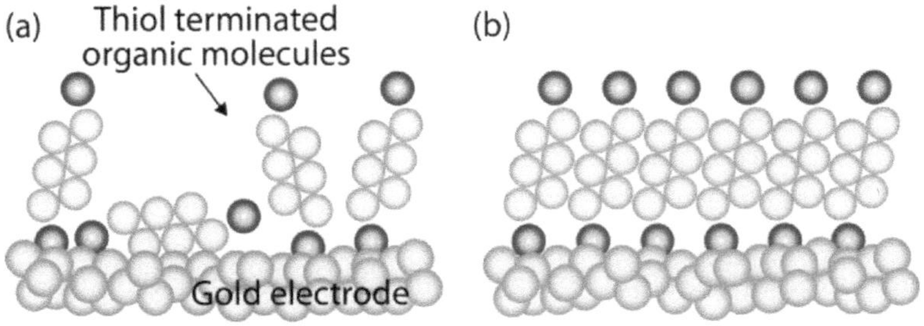

Figure 10.6 Schematic of self-assembly process (a) before and (b) after an ordered monolayer of organic molecules is formed (Adapted with permission from Lee et al. [16]).

surface, as shown in Fig. 10.6. This is usually accomplished by exposing a metal surface to a solution containing organic molecules terminated with reactive chemical groups, such as thiols (–SH) or amines (–NH_2). The organic molecules chemically bind to the metal surface through the formation of chemical bonds between the metal and the reactive end groups [54]. Given sufficient time, the organic molecules spontaneously self-assembled into an ordered monolayer of molecules (Fig. 10.6b) [54]. These self-assembled organic molecules can readily be used to form metal–molecule–metal junctions by using the STM-BJ or CP-AFM techniques.

10.6.2 *Scanning Tunneling Microscope Break Junction (STM-BJ) Technique*

After a substrate of self-assembled organic molecules is prepared, it is placed in the close proximity of a sharp STM tip (Fig. 10.7a), and the STM tip is slowly driven toward the substrate as the first step of the technique. During this approach, a small voltage bias is applied between the STM tip and the substrate, and the current through the STM tip is continuously monitored. When the measured current reaches a large predetermined value, it implies that a large number of metal (substrate)–molecule–metal (STM tip) junctions are formed due to the chemical interaction between the reactive end groups of molecules and the STM tip. Then the STM tip is slowly withdrawn from the substrate, and the molecules that are trapped between the metal electrodes (the substrate and the STM tip) start breaking away one at a time until only a single molecule remains between the metal electrodes (Fig. 10.7b). Figure 10.7c shows the conductance G of a molecular junction during this process of withdrawal It can be seen that the conductance decreases in discrete steps, indicating that the molecules are breaking off one at a time. Each step corresponds to the breaking of one or more molecules. Eventually, when the last molecule trapped between the electrodes also breaks away, the monitored electrical current falls below the detection limit [42, 43].

The withdrawal process can be stopped when only a single molecule is trapped between the metal electrodes. The amount of time that this single-molecule junction can be stably maintained largely depends on the temperature and the stability of the

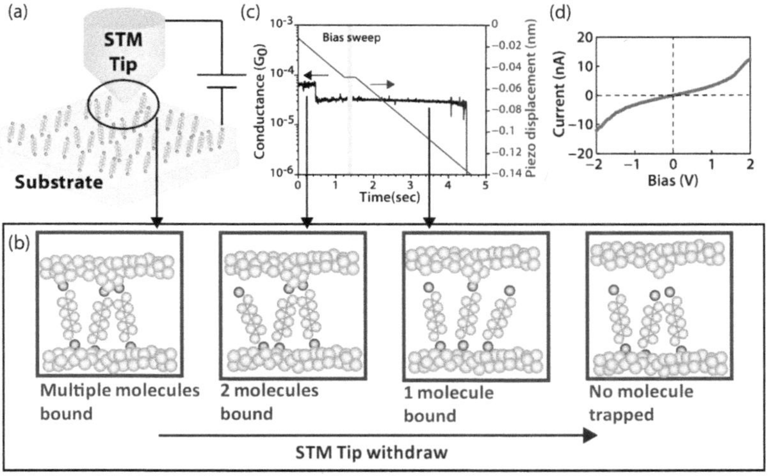

Figure 10.7 The schematic of the STM-BJ technique. (a) A sharp STM tip is placed in the close proximity of a self-assembled monolayer of organic molecules. The approach, withdraw and hold sequence of STM tip is carefully controlled to form metal–single-molecule–metal junctions. (b) Schematics depict molecules that were trapped between the metal electrodes break off one at a time during the withdraw process. (c) A representative conductance graph shows steps that imply one or more molecules break off from the electrodes. In the green shaded area, the I–V characteristics of a single-molecule junction were measured. (d) A representative I–V curve obtained from the voltage sweeping of (c) (Panels reproduced with permission from Lee et al. [45]).

experimental system and is influenced by vibrations and thermal drift [45]. In addition to measuring the electrical conductance of molecular junctions, it is also possible in this setup to measure their current–voltage (I–V) characteristics, which are essential for performing additional spectroscopic studies [45]. To obtain the I–V characteristics of single-molecule junctions, the withdrawal process is stopped when only a single molecule is trapped between the metal electrodes. Subsequently, a voltage bias is swept linearly while monitoring the electrical current flowing through the molecular junction. This procedure is described in Fig. 10.7c (green area), and the I–V characteristic obtained in one such experiment is shown in Fig. 10.7d.

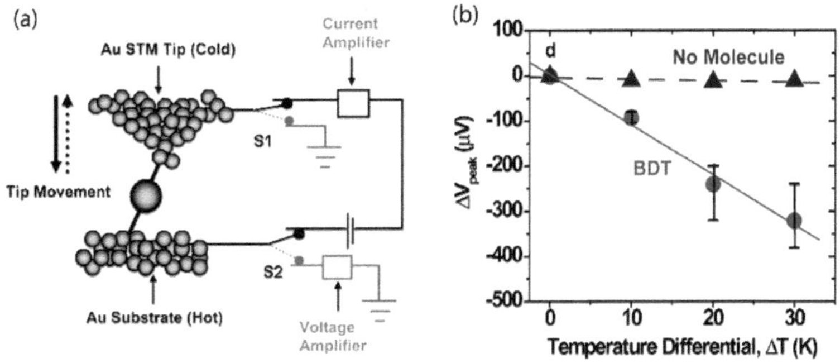

Figure 10.8 The modified STM-BJ technique to measure the thermoelectric properties of molecular junctions. (a) The experimental setup for measuring the thermoelectric voltage of a molecular junction ΔV induced by the temperature differential between the STM tip and the substrate ΔT. Establishing temperature differential across the molecular junction requires careful design of experiments. (b) Measured results obtained in the experimental platform of (a) for 1,4-benzenedithiol (BDT) molecular junctions. The Seebeck coefficient of such junctions turns out to be $+ 8.7 \pm 2.1$ μV/K (Adapted with permission from Reddy et al. [2]).

Measuring thermoelectric properties of molecular junctions includes more subtle issues. This is because establishing a temperature differential ΔT across single or multiple molecules is essential for thermoelectric studies. A technique to study thermoelectric effects in molecular junctions has been developed recently by Reddy et al. [2] with a modification of the STM-BJ technique. The key to making this measurement is to establish a temperature differential ΔT, across the molecular junction. A single-molecule junction formed by using the STM-BJ technique is described in Fig. 10.8a. The noticeable difference of this experimental technique compared with the conventional STM-BJ technique is that the substrate is heated to $T + \Delta T$, while the STM tip is kept at room temperature T by putting it in contact with a large thermal reservoir. This creates a tip-substrate temperature differential, ΔT, that can be achieved only due to the much smaller thermal conductance ($\sim$10–100 pW/K) of molecular junctions than that between the STM tip and the thermal reservoir with which it is in contact [2, 55]. In this configuration, the temperature of the STM tip remains at the reservoir temperature because the heat flow from the hot substrate to the cold STM tip is negligible.

The experimental procedure to measure the Seebeck coefficient of molecular junctions involves, first, the trapping process described above. After molecules are trapped between the metal electrodes, the current amplifier, which was used to monitor the current, is disconnected and a custom-built high input-impedance voltage amplifier is connected as shown in Fig. 10.8a, to measure the induced tip-substrate thermoelectric voltage. The STM tip is then slowly withdrawn until all the molecules trapped between the electrodes break off. During this process the output voltage ΔV is continuously monitored while the STM tip is grounded [2]. The results obtained in a large number of such measurements for 1,4- benzenedithiol (BDT) molecules trapped between an Au STM tip and the Au substrate are shown in Fig. 10.8b for temperature differentials from 0 K to 30 K. The output voltage ΔV, which is induced by the temperature differential ΔT, is a measure of the Seebeck coefficient S_{junc} of the molecular junction, which is obtained from Eq. 10.26 to be $+8.7 \pm 2.1\ \mu$V/K. Here, the voltage drop across the molecular junction is obtained from ΔV by accounting for the thermoelectric voltage drop in the leads where temperature differentials may exist (see supporting information of Ref. [2] for more details).

The sign of the Seebeck coefficient indicates whether the charge transport of a molecular junction is HOMO- or LUMO-dominated according to the Landauer theory (Eq. 10.34), an observation that can also be made based on the Lorentzian transmission model presented earlier in Section 10.5.1. Looking more closely at the expression for thermopower S (Eq. 10.39), if $(\mu - E_m)^2 \gg (\Gamma_{m,1} + \Gamma_{m,2})^2$, a condition known as weak coupling, then the thermopower of a single Lorentzian transmission function simplifies to:

$$\approx \frac{\pi^2 k_B^2 T}{3|e|} \frac{2}{(\mu - E_m)} \tag{10.46}$$

Hence, if S is positive, then the chemical potential is higher in energy than the nearest molecular orbital ($\mu > E_m$), implying HOMO-dominated transport. If S is negative, then the chemical potential is lower in energy than the nearest molecular orbital ($\mu < E_m$), implying LUMO-dominated transport.

The first experiments on the thermometric properties of molecular junctions focused on understanding the length dependence of

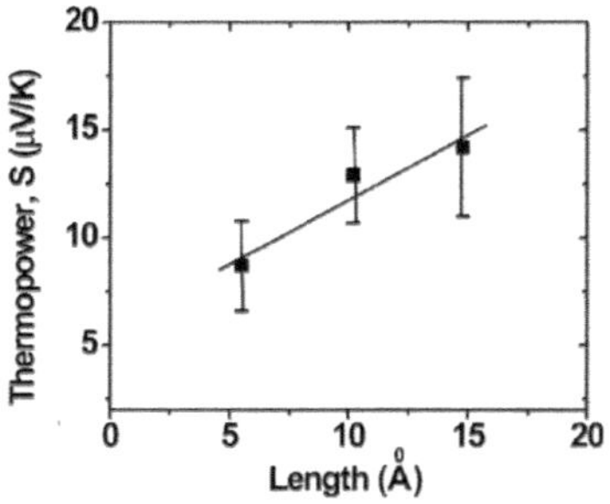

Figure 10.9 Length dependence of the Seebeck coefficient of Au-aromatic dithiol-Au junctions (Adapted with permission from Reddy et al. [2]).

thermoelectric properties and the nature of the dominant charge carriers. For example, Reddy et al. [2] showed that the charge transport in a series of thiol-terminated aromatic molecular junctions (benzenedithiol, dibenzenedithiol, and tribenzenedithiol) created from Au electrodes is HOMO dominated (as shown by their positive Seebeck coefficients, see Fig. 10.9). These initial measurements also suggested that the Seebeck coefficient of molecular junctions increases approximately linearly with length as opposed to the exponential length dependence seen in the electrical conductance of molecular junctions.

Subsequent experiments highlighted that the magnitude of the thermopower S could be predictably tuned based on the chemistry of the molecule primarily by varying end groups or by introducing substituents [48, 50]. For example, the BDT molecules were modified by the substitution of electron-donating or electron-withdrawing groups for the hydrogen atoms on the benzene ring. In free standing BDT, substitution by electron-withdrawing groups reduces the electron density on the π-dominated HOMO, and hence stabilizes it to a lower energy. In the STM-BJ-trapped BDT, this same substitution moved the HOMO to lower energy and further from the chemical potential, resulting in an observed reduction in S, as shown in Fig. 10.10.

An opposite effect was observed with the substitution of electron-donating groups, which increased the electron density of the HOMO and moved it nearer to the chemical potential, therein increasing S. Electrical conductance measurements of similar molecular junctions were consistent with these results, as electron-

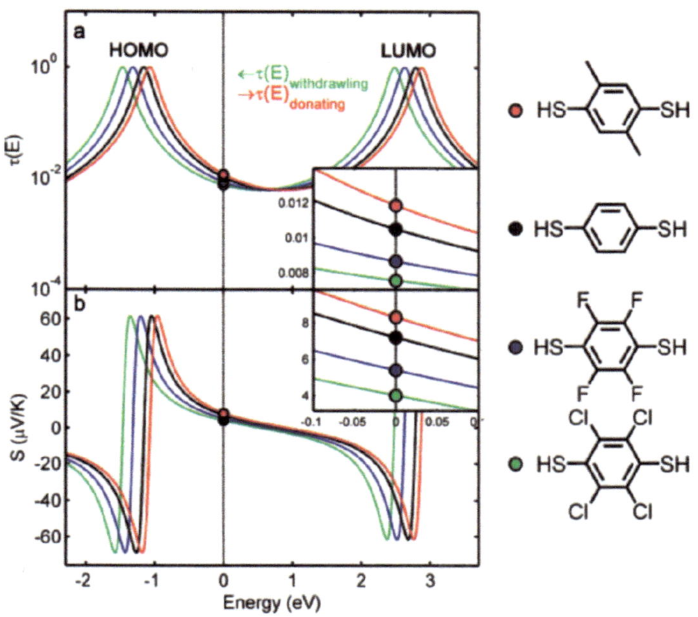

Figure 10.10 Through substitution of the hydrogens on the benzene ring of BDT (black) with electron-donating groups (red) and electron-withdrawing groups (blue and green), predictable shifts in the HOMO resulted in increased and decreased thermopower (Adapted with permission from Baheti et al. [48])

withdrawing groups resulted in lower conductance because the HOMO moved further from the chemical potential [56]. This result, summarized by Fig. 10.10, suggests the possibility of concurrently increasing S and G_e in molecular junctions, which is very hard to accomplish in bulk materials. Additional experiments [48] showed that it is possible to change the sign of S for molecular junctions by changing the end group from thiol (–SH) to cyanide (–CN).

A variation of the Seebeck coefficient measurement based on the STM-BJ technique has been developed recently by Widawsky et al. [51] and Evangeli et al. [57] to successfully study the thermoelectric properties of a variety of molecular junctions. In this technique, instead of measuring the open-circuit voltage, the short-circuit electrical current resulting from the thermoelectric voltage is obtained by electrically shorting the tip and the sample while applying no external voltage bias. These experiments enabled a concurrent measurement of the electrical conductance and Seebeck coefficient to junctions and revealed several interesting effects. For example, Widawsky et al. [51] studied the electrical conductance and Seebeck coefficient of pyridine–Au-linked junctions and showed that transport in them is LUMO, whereas transport in amine-Au-linked junction is HOMO dominated. In addition, recent experiments have focused on probing the thermoelectric properties of molecular junctions that are connected via the Au–C bond. Such junctions were found to have significantly increased electrical conductances, in comparison to junctions having Au–SH or Au–NH_2, bonds while the Seebeck coefficients were not appreciably smaller. This resulted in a significant increase in the power factor S^2G_e (Eq. 10.2) that is desirable for thermoelectric energy conversion.

In this context, recall that for an improved ZT, it is desirable to improve both the Seebeck coefficient and the electrical conductance of the junction, which can in principle, be accomplished by tuning the alignment between the chemical potential and a molecular orbital. Toward this goal, some of the recent measurements have focused on probing thermoelectric phenomena in fullerene junctions [57, 58]. Earlier measurements of conductance in fullerene C_{60} molecules showed that their conductances were orders of magnitude higher than the aromatic molecules that were an initial focus of single-molecule thermopower measurements [59–61]. Furthermore, C_{60} has LUMO-dominated electronic transport in bulk molecular solids, and hence should demonstrate LUMO-dominated transport in molecular junctions. Experimental measurements of thermopower in Au–C_{60}–Au junctions confirmed this hypothesis, finding negative values of thermopower $S_{\text{Au}-\text{C}_{60}-\text{Au}} = -14.5 \pm 1.2$ μV/K [58]. Its higher magnitude, relative to the thermopower of aromatics, also suggests improved alignment of the chemical potential with the nearest molecular orbital. Similar trends were observed in C_{70} and PCBM, which is a soluble fullerene derivative. Recent work by Evangeli et al. [57] has probed both the conductance and the Seebeck coefficient of Au–C_{60}–Au junctions and Au–$(C_{60})_2$–Au junctions, where $(C_{60})_2$ refers to a C_{60} dimer. In this work, it was found that the conductance and the Seebeck coefficient of Au–C_{60}–Au junctions were $\sim 0.1G_0$ and ~ -18 μV/K to -23 μV/K, respectively. Interestingly, it was found that the Seebeck coefficient

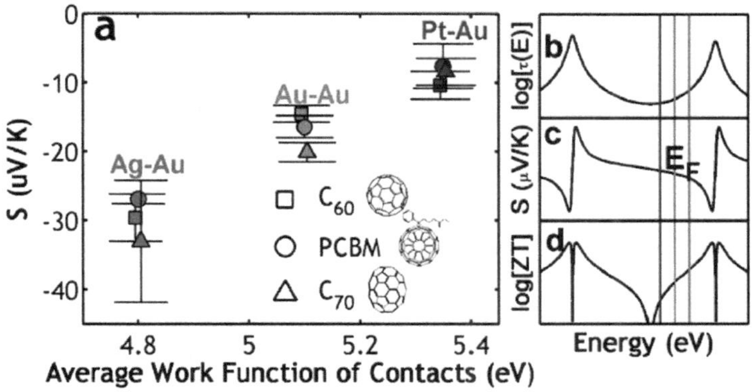

Figure 10.11 (a) Thermopower of metal–fullerene–metal junctions increases as the average work function of the contact metals decreases. (b–c) A Lorentzian depiction of $\tau(E)$ and $S(E)$ suggests that increased thermopower results from improved alignment between the chemical potential E_F of the contacts and the LUMO of the molecule. The position of the chemical potential and the points in (a) are color coded. (d) The Lorentzian depiction also suggests that ZT can be improved by aligning the chemical potential with the LUMO (Adapted with permission from Yee et al. [58]).

of the dimers (Au–$(C_{60})_2$–Au junctions) was higher at ~ -25 μV/K to -50 μV/K, while the electrical conductance was $\sim 0.002G_0$. These reported Seebeck coefficients are among the highest values observed in molecular junctions.

An additional benefit of working with LUMO-dominated molecular junctions is the availability of air-stable contacts with lower work functions than Au that can be used to adjust the position of the chemical potential with respect to the LUMO. For example, Ag has a lower work function than Au, and measurements of Ag–C_{60}–Au junctions yielded a negative thermopower with higher magnitude than Au–C_{60}–Au junctions. This effect is summarized in Fig. 10.11a, which shows the thermopower of C_{60}, PCBM, and C_{70} versus the average work function of the contacts. A clear trend confirms that reducing the work function of the contacts increases the thermopower of the junction. A Lorentzian depiction of $\tau(E)$ and $S(E)$, shown in Fig. 10.11b,c demonstrates that moving the chemical potential to lower energies makes it more well aligned with the LUMO of the fullerene molecule. Direct measurements of conductance in these junctions were too variable to yield confirmation of concurrent increases in G_e. Nonetheless, the Lorentzian picture of ZT, shown in Fig. 10.11d, suggests that this is a step in the right direction.

10.6.3 *Contact Probe Atomic Force Microscope Technique*

In the CP-AFM technique, the top surface of a molecule-coated metal substrate is placed in mechanical contact with a metal-coated AFM tip as shown in Fig. 10.12a. After this process, ~10–100 molecules are trapped between the metal electrodes forming multiple metal–molecule–metal junctions; the number of molecules trapped between the electrodes is determined by the radius of the AFM tip [47]. One advantage of the CP-AFM technique over the STM-BJ technique is that the metal coating on the AFM tip can be chosen to be different from the substrate. Therefore, it is possible to study the effect of chemical bonding on the charge transport through molecular junctions. Further, this technique is mechanically stable to hold molecular junctions longer than the STM-BJ technique. When a voltage bias is linearly swept across a group of molecular junctions, studies on the current–voltage (I–V) characteristics of molecular junctions can be performed (similar to the I–V studies performed

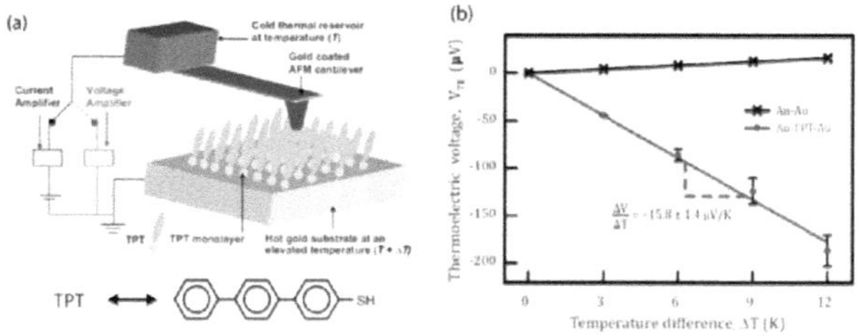

Figure 10.12 The modified CP-AFM technique to measure the Seebeck coefficient of molecular junctions. (a) The experimental setup for measuring the thermoelectric voltage of a group of molecular junctions (~10–100 Au–TPT–Au junctions) ΔV induced by the temperature differential across the molecular junctions ΔT. (b) The measured thermoelectric voltages show linear increase as the temperature differentials increase. The Seebeck coefficient of Au–TPT–Au junctions is observed to be $+16.9 \pm 1.4$ μV/K, which indicates the HOMO–mediated transport (Adapted with permission from Tan et al. [53]).

using the STM-BJ technique, as described earlier in Section 10.6.1). Given the wide variety of molecules that can be self-assembled on metal surfaces, the CP-AFM technique has been extensively used to understand the effect of: (1) the length of molecules and (2) the reactive end groups of molecules on the charge transport of molecular junctions [46, 47, 62].

To measure the Seebeck coefficient of molecular junctions with the CP-AFM technique, Tan et al. [52, 53] made modifications on the conventional CP-AFM technique (Fig. 10.12a). These modifications include an electrical heater attached to the substrate on which the molecules are self-assembled, which makes it possible to raise the substrate temperature to be $T + \Delta T$, and a short (~125 μm long, 35 μm wide and 1 μm thick) Si cantilever coated with Au, instead of a silicon nitride (Si_3N_4) cantilever employed in the conventional CP-AFM technique, which is attached to a thermal reservoir at a temperature T. Given the large thermal conductivity of silicon [63] (~150 W/m.K) and the relatively poor thermal conductivity of the surrounding air (~0.024 W/m.K), thermal modeling [64] suggests that the temperature of the metal-coated cantilever tip, which is in contact with molecules, must be in between T and $T + 0.05\Delta T$. This implies that at least 95% of the temperature differential occurs across the molecules trapped in between the metal electrodes. Subsequently, the thermoelectric voltage of molecular junctions was measured by connecting a custom-built high input-impedance voltage amplifier between the AFM cantilever and the substrate. The measured thermoelectric voltages ΔV induced by the temperature differentials ΔT are shown in Fig. 10.12b. The measured thermoelectric voltages were found to increase linearly when the temperature differentials established across the molecular junctions were increased from 0 K to 12 K in steps of 3 K. These results suggest that the Seebeck coefficient of Au–1,1′,4′,1″-terphenyl-4-thiol (TPT)–Au junctions is $+16.9 \pm 1.4$ μV/K. Control experiments were also performed on Au–Au junctions, and the measured Seebeck coefficient of Au–Au junctions varied between 0.1 μV/K for low contact resistances (mΩ) and a maximum of ~1.3 μV/K (for contact resistance of ~100 Ω). These small values clearly demonstrate that the measured thermoelectric voltages (Fig. 10.12b) originate from the TPT molecules trapped between the electrodes.

10.7 Experimental Techniques for Probing the Heat Dissipation and Heat Transport Properties of Single-Molecule Junctions

Recently, Lee et al. [45] have developed a novel experimental technique to probe heat dissipation in atomic-scale junctions—single-molecule junctions and Au–Au junctions. Using custom-fabricated nanoscale-thermocouple-integrated scanning tunneling probes (NTISTP) and a modulation scheme, the authors have shown that the heat dissipation properties of atomic-scale junctions are closely related to the electronic transmission characteristics, which can be understood in the context of the Landauer theory (see Section 10.2). Specifically, their studies showed that the heat dissipation in the electrodes of molecular junctions (Au–1,4-benzenediisonitrile (BDNC)–Au and Au–1,4-benzenediamine (BDA)–Au junctions), which have relatively strong energy-dependent transmission, is asymmetric and bias polarity dependent. In contrast, symmetric heat dissipation was observed in experiments conducted on Au–Au atomic junctions, which have relatively energy-independent transmission.

The study of heat transport properties of single-molecule junctions is experimentally challenging since the expected heat current through the molecular junctions is extremely small, 10–100 pW, for an applied temperature bias of 1 K. None of the currently available single-molecule measurement techniques has the desired resolution to perform this study. Until now, a few pioneering experimental studies have probed heat transport in monolayers of molecules and polymer fibers with nanoscale diameters [33, 65]. However, probing heat transport at the atomic scale has remained out of reach despite several interesting computational predictions [66, 67].

To overcome this limitation and achieve thermal transport measurements through a variety of single-molecule junctions and single-polymer chains, it is required to develop novel experimental techniques that enable detection of heat currents with picowatt resolution since the computationally predicted thermal conductance of single-molecule junctions and single-polymer chains is in the range of 10–100 pW/K. Further, these techniques should demonstrate three essential capabilities: (1) trap a single molecule between the metal electrodes to form metal-single-molecule–metal junctions, (2) establish known temperature differentials across the electrodes that are separated by a few nanometers, and (3) measure the small heat currents resulting from the temperature differentials established across the electrodes. All these capabilities are currently available only on individual experimental platforms. For example, the STM-BJ technique described earlier has the capability of accomplishing the first two requirements, while the recently developed [68, 69] calorimetric techniques have demonstrated the ability of measuring heat currents with <10 pW resolution. If a novel combination of these techniques is developed, the experimental platform can potentially be leveraged to probe interesting thermal transport properties at the single-molecule/single-polymer chain level.

10.8 Summary

- Molecular junctions composed of organic–inorganic interfaces can be exploited for thermoelectric energy conversion due to their unique transport properties. Thermoelectric devices directly convert heat into electricity and have great potential for future use in scavenging waste heat, which is a byproduct of conventional energy devices
- The efficiency of a thermoelectric device is related to its material properties through the thermoelectric figure of merit $ZT = (S^2 \sigma T)/\kappa$. In bulk 3D systems, it is difficult to increase both S and σ simultaneously. Recently, breakthroughs have been made to reduce κ by reducing phonon contributions through nanostructuring. Single-level charge transport gives the highest ZT and this can theoretically be achieved in molecular junctions. The alignment of the discrete energy states in the molecules with respect to the continuous states in the contacts is the key factor affecting transport properties across the junction.
- The electric current through a 1D chain of atoms between two electrodes L and R, as given by the Landauer formalism, is $I_{L\to R} = \int_{-\infty}^{\infty} (f_L(E) - f_R(E))v(E)eD(E)\,dE$ and the

quantum of electronic conductance at zero temperature is $G_e^o = 2e^2/h$.

- The thermal energy flux between the electrodes due to phonons is given by $J_{L\to R,p} = \sum_i \int_0^\infty \frac{dk}{2\pi}\hbar\omega_i(k)\, v_{g_i}(k)$ $(n_L - n_R)$ and the quantum of thermal conductance due to phonons for a single polarization is $G_{Th,p}^0 = k_B^2\pi^2 T/3h$. The quantum of thermal conductance due to electrons is given by $G_{Th,e}^0 = 2G_{Th,p}^0$ due to electron spin degeneracy.
- In a molecular junction, the discrete energy states in the molecule spread after interaction with the contacts. To model transport across these junctions, a transmission function $\tau_e(E)$, ranging in value between 0 and 1, must be introduced into the Landauer formalism to model the nonideality of the junction. This modifies the expressions for the transport properties of an ideal junction. For example, in this modified regime, the quantum of electronic conductance becomes $G_e = (2e^2/h)\tau_e(E = \mu)$. Similar changes can be made for the expressions for thermal conductance due to electron and phonon contributions
- Thermopower or Seebeck coefficient S of a thermoelectric device is given by $S = -\Delta V/\Delta T$. The relationship between S and $\tau_e(E)$ can be derived by setting $I_{L\to R} = 0$ and further simplified to $S = -\frac{\pi^2 k_B^2 T_L}{3|e|}\left(\frac{1}{\tau_e(E)}\frac{\partial\tau_e(E)}{\partial E}\right)\Big|_{E=\mu}$. A thermoelectric current can be created only when an asymmetry in the transmission function exists in the vicinity of the chemical potential.
- Accurate, quantitative representation of the transmission function is essential for predicting the transport properties. In case of charge transport, Dirac-delta functions and Lorentzians are most common and efforts have been made to modify these to ensure maximum power output and minimize opposing flow of thermally excited electrons.
- For phonon transport, the Acoustic Mismatch Model (AMM) and Diffuse Mismatch Model (DMM) have been used for solid–solid crystal interfaces and have also been recently adapted for a simple SAM interface. Computational studies using molecular dynamics and Green's function approaches

have also been employed to model transmission functions in realistic systems.

- For a weakly coupled molecular junction with an assumed Lorentzian electronic transmission function, $Z_i T = \frac{4\pi^2 k_B^2 T^2}{3(\mu - E_i)^2(1+R)}$ for the ith molecular orbital. It can be related to the maximum possible efficiency for device as $\eta_{TE} = \frac{T_H - T_C}{T_H} \frac{\sqrt{1+ZT_m}-1}{\sqrt{1+ZT_m}+T_C/T_H}$ where $T_m = (T_H + T_C)/2$. η_{TE} is limited by the Carnot efficiency $\eta_c = 1 - T_C/T_H$. Computational efforts have been made to predict the value of ZT in realistic molecular junctions.
- In recent years, experimental techniques have been developed to study the transport properties of molecular junctions. The STM-BJ and CP-AFM techniques are most widespread in measuring the electronic conductance and thermopower of a molecular junction.
- A molecule–metal interface is created by exposing a metal surface to a solution containing organic molecules. This leads to the formation of an ordered monolayer of the molecules on the surface through self-assembly. This is then used in both STM-BJ and CP-AFM experiments to obtain single-molecule junctions.
- In STM-BJ, a sharp STM tip is brought onto the surface of the molecule and withdrawn slowly. During this process, the number of molecules attached to both the tip and metal surface decreases till only a single molecule remains. The I-V characteristics of this junction can be studied by applying a voltage V and measuring the current I. For measuring the thermopower a temperature differential ΔT is set up across the junction and the open-circuit voltage is measured. It can be shown that the sign of S can be used to determine whether the transport is HOMO or LUMO dominated. This can also be seen while deriving the expression for S using a simple Lorentzian transmission function.
- In CP-AFM, the molecular layer is brought in contact with a metal-coated AFM tip. A large number of molecules, which depends on the diameter of the tip, get attached to it. To measure S, the metal substrate is attached to a heater to create a temperature differential across the junction and

the open-circuit thermoelectric voltage is measured. This method is more stable in holding molecular junctions for a longer time than STM-BJ.

- Heat dissipation in molecular junction is also closely related to the electronic transmission characteristics. It is shown to be asymmetric when the transmission function is strongly energy dependent and symmetric when not.

Problems

1. Derive an expression for thermopower S for a transmission function having a single Gaussian function centered at energy level E_m. Assume the standard deviation $\sigma = 1/\sqrt{2\pi}$. Compare this to the expression given by Eq. 10.39 for a single energy level $E_m = 1$, assuming $\Gamma_m = 2\Gamma_{m,1} = 2\Gamma_{m,2} = 2/\pi$ and $\frac{\pi^2 k_B^2 T}{3|e|} = 1$, by plotting all on the same plot as a function of E. What is the value of S predicted by these expressions for $\mu = 1.5$?
2. For molecular junctions consisting of alkane molecules connected to two reservoirs, electron transport takes place through tunneling. How will G_e and S vary with molecular length L? Hint: Derive and plot the expressions of G_e and S as functions of L assuming that the transmission function $\tau(E)$ can be represented by an electron tunneling barrier potential where the barrier height is E_m, barrier width equals L, and $\mu < E_m$. You can assume $\frac{\pi^2 k_B^2 T}{3|e|} = 1$, $\frac{2e^2}{\hbar} = 1$, $E_m = 1.5$, and $\sqrt{2m/\hbar^2} = 1$.
3. From a qualitative standpoint, would the DMM (Eq. 10.42— only elastic scattering) predict a higher or lower phonon transmission as the density of states of the two substrates in contact increase in mismatch? Give reasons.
4. Derive expressions for the transmission function τ_{el} across an interface using the DMM (assuming only one polarization), given by Eq. 10.42, for the following kinds of dispersions:

   a. Debye: $\omega = v_g k$
   b. Sinusoidal: $\omega = 2v_g \sin(k/2)$

Plot both expressions for τ_{el} on the same plot as a function of ω for $v_{g,L} = 2$, $v_{g,R} = 4$, and $0 \leq k \leq \pi$.

5. Suppose you are setting up an experiment to increase thermopower in a molecular junction by tuning its electronic structure. Can a gate electrode be used to control the S of a molecular junction? How would this effect the measured value of S? What requirements should be satisfied by your experimental platform?

References

1. Briseno, A.L., et al., Patterning organic single-crystal transistor arrays. *Nature*, 2006. **444**(7121): pp. 913–917.
2. Reddy, P., et al., Thermoelectricity in molecular junctions. *Science*, 2007. **315**(5818): pp. 1568–1571.
3. *Energy in Sweden: Facts and Figues 2010*. 2010: National Energy.
4. IEA (International Atomic Agency) *Key World Energy Statistics*. 2012.
5. Seebeck, T.J., Magnetic polarization of metals and minerals. *Abhandlungen der Deutschen Akademie der Wissenschaften zu* Berlin, 1822. **265**
6. Cutler, M. and N. Mott, Observation of anderson Localization in an electron gas. *Physical Review*, 1969. **181**(3): pp. 1336–1340.
7. Jonson, M. and G. Mahan, Electron-phonon contribution to the thermopower of metals. *Physical Review B*, 1990. **42**(15): pp. 9350–9356.
8. Heremans, J.P., et al., Enhancement of thermoelectric efficiency in PbTe by distortion of the electronic density of states. *Science*, 2008. **321**(5888): pp. 554–557.
9. Harman, T.C., Quantum Dot superlattice thermoelectric materials and devices. *Science*, 2002. **297**(5590): pp. 2229–2232.
10. Hsu, K.F., Cubic AgPbmSbTe2+m: Bulk thermoelectric materials with high figure of merit. *Science*, 2004. **303**(5659): pp. 818–821.
11. Venkatasubramanian, R., et al., Thin-film thermoelectric devices with high room-temperature figures of merit. *Nature*, 2001. **413**(6856): pp. 597–602.
12. Poudel, B., et al., High-thermoelectric performance of nanostructured bismuth antimony telluride bulk alloys. *Science*, 2008. **320**(5876): pp. 634–638.

13. Kittel, C., and P. McEuen, *Introduction to Solid State Physics*. Vol. 19. John Wiley.
14. Mahan, G.D., and J.O. Sofo, The best thermoelectric. *Proceedings of the National Academy of Sciences*, 1996. **93**(15): pp. 7436–7439.
15. Wang, R.Y., R.A. Segalman, and A. Majumdar, Room temperature thermal conductance of alkanedithiol self-assembled monolayers. *Applied Physical Letters*, 2006. **89**(173113): pp. 1–4.
16. Lee, W., B. Song, and P. Reddy, Measurement of thermoelectric and thermal transport properties of single-molecule junctions. *Annual Review of Heat Transfer*, 2013. **16**(1).
17. Butcher, P.N., Thermal and electrical transport formalism for electronic microstructures with many terminals. *Journal of Physics: Condensed Matter*, 1990. **2**(22): pp. 4869–4878.
18. Malen, J.A., et al., Fundamentals of energy transport, energy conversion, and thermal properties in organic-inorganic heterojunctions. *Chemical Physics Letters*, 2010. **491**(4): pp. 109–122.
19. Humphrey, T.E. and H. Linke, Reversible thermoelectric nanomaterials. *Physical Review Letters*, 2005. **94**(9): pp. 096601 1–4.
20. Nakpathomkun, N., H.Q. Xu, and H. Linke, Thermoelectric efficiency at maximum power in low-dimensional systems. *Physical Review B*, 2010. **82**(23): pp. 235428 1–9.
21. Karlström, O., et al., Increasing thermoelectric performance using coherent transport. *Physical Review B*, 2011. **84**(11): pp. 113415 1–4.
22. Curzon, F.L., Efficiency of a Carnot engine at maximum power output. *American Journal of Physics*, 1975. **43**(1): pp. 22–24
23. Datta, S., *Quantum Transport: Atom to Transistor*. Vol. 14. Cambridge University Press.
24. Little, W.A., The transport of heat between dissimilar solids at low temperatures. *Canadian Journal of Physics*, 1959. **37**(3): pp. 334–349.
25. Swartz, E.T. and R.O. Pohl, Thermal boundary resistance. *Reviews of Modern Physics*, 1989. **61**(3): pp. 605–668.
26. Duda, J.C., et al., Assessment and prediction of thermal transport at solid-self-assembled monolayer junctions. *Journal of Chemical Physics*, 2011. **134**(094704): pp. 1–6.
27. Fermi, E., J. Pasta, and S. Ulam, Studies of nonlinear problems. 1955: *Los Alamos Scientific Laboratory Report No. LA-1940.*
28. Casher, A., Heat flow in regular and disordered harmonic chains. *Journal of Mathematical Physics*, 1971. **12**(8): pp. 1701–1711.

29. Lepri, S., R. Livi, and A. Politi, Heat conduction in chains of nonlinear oscillators. *Physical Review Letters*, 1997. **78**(10): pp. 1896–1899.

30. Lepri, S., R. Livi, and A. Politi, On the anomalous thermal conductivity of one-dimensional lattices. *Europhysics Letters*, 1998. **43**(3): pp. 271–276.

31. Segal, D., A. Nitzan, and P. Hänggi, Thermal conductance through molecular wires. *Journal of Chemical Physics*, 2003. **119**(13): pp. 6840 1–16.

32. Henry, A. and G. Chen, High thermal conductivity of single polyethylene chains using molecular dynamics simulations. *Physical Review Letters*, 2008. **101**(23): pp. 1–4.

33. Shen, S., et al., Polyethylene nanofibres with very high thermal conductivities. *Nature Nanotechnology*, 2010. **5**(4): pp. 251–255.

34. Sasikumar, K. and P. Keblinski, Effect of chain conformation in the phonon transport across a Si-polyethylene single-molecule covalent junction. *Journal of Applied Physics*, 2011. **109**(11): pp. 114307 1–6.

35. Murphy, P., S. Mukerjee, and J. Moore, Optimal thermoelectric figure of merit of a molecular junction. *Physical Review B*, 2008. **78**(16): pp. 161406 1–4.

36. Christopher, M.F., et al., Conformation dependence of molecular conductance: Chemistry versus geometry. *Journal of Physics: Condensed Matter*, 2008. **20**(2): p. 022203.

37. Fano, U., Effects of configuration interaction on intensities and phase shifts. *Physical Review*, 1961. **124**(6): pp. 1866–1878.

38. Ke, S.-H., et al., Thermopower of molecular junctions: An ab initio study. *Nano Letters*, 2009. **9**(3): pp. 1011–1014.

39. Park, H., et al., Nanomechanical oscillations in a single-C-60 transistor. *Nature*, 2000. **407**(6800): pp. 57–60.

40. Tsutsui, M., M. Taniguchi, and T. Kawai, Local heating in metal-molecule-metal junctions. *Nano Letters*, 2008. **8**(10): pp. 3293–3297.

41. Venkataraman, L., et al., Dependence of single-molecule junction conductance on molecular conformation. *Nature*, 2006. **442**(7105): pp. 904–907.

42. Xu, B.Q. and N.J.J. Tao, Measurement of single-molecule resistance by repeated formation of molecular junctions. *Science*, 2003. **301**(5637): pp. 1221–1223.

43. Jang, S.Y., et al., Interpretation of stochastic events in single molecule conductance measurements. *Nano Letters*, 2006. **6**(10): pp. 2362–2367.

44. Mishchenko, A., et al., Influence of conformation on conductance of biphenyl-dithiol single-molecule contacts. *Nano Letters*, 2010. **10**(1): pp. 156–163.
45. Lee, W. and P. Reddy, Creation of stable molecular junctions with a custom-designed scanning tunneling microscope. *Nanotechnology*, 2011. **22**(48): p. 1–8.
46. Beebe, J.M., et al., Contact resistance in metal-molecule-metal junctions based on aliphatic SAMs: Effects of surface linker and metal work function. *Journal of the American Chemical Society*, 2002. **124**(38): pp. 11268–11269.
47. Choi, S.H., B. Kim, and C.D. Frisbie, Electrical resistance of long conjugated molecular wires. *Science*, 2008. **320**(5882): pp. 1482–1486.
48. Baheti, K., et al., Probing the chemistry of molecular heterojunctions using thermoelectricity. *Nano Letters*, 2008. **8**(2): pp. 715–719.
49. Malen, J.A., et al., The nature of transport variations in molecular heterojunction electronics. *Nano Letters*, 2009. **9**(10): pp. 3406–3412.
50. Malen, J.A., et al., Identifying the length dependence of orbital alignment and contact coupling in molecular heterojunctions. *Nano Letters*, 2009. **9**(3): pp. 1164–1169.
51. Widawsky, J.R., et al., Simultaneous determination of conductance and thermopower of single molecule junctions. *Nano Letters*, 2012. **12**(1): pp. 354–358.
52. Tan, A., et al., Effect of length and contact chemistry on the electronic structure and thermoelectric properties of molecular junctions. *Journal of the American Chemical Society*, 2011. **133**(23): pp. 8838–8841.
53. Tan, A., S. Sadat, and P. Reddy, Measurement of thermopower and current-voltage characteristics of molecular junctions to identify orbital alignment. *Applied Physics Letters*, 2010. **96**(1): pp. 013110 1–4.
54. Love, J.C., et al., Self-assembled monolayers of thiolates on metals as a form of nanotechnology. *Chemical Reviews*, 2005. **105**(4): pp. 1103–1169.
55. Shi, L. and A. Majumdar, Thermal transport mechanisms at nanoscale point contacts. *Journal of Heat Transfer-Transactions of the ASME*, 2002. **124**(2): pp. 329–337.
56. Venkataraman, L., et al., Electronics and chemistry: Varying single-molecule junction conductance using chemical substituents. *Nano Letters*, 2007. **7**(2): pp. 502–506.

57. Evangeli, C., et al., Engineering the thermopower of C60 molecular junctions. *Nano Letters*, 2013. **13**(5): pp. 2141–2145.

58. Yee, S.K., et al., Thermoelectricity in fullerene-metal heterojunctions. *Nano Letters*, 2011. **11**(10): pp. 4089–4094.

59. Bohler, T., A. Edtbauer, and E. Scheer, Conductance of individual C-60 molecules measured with controllable gold electrodes. *Physical Review B*, 2007. **76**(12): p. 125432.

60. Kiguchi, M., Electrical conductance of single C-60 and benzene molecules bridging between Pt electrode. *Applied Physics Letters*, 2009. **95**(7).

61. Neel, N., et al., Controlled contact to a C-60 molecule. *Physical Review Letters*, 2007. **98**(6).

62. Engelkes, V.B., J.M. Beebe, and C.D. Frisbie, Length-dependent transport in molecular junctions based on SAMs of alkanethiols and alkanedithiols: Effect of metal work function and applied bias on tunneling efficiency and contact resistance. *Journal of the American Chemical Society*, 2004. **126**(43): pp. 14287–14296.

63. Incropera, F.P., et al., *Fundamentals of Heat and Mass Transfer* 6th ed. 2007 John Wiley.

64. Xue, Y.Q. and M.A. Ratner, Microscopic study of electrical transport through individual molecules with metallic contacts. I. Band lineup, voltage drop, and high-field transport. *Physical Review B*, 2003. **68**(11): pp. 115406 1–18.

65. Losego, M.D., et al., Effects of chemical bonding on heat transport across interfaces. *Nature Materials*, 2012. **11**(6): pp. 502–506.

66. Lepri, S., R. Livi, and A. Politi, Thermal conduction in classical low-dimensional lattices. *Physics Reports-Review Section of Physics Letters*, 2003. **377**(1): pp. 1–80.

67. Li, N.B., et al., Colloquium: Phononics: Manipulating heat flow with electronic analogs and beyond. *Reviews of Modern Physics*, 2012. **84**(3): pp. 1045–1066.

68. Sadat, S., et al., Room temperature picowatt-resolution calorimetry. *Applied Physics Letters*, 2011. **99**(4): pp. 043106 1–3.

69. Sadat, S., E. Meyhofer, and P. Reddy, Resistance thermometry-based picowatt-resolution heat-flow calorimeter. *Applied Physics Letters*, 2013. **102**(16).

Chapter 11

Interference Effects in Single-Molecule Transport

Gemma C. Solomon
Nano-Science Center and Department of Chemistry, University of Copenhagen, Universitetsparken 5, 2100 Copenhagen Ø, Denmark
gsolomon@nano.ku.dk

Destructive interference effects in single-molecule transport have attracted attention in recent years as a way to introduce a range of electronic functionalities into molecular electronics. In this chapter, we provide an introduction to this area and address may of the topics that constitute background information for understanding recent publications. We discuss why interference effects might be useful, their experimental and theoretical signatures, the range of chemical systems they occur in, and possible applications.

11.1 Introduction

In molecular electronics, chemists dream of having electrical functionality controlled by careful molecular design. With the change of a substituent, a molecule would change from, e.g., a wire to some other functionality, maybe a rectifier or switch.

Handbook of Single-Molecule Electronics
Edited by Kasper Moth-Poulsen

ISBN 978-981-4463-38-6 (Hardcover), 978-981-4463-39-3 (eBook)
www.panstanford.com

Nature can seemingly exert this level of fine control, with biological systems exhibiting rates of electron transfer ranging over orders of magnitude as a consequence of subtle changes in the natural system. Replicating this mastery in synthetic systems has, however, proved challenging.

A great deal of understanding has been developed over the years from studies on electron transfer and transport in synthetic systems. Experimental techniques have emerged that allow single molecules to be reproducibly bound in conducting junctions and their conductance determined. Like all single-molecule measurements, there is necessarily some variation from measurement to measurement and also in time. Statistical techniques, where the conductance can be determined as an average over hundreds to thousands of measurements, allow this difficulty to be overcome and the conductance of molecular wires can be obtained. A molecular wire is a basic functionality in any vision for molecular electronics; however, more complex functionalities also need to be achieved, e.g., switches, rectifiers, dielectric layers, and logic elements. In all these cases, it is not sufficient to simply maximize the charge transport through the molecule. It may be that an "on" state must be complemented with an easily accessible "off" state, with the two states being activated or deactivated in response to some external stimuli. Or the conductance might need to be switched off while the molecular polarizability is maintained. Or for an application such as thermoelectrics, the energy dependence of the charge and heat transport properties needs to be tailored in a precise fashion.

This chapter proceeds as follows: first we outline why interference effects might lead to useful functionality in molecules and some explanation of the interest in interference effects in single-molecule transport. Then we describe some of the signatures of interference in theory and experiment and illustrate the range of chemical systems that exhibit destructive interference. We also summarize the proposals for how to predict when interference effects will dominate. We then address the question of "what is interfering with what" and provide a suggestion of how to think about these processes at the single-particle level. From this base, we then consider how interference effects can be tuned and how they might be applied to more advanced single-molecule electronic components.

At the outset, we should also say what is not included in this chapter. This chapter is not a conclusive review of interference effects in molecules, but rather an explanation/introduction to some of the important concepts. The discussion is dominated by the theoretical description of these effects, as this dominates the literature to date. This theoretical discussion is undertaken at the simplest possible (single-particle) level, and consequently some processes are ignored. More specifically, electron–electron interactions, correlation, inelastic scattering, Coulomb blockade, spin-systems, and Kondo physics may all have a bearing on interference effects, but they are all neglected in this chapter. In many cases, these effects may only induce quantitative changes from the single-particle picture; however, this will not always be the case and care is required in this regard.

We hope that this chapter will provide an introduction to some of the things that are not explained in articles about interference effects and thus constitute a starting point for someone jumping into the field.

11.2 Why Interference?

In answer to a question such as this, there is always the trivial response: why not? Indeed, a large part of scientific discovery is about simply observing things as they are, not driven by applications or technology. This, however, is not the only kind of research, and we often study the world with a purpose or application in mind, seeking to optimize our system for that purpose. In the case of molecular electronics, we look for electronic functionality driven by chemical functionality.

How can we achieve some desired electrical function from a single molecule? Like other physical properties, we can imagine that different chemical structures will have different charge transport properties, so we can vary our molecule until we find the desired functionality, and this is correct. The structure–function relationships that govern molecular conductance are not so well mapped out as other types of physical properties, so it might take some time to find the molecule we want, but the principle is the same. With chemical modifications, the conductance of a molecule can be increased. This increase is not generally so much as to make a molecule as conductive as a metal wire, but in principle it can be. Part of the challenge is that we need to control the properties of the molecule bound in the conducting junction, not just the isolated molecule, but this is a topic for a different chapter. Suffice to say that in the case of interference effects, we can understand the properties of the molecule in the junction by understanding the properties of the molecule alone.

What if we want more than just a molecular wire, e.g., we want to design a molecular switch. Now, we need two distinct conductance states and the possibility to switch between them. If our molecule has a single conducting channel, the conductance for our ideal switch is shown in Fig. 11.1. In the "on" state, the molecule will transmit perfectly with a conductance of 1G0, and in the "off" state, it will be a perfect insulator. Most molecules do not even get close to 1G0 at low bias, so more likely we are considering the difference between a slightly low "on" state and a very low "off" state.

The challenges for the molecular design are as follows:

- What are the chemical/electrical changes we need to make in the molecule to induce these conductance changes?

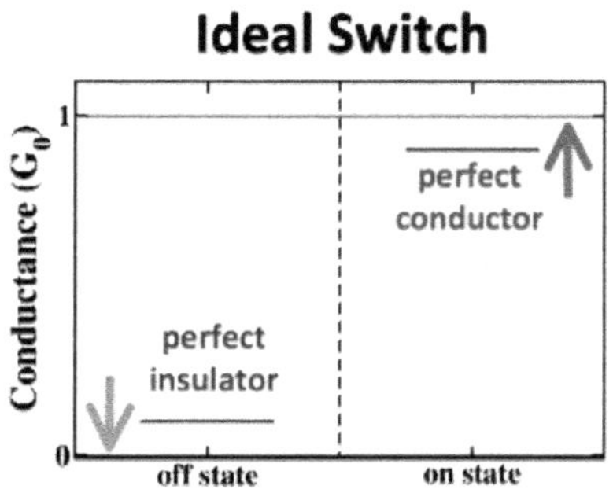

Figure 11.1 An ideal one-channel molecular switch: a perfect insulator in the "off" state and a perfect conductor in the "on" state. Reprinted with permission from Ref. [1]. Copyright (2013) American Chemical Society.

- What stimuli do we need to induce these changes in the molecule?
- Is this process reversible or a one-time switch?
- What is the time scale of the switching event?

A range of answers can be, and have been, proposed for these questions, but there is one inescapable truth: as long as nuclear motion is involved, the kinetics of that process (slow) will always limit the switching rate. This is the same whether we consider a chemical reaction, reduction/oxidation, or conformational change. A slow switching rate will not always be a disadvantage, nor will an irreversible process, but if we want fast, reversible switching, we need to use a different strategy.

Another approach is the use of destructive interference effects. In some cases, this approach constitutes a subset of the processes described above since destructive interference can be switched on or off with chemical or redox reactions. The difference is that interference effects can also be tuned by electronic processes alone and, therefore, have the potential for rapid switching without mass transport limiting the kinetics.

Figure 11.2 illustrates how interference can manifest in an idealized way. In these plots, we illustrate the transmission relative to some imaginary Fermi energy of our (unbiased) electrodes. Throughout the chapter, we will consider the two electrodes to be made of the same material, and thus they would have a common Fermi energy in the absence of an applied bias. We will use this energy as our reference energy. In the case of atomistic calculations, we will consider gold electrodes, putting our energy reference around -5 eV.

For our idealized system, a destructive interference feature in the electronic transmission creates an energy window where a system that would, otherwise, behave like a perfect conductor instead has the properties of a perfect insulator (Fig. 11.2a). This is exactly what we want for our switch; we need to only manipulate the interference feature, and we can tune the system between a conductor and an insulator. Some ways to achieve this in molecules will be discussed later, but for now it is sufficient to understand that this is the feature we are looking for.

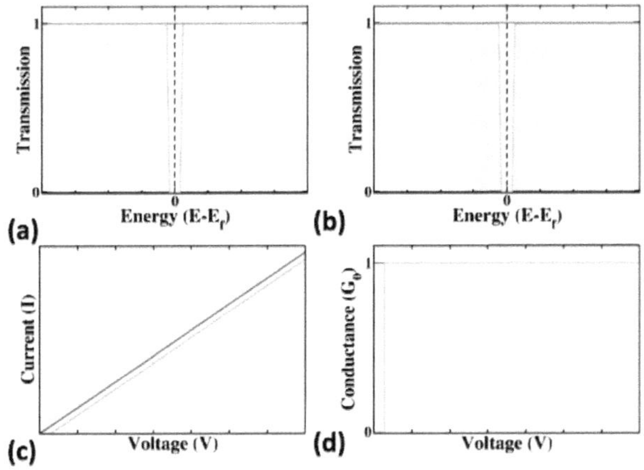

Figure 11.2 An illustrative view of interference effects. (a) Electronic transmission through three systems: a perfect one-channel conductor (red line), a perfect insulator (blue line), and an idealized system with destructive interference (green). The system with destructive interference is a perfect conductor outside a small energy window in which it is an insulator. (b) The current through these systems can be calculated by integrating the transmission over a window controlled by the applied bias leading to the current (c) and conductance (d) as a function of bias for the three systems. Reprinted with permission from Ref. [1]. Copyright (2013) American Chemical Society.

11.3 Interest in Interference Effects

Whether one considers interference effects in single-molecule transport novel or interesting depends on perspective. A physicist can regard these effects as well-known phenomena from coupled quantum dots, interferometers, or waveguides repackaged for the case of molecular systems. An organic chemist examining the types of chemical structures may be unsurprised that interference effects occur in these systems on the basis of the intuition they have from studying reactivity at different positions in conjugated molecules. A researcher familiar with the huge body of work on photoinduced electron transfer in donor–bridge–acceptor systems may again be unsurprised because of having a feeling for the rates of electron or hole transfer in similar bridges.

In some sense, all these sentiments are correct. On carefully considering all the understanding in these fields, one would indeed find it unsurprising in retrospect that interference effects result in the transport properties they do in molecules. This is, however, different from predicting the full range of transport properties we see in molecules with interference effects. One should not confuse what is unsurprising in retrospect with what is obvious beforehand.

Molecules in transport junctions are their own special case. Quantum dots, interferometers, and waveguides give us no sense of the molecules that will exhibit destructive interference effects. Further, the coherence lengths and distances over which coherence must be maintained for interference are very different in these systems. Organic chemists have a phenomenal sense of molecules from their feeling for chemical reactions, but there is no simple one-to-one mapping for reactivity and conductance. Intramolecular electron transfer is a close cousin of electron transport; however, the properties of the donor and the acceptor play a different role from metallic leads in the rate of electron transfer, and we cannot see the full energy range of the coupling that becomes accessible when a bias is applied across a molecule.

For these reasons, we will outline some of the concepts related to interference effects in molecules coupled to conducting leads without much reference to these related fields. An interested reader should certainly branch out and get inspiration from all the prior studies on condensed matter physics, organic chemistry, and electron transfer, as the process is thoroughly enjoyable and stimulating.

The origins of molecular electronics can be traced back to the work of Aviram and Ratner in 1974 [2] with their proposal for a single-molecule rectifier. The first studies on interference effects came not long after; the work of Joachim et al. [3, 4] showed that different substitution patterns around a central benzene ring will result in dramatically different transport properties. This group has continued to work with interference effects and has proposed a remarkable range of functional molecular electronic devices utilizing interference to achieve all kinds of electronic functionality [5–8].

In recent years, interest in interference effects in cyclic and acyclic molecular systems, both in theory [9–18] and experiment

[19–22], has increased. Cross-conjugation in both cyclic and acyclic systems has been investigated with studies addressing the nature, stability, amenability to control, and application of interference effects in these systems. At the same time, interest in the use of interference in cyclic systems for transistor or thermoelectric applications has developed.

11.4 Signatures of Interference

11.4.1 *In a Model-System Calculation*

Before we discuss the types of systems that exhibit interference effects, we should find out what are the signatures of interference, in both theory and experiment. The simplest case is a model-system calculation, typically a Hückel or tight-binding model representing the π-system of a conjugated molecule. In these cases, destructive interference is evident as sharp dips in the electronic transmission on a logarithmic scale. As a side note, when we consider transmission in this chapter, we consider coherent tunneling as calculated at the so-called Landauer level. The minimum is controlled only by the energy resolution of the plot for numerical calculations and can be shown to be explicitly equal to zero with a functional form for the transmission. Figure 11.3 illustrates these features as they appear for *ortho*- and *meta*-substituted models of a benzene ring coupled between two electrodes.

11.4.2 *In a More Realistic Calculation*

Moving from a model-system calculation to an atomistic calculation for a conjugated molecule means that the story is not so simple. In addition to the π-system, the σ-system is now included, and it can give a significant contribution to the transmission. This statement is, somewhat, counter-intuitive, as the starting assumption would always be that the π-system dominates the transport properties in conjugated molecules. This is because we have effectively killed the contribution of the π-system through interference, so that the σ-system can come to dominate in these cases.

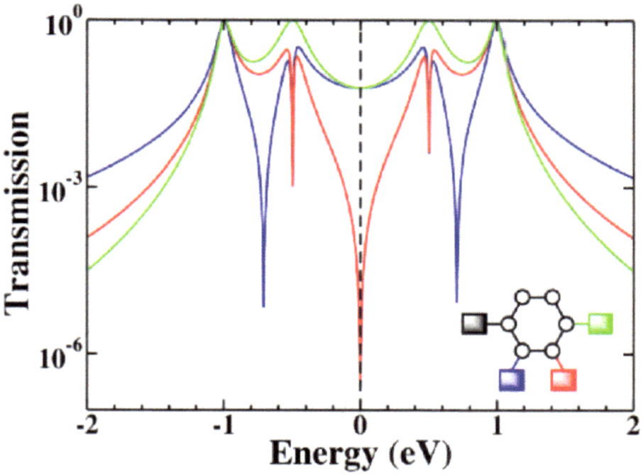

Figure 11.3 The electronic transmission as a function of energy (relative to a common electrode Fermi energy) for the Hückel model calculation for the π-system of a benzene ring. In all cases, the black box represents the left electrode, while the right electrode varies to illustrate the effects associated with changing substitution between *para* (green), *meta* (red), and *ortho* (blue). A number of interference features are evident as sharp dips in the transmission. Reprinted with permission from Ref. [23]. Copyright (2013) American Institute of Physics.

Figure 11.4 illustrates how this manifests in some small molecules: The sharp dip in the transmission is no longer evident in the total transmission, but only in the π-component. In these systems, the current is significantly reduced, but no characteristic line shape is visible in the current as a function of voltage, to indicate an interference effect. An important difference between the π-system transport and the σ-system transport is how fast the σ-system transport decreases with increasing molecular length. A modest increase in system size will result in a significant decrease in the σ-transmission, and consequently more of the interference dip will be visible in the total transmission. In these cases, the line shape of the current and/or conductance might reveal the interference dip.

Not all molecules allow for the symmetry separation that revealed the interference dip in the transmission as shown in Fig. 11.4. In many cases, the molecule will not have a plane of symmetry, or the interference could simply be in some component of the σ-system transport. In these cases, we need another signature in our

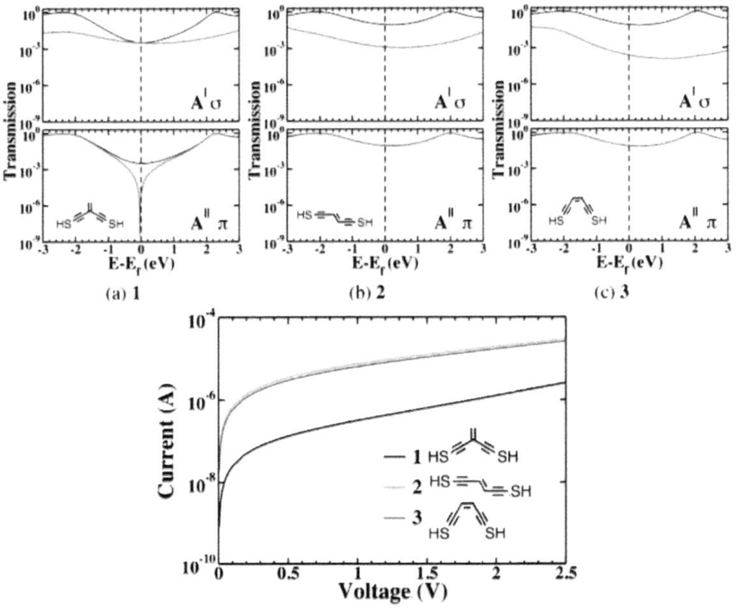

Figure 11.4 The electronic transmission (top) as a function of energy (relative to a common electrode Fermi energy) for a series of small molecules bound between Au electrodes using gDFTB. In each case, the red curve illustrates the individual σ- or π-components of the transmission, and the black curve is the total transmission. In the cross-conjugated species (1), an interference feature is evident in π-transmission, which means that the σ-transmission dominates at the Fermi energy. The interference effect results in a dramatic reduction in the current through the system, as shown below. Reprinted with permission from Ref. [14] Copyright (2013) American Chemical Society.

calculations to distinguish between low levels of transport induced by destructive interference, and generically low transport. The local transport through a molecule can provide us with another signature of interference: ring-current reversal about the interference feature. Figure 11.5 shows the local transmission through the small cross-conjugated system introduced in Fig. 11.4. Around the energy of the interference feature, a ring-current reversal is evident in the local transmission (top); this is seen in both cyclic and acyclic systems and was predicted by other methods [24].

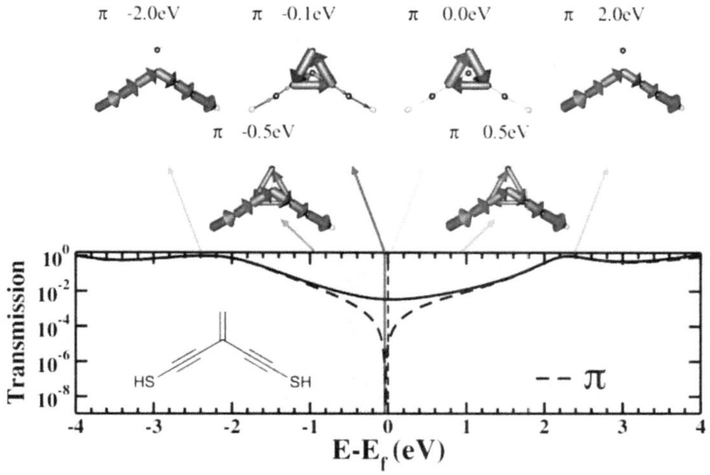

Figure 11.5 The local π-transmission through a small cross-conjugated molecule (top) at a selection of energies relative to the Au Fermi energy. The π-system transmission (dashed) and total transmission (solid) are shown below. At the energy of the interference feature, a ring-current reversal is evident in the local transmission. Reproduced from Ref. [25]. Copyright (2010) Nature Publishing Group.

The advantage of using local current analysis as a diagnostic for interference effects comes in systems where a sharp dip cannot be distinguished. For example, the conductance of saturated systems can vary significantly with conformation, but sharp dips are not observed in the transmission. Figure 11.6 shows an example of this kind. When the central dihedral of 1,8-octanedithiol is varied from 180° to 0°, a marked decrease in the transmission is visible around the Fermi energy (Fig. 11.6a). A sharp interference feature is not evident, but Fig. 11.6b,c illustrates that a ring-current reversal can be observed about the transmission minimum of the modified structures.

11.4.3 *In Experimental Measurements*

All signatures of interference described thus far are purely theoretical observations. The forms of the electronic transmission and the local transmission are not experimental observables, so we are

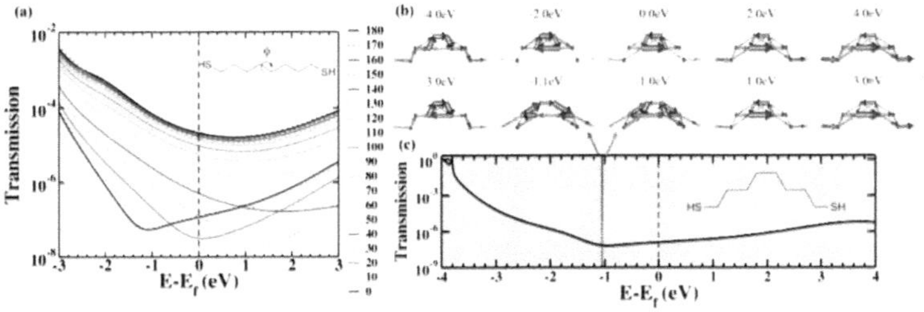

Figure 11.6 The transmission (a) through 1,8-octanedithiol as the central dihedral is varied relative to the Au Fermi energy. The local transmission (b) for the 0° system shows a ring-current reversal about the transmission minimum (c). Reproduced from Ref. [25]. Copyright (2010) Nature Publishing Group.

faced with the question: How can interference effects be actually observed? The first part of this challenge is to know whether there is even a molecule bound in the junction since the conductance of molecules with dominant destructive interference is very low. What is the difference between the signatures of a bound molecule with conductance below a measurement threshold and an open junction without a molecule bound? An answer to this question was found by analyzing the breaking force as well as the conductance, as shown in Fig. 11.7. While the conductance versus distance histogram is not so clear for the *meta*-substituted system due to the very low levels of transport, the breaking force clearly matches that of the more highly conducting *para*-substituted system, indicating that the molecule is indeed bound.

While the breaking force is clearly a good indication that a weakly conducting molecule is bound in the junction, it is not a measurement that can be made in all experimental setups; nor it is a signature of interference alone. In general, the only signature of interference in the conductance is a very low level of conductance relative to other conjugated molecules of the same length. This is all that can be observed in the conductance if:

(a) There is a sharp interference feature in the transmission, but it does not lie exactly at the Fermi energy; or
(b) There is significant σ-transport, inelastic transport, or some alternate pathway in addition to the (π) component with the interference.

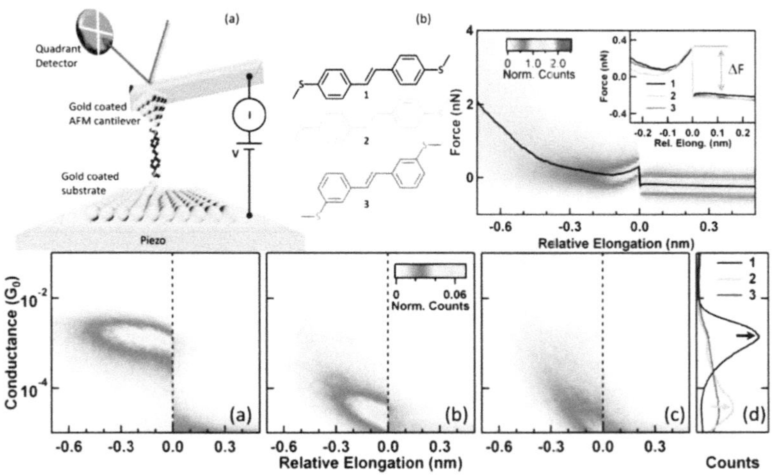

Figure 11.7 Top: A diagram of experimental setup and the three molecules measured (left). The force as a function of elongation for *X* with all three molecules together in the inset, showing the same breaking force for all three junctions. Bottom: The conductance versus distance histograms show very different conductance signatures, illustrating the changing nature of the conjugation of the bridge, destructive interference from the two *meta*-substituted rings in three results in a dramatic reduction in the conductance. Reprinted with permission from Ref. [19]. Copyright (2013) American Chemical Society.

However, in a special case, a sharp feature in the transmission may lead to a sharp feature in the conductance: where the interference feature lies very close to the electrode Fermi energy. While this might seem to be a very unlikely scenario, it was, in fact, observed in recent experiments. Figure 11.8 illustrates the differing line shapes observed in the conductance versus bias depending on whether the interference feature is well aligned with the Fermi energy (sharp dip).

Clearly, a sharp dip in the conductance is a very nice signature of interference; however, we can still have destructive interference and an absence of this feature. So is there another experimental signature we can look for? Essentially, we need something else that tells us qualitatively what kind of low-level transport we have. Is it low transport because a component of the transmission is dropping rapidly to zero (with a proximate interference effect) or

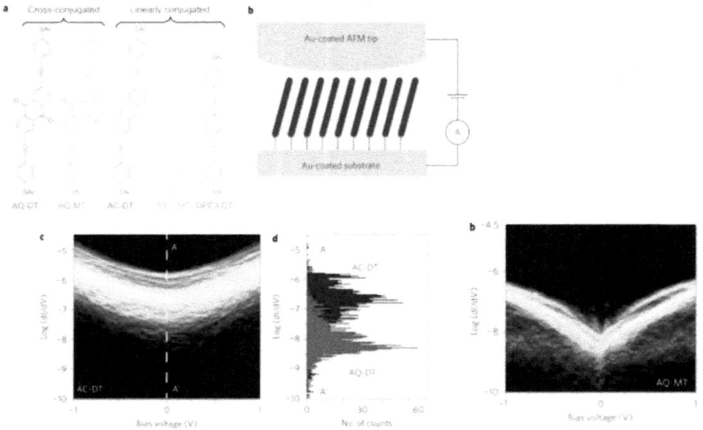

Figure 11.8 Top (left to right): A selection of molecules studied; an illustration of the experimental setup. Bottom (left to right): A molecule without effects where a sharp dip is not evident; the dramatically different magnitude of the conductance when interference effects are evident; the sharp dip in conductance, which is a signature of interference effects when the molecule is only substituted with a single thiol binding group. Reprinted with permission from Macmillan Publishers Ltd: Nature Nanotechnology [26], copyright 2013.

low-flat transmission in the middle of wide gap? The thermoelectric response is another experimental observable that is sensitive to the gradient of the transmission. It has been shown [27] that the thermoelectric response scales with the gradient of the transmission, meaning that it is an interesting candidate for a measurement that can signal that the low transport is, in fact, due to interference. One proposal, as shown in Fig. 11.9, is that in molecules where multiple degenerate interference features lead to a higher-order feature, the line shape of the transmission around the interference feature will lead to very high levels of thermoelectric response. Just like it is not possible to measure force and conductance together in all experimental setups, it is certainly not possible to measure conductance and thermoelectric response together in all setups. This measure simply represents another possible signature of when interference effects dominate.

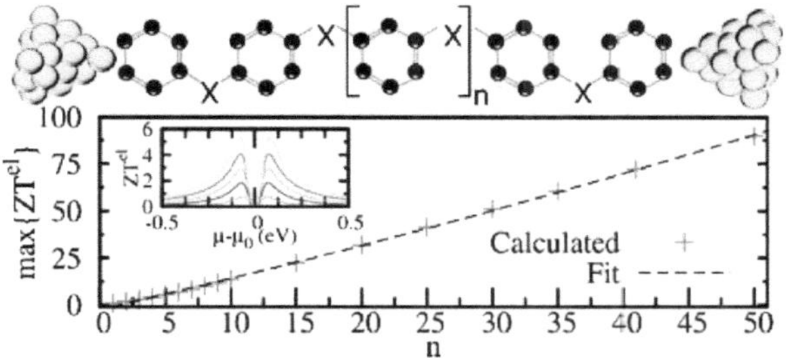

Figure 11.9 In molecules where multiple units with interference effects are connected in series, higher-order nodes (degenerate nodes) are possible. In this case, the thermoelectric response can be shown to scale with the order of the node. Reprinted with permission from Ref. [28]. Copyright (2013) American Chemical Society.

11.5 Range of Chemical Systems

Part of the appeal of interference effects in molecular systems comes from the range of chemical systems that exhibit these effects. While being by no means ubiquitous, interference effects dominate in molecules with varying topologies. Figure 11.10 illustrates some examples of different systems that have been studied as quantum interference units (QIU). Some characteristics are necessarily common to all these functionalities despite their apparent differences, and we will discuss this further below. In general, they can be cyclic or acyclic, large or small, with heteroatoms and/or electron-donating/withdrawing groups, and interference effects can still be evident.

11.5.1 *Predicting Interference Effects*

A list of molecular subunits responsible for interference effects will never be complete, as small chemical modifications result in distinct chemical species that might retain the physical properties of the parent compound. More useful are strategies for predicting when interference effects will dominate. The challenge here is to take a given molecule (possibly even before synthesis has been undertaken) and predict whether interference effects will dominate at low bias. Ideally, this can be achieved without resorting to any

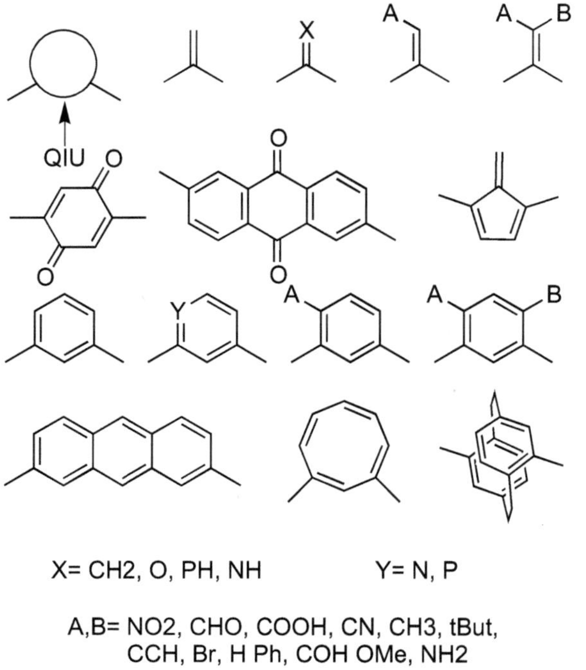

Figure 11.10 Some examples of molecular subunits (QIU) that will all result in interference effects in the transmission. All cases have two bonds without terminal substituents, indicating the positions at which the molecule should be connected (either directly or indirectly) to the electrodes.

type of calculation, or failing that it should be a relatively simple calculation. We will present here a summary of a whole range of proposed predictive tools of this kind. The first three methods do not require any calculation to be performed, whereas the last three do. While the approaches might seem quite different on the surface, underlying commonalities link them. In some cases, we could also use one method to predict the signatures of the other, but in other cases, the link underlying the predictive strategies is not so clear.

For the sake of illustration, we consider a *meta*-substituted benzene ring as shown in Fig. 11.11. In each case, we briefly describe the strategy in general and then apply it to the *meta*-substituted benzene system. The rules state that a system will be dominated by destructive interference:

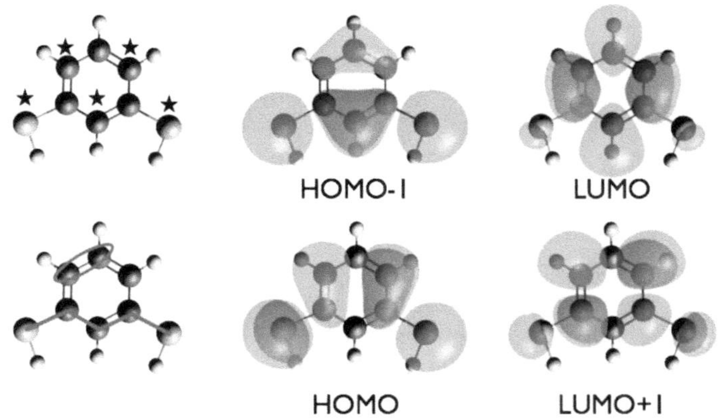

Figure 11.11 Some of the strategies applied to *meta*-substituted benzene (left) and a selection of the molecular orbitals (right). Irrespective of the method used, we can (fortunately) predict that interference effects will dominate.

(1) *If, after removing all atoms in the shortest path between the electrodes, the atoms that remain cannot be grouped into neighboring pairs* [29]. This rule has a convenient graphical representation as illustrated in the lower left of Fig. 11.11 of our system in question. We can rule out the atoms in the shortest path, and then we are left with one pair of atoms and one unpaired atom. Therefore, we predict interference.

(2) *If, in a cyclic system, the path length between the two paths differs by* $4m + 2$ *atoms, where* $m = 0, 1, 2, \ldots$ [30].
In our system in question, we have two paths going one way or the other around the benzene ring. *Meta*-substitution results in the path length differing by two atoms, thus indicating a system with interference.

(3) *If, when a star is placed on every second atom in a bipartite system, the electrodes bind to two atoms of the same type (either starred or unstarred)* [14, 31].
The starring for our example is shown in the top left of Fig. 11.11. Whether we designate the sulfur atoms starred or unstarred, we end up with connections to two atoms of the same type, thus indicating interference.

(4) *If the sign of the product of the molecular orbital coefficients of the two binding sites on the highest occupied molecular orbital (HOMO) is the same as the product of the same sites on the lowest unoccupied molecular orbital (LUMO)* [32–34].
This is the first of the approaches that requires some calculation (or at least knowledge of the form of the molecular orbitals for the molecule in question). The orbitals shown in Fig. 11.11 are not the simplest to interpret because they are the orbitals with the terminal sulfur atoms included (not the benzene ring alone). A further difficulty is that if we simply consider benzene, we are confronted with a degenerate pair of orbitals for the HOMO and another degenerate pair for the LUMO. So which combinations do we consider? It turns out that we can see interference between pairs of pairs, but in the case as plotted above, it is not the HOMO with the LUMO, but the HOMO-1 with the LUMO and the HOMO with the LUMO+1.

(5) *If there is an increase in the number of non-spanning nodes in the wave function between the HOMO and the LUMO, where a non-spanning node is defined as a node that does not span the path between the two electrodes* [35]. *This definition can be qualified in the case of cyclic systems to be a node that does not span the shortest path between the electrodes* [36].
Here we can see that the number of nodes in the longer path around the ring increases as we go from HOMO to LUMO.

(6) *If, when two independent parts of the molecular orbitals are constructed for a bipartite system, one component vanishes* [37].
This again requires a calculation, and perhaps a less familiar one that the molecular orbitals. An interested reader should see Ref. [37] where the case of *meta*-substituted benzene is reproduced.

All these strategies need to be used with a degree of caution with large molecules or systems with hetero atoms (N, O, P, etc.), which can shift the position of the interference feature. There will be a significant difference in the transport properties if an interference feature is shifted even 0.5 eV away from the Fermi energy, and this position will vary with the choice of electrode material and the charge transfer between the electrode and the molecule.

Another potential pitfall of methods 3 and 4 is that they will fail for even numbers of units with interference features connected in series, e.g., two *meta*-substituted benzene rings together. In this case, the HOMO and LUMO and the starring will not indicate that there is interference when in fact there is. Similarly, method 2 describes cyclic systems without side groups well but may fail to correctly predict interference effects when a cyclic system has side groups that form part of the conjugated system.

With these caveats in mind, they are all fantastic shortcuts for predicting interference effects in molecules, although our personal preference is method 1 for its simplicity and generality (cyclic or acyclic systems, odd- or even-membered rings, and no calculation required). If different methods give different predictions, we would favor method 1 as the most reliable.

11.6 What is Interfering with What?

So far, we have not addressed the question of how or why this destructive interference occurs in some molecules. In part, the question "what is interfering with what" is necessary if we want to understand why this is observed in only some molecules, if we should also see constructive interference or how we can disrupt/perturb interference effects. The answer is not entirely straightforward for molecules, and the literature has different answers to this question. Here we will explain a suggestion that we find a useful picture for chemical systems.

When we first try to approach this question, an obvious starting point is the descriptions used for interference effects in other types of systems. The classic picture of interference has to be a two-slit experiment. Two slits yield two paths in space for a wave to pass through, and as the waves diffract from the slits, an interference pattern is set up. In cyclic molecules, it is tempting to apply the same picture and argue that different segments of the molecule constitute different paths in space; however, in acyclic molecules, the analogue is not so clear. Alternatively, in acyclic molecules, it is tempting to apply the model of a stub-resonator or some bound state (on a side arm) that interacts with a continuum to explain the interference; this is not so useful for other topologies. When it comes to more complex systems (e.g., cyclophanes), these analogous arguments from physics become even more strained and we must look elsewhere.

Molecules have their own character, and at the single-particle level it is frequently characterized by molecular orbitals. When a molecule is bound in a conducting junction, the orbital character of the isolated system is perturbed but, we would argue, not destroyed, and we can instead look at the eigenvalues of the Green's functions used in transport calculations. We can take a step even further and decompose the transmission into components, which, broadly speaking, come from each orbital. These components are summed and then squared to give us the total transmission. A more rigorous description of this decomposition is given elsewhere [23]. Figure 11.12 shows this decomposition for our example of the *meta*-substituted benzene ring.

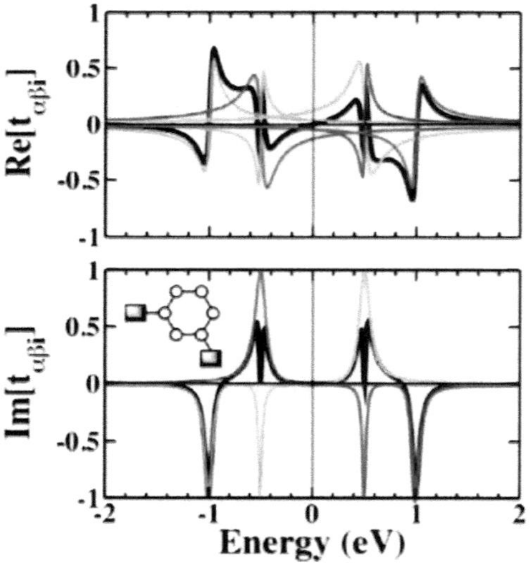

Figure 11.12 Real and imaginary components of the transmission separated into the contributions from each orbital. In this picture, it is clear that, in the middle of the gap, contributions from the HOMO are cancelled by those from the LUMO+1 and contributions from the LUMO are cancelled by those from the HOMO-1. Reprinted with permission from Ref. [23] Copyright (2013), American Institute of Physics.

With this type of decomposition in mind, we would argue that the question of "what is interfering with what" can be answered. We would answer that it is not contributions from different paths through space, as we think of in the case of the two-slit experiment, rather contributions from overlapping paths through space but separated in energy, as this is perhaps a more accurate description of two orbitals (in many cases, the HOMO and the LUMO of the molecule). For many molecules, a decomposition of the type shown in Fig. 11.12 will suggest that an interference feature appears near the Fermi energy because there is interference between contributions coming from the tails of the occupied orbitals (often dominated by the HOMO) and tails from the unoccupied orbitals (often dominated by the LUMO).

11.7 (How) Can We Use Interference Effects?

The applications of interference effects are limited only by our imagination. Essentially, interference effects give us the opportunity to "switch-off" electronic communication between different parts of a molecule. This switching can be temporary or permanent, involve nuclear motion or be purely electronic, and can be tuned to only partially switch of transport if required.

11.7.1 *Switching*

A molecular switch was the first functionality we introduced to motivate our discussion on interference, so now we need to address how this functionality can be achieved. So far we have not shown how interference effects can be perturbed, but the idea that they arise from interference between contributions from differing orbitals gives us a clue of how to proceed.

First, we have to consider what we need to achieve in terms of transmission. Figure 11.13 illustrates some of the possibilities. From our initial system with a sharp interference feature at the Fermi energy, we can either shift the interference feature (using a bias across the molecule or a chemical modification) or we can destroy the interference (most likely through a chemical modification). In

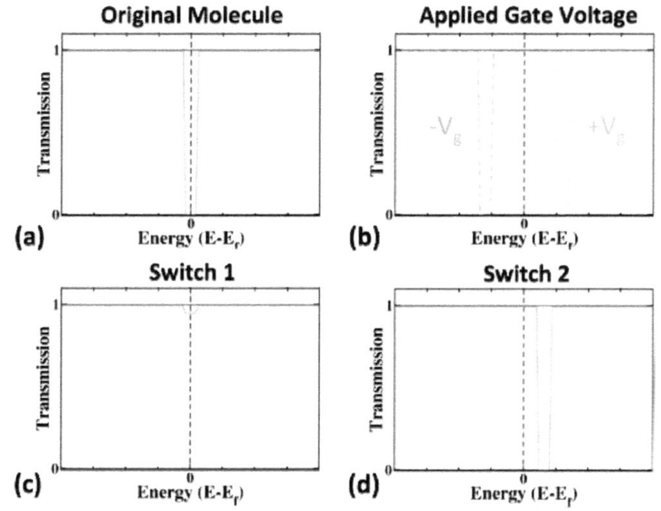

Figure 11.13 Perturbing interference features to create a molecular switch: (a) the original system; (b) perturbed by a gate, the interference can be shifted away from the Fermi energy; (c) chemical modification, the interference feature is removed; (d) chemical modification, the interference feature is shifted. Reprinted with permission from Ref. [1]. Copyright (2013) American Chemical Society.

either of these cases, the transmission around the Fermi energy will increase and our switch will be "on."

Depending on how the switching is achieved, the process can be reversible or irreversible, fast or slow. If fast, reversible switching is required, then we need to avoid nuclear motion and switching via a bias is optimal. Figure 11.13 illustrates this for a gate bias that can shift the interference feature in either direction, but in some cases this is also possible with the source–drain bias [1].

If the speed of the switching event is not a material consideration, or if permanent switching is desirable, e.g., in the case of a single-use sensor, then a chemical reaction may be optimal. In some cases, a large change in the molecule is not required; e.g., Fig. 11.14 illustrates two systems, one dominated by interference and one without interference effects. We can imagine a situation where we switch reversibly or irreversibly between these two systems, starting from either one or the other, depending on the purpose.

Figure 11.14 When this system is connected to two electrodes through the R and R′ substituents, it can be easily modified from a system with interference near the Fermi energy (left) to a system without (right), switching from "off" to "on" in this process.

When we consider the picture we have of these interference effects from our previous discussion, we can say that reactions that change the form of the molecular orbitals, the position of the nodes, and the relative phase on the ends of the system will obviously change whether or not interference is observed anywhere near the Fermi energy. Changes that merely shift the orbitals slightly, but do not change their form, may shift the interference features away from the Fermi energy or not (if multiple orbitals move).

11.7.2 *Chemical Control*

Using a combination of the effects of substituents and bias, it is possible to achieve other types of electrical functionality, e.g., negative differential resistance and rectification [1]. Up to this point, we have largely considered the situation where the interference feature lies very close to the Fermi energy, but there are situations where this is not desirable. The beauty of chemistry is that with the right substituents, we can tune the position interference features as shown in Fig. 11.15.

Clearly, electron-donating and -withdrawing substituents can be used to tune the position of this interference feature with relative ease; however, this acyclic cross-conjugated system is only one of many topologies that exhibit interference effects. For the broader range of systems shown in Fig. 11.10, much work has to be done to see how all these interference effects can be tuned and whether they are amenable to chemical modification in the same fashion. The challenge for chemists is that we need to understand not just how

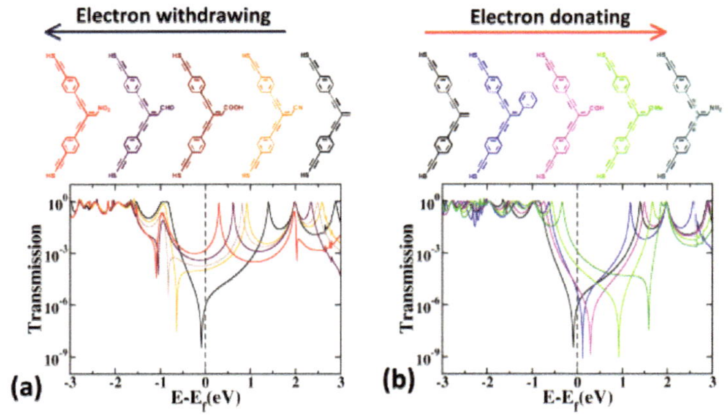

Figure 11.15 Electron-withdrawing or -donating substituents can both shift the position of the interference feature in a cross-conjugated molecule. With a careful choice of substituent, the interference feature can be tuned across a large (over 2 eV) range. Reprinted with permission from Ref. [1]. Copyright (2013) American Chemical Society.

these substituents might tune the zero bias characteristics, but how we can use substituents to control the properties under bias.

11.8 Conclusion

As our understanding of interference effects in molecules grows, so too can our strategies for utilizing them expand. Here we have provided an introduction to some of the basic concepts and the ideas that underlie the research focused on applications, but still much work has to be done. At this stage, our understanding is still relatively limited, both because of the physics we neglect in this coherent single-particle picture and our limited understanding of how to control interference chemically, essentially a lack of insight into the structure–function relationships at work.

What is clear, however, is that these types of systems have a huge potential. The ability to take a fully conjugated molecule, which would generally be a good conductor, and with a small modification turn it into an insulator is a serious asset. With a careful approach, much can be learned from these molecules, and who knows what applications we might find in the process.

Problems

1. Consider the acyclic cross-conjugated molecule from Fig. 11.4 as shown below. Use the predictive approaches 1 and 3 (grouping of atoms not in the shortest path and starring) from 11.5.1 to show that these methods predict interference for this system.

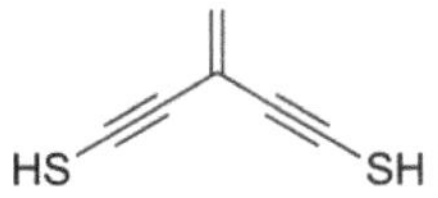

Figure 11.16 A small acyclic cross-conjugated molecule.

2. Consider the cyclic system shown in Fig. 11.17.

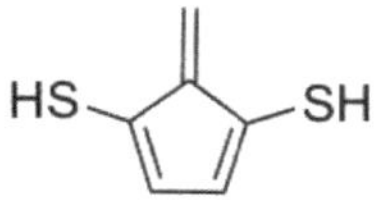

Figure 11.17 A small cyclic molecule.

   (a) Show that predictive method 1 again predicts interference
   (b) Explain why we cannot use method 3 in this case.
   (c) What does method 2 predict for this system? Is it correct to use this method for this molecule?

3. Consider the three molecules shown in Fig. 11.18.

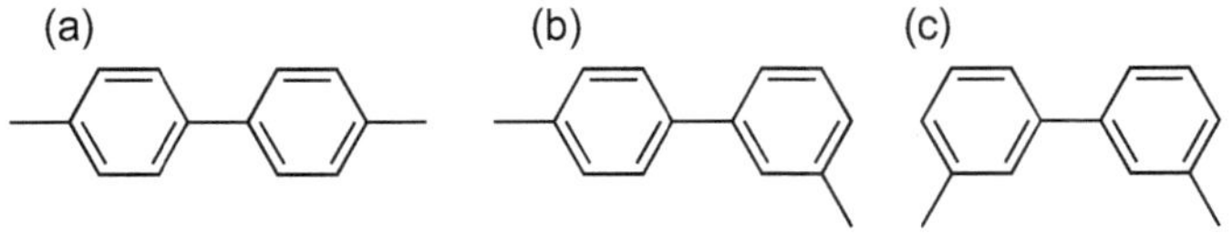

Figure 11.18 Three substitution patterns on a biphenyl system.

(a) Which of these molecules is predicted to have interference using method 1?
(b) What about method 2?
(c) Method 3?

(d) The molecular orbitals for biphenyl are shown below. What does method 4 predict?
(e) Which method should we rely on?

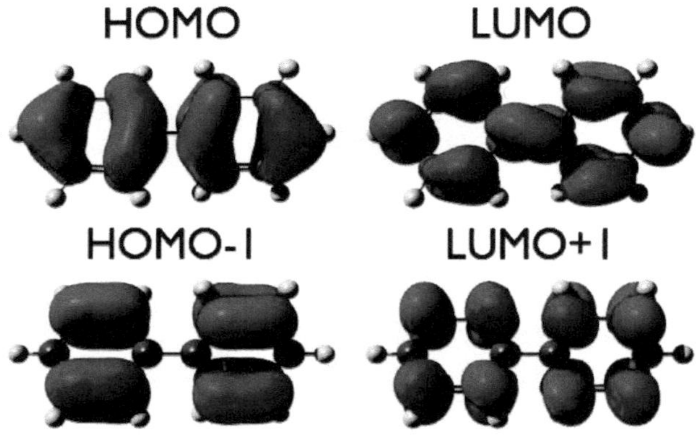

Figure 11.19 Selected molecular orbitals for the biphenyl system.

References

1. Andrews, D.Q., et al., Single molecule electronics: Increasing dynamic range and switching speed using cross-conjugated species. *Journal of the American Chemical Society*, 2008. **130**(51): pp. 17309–17319.
2. Aviram, A., and M.A. Ratner, Molecular rectifiers. *Chemical Physics Letters*, 1974. **29**(2): pp. 277–283.
3. Patoux, C., et al., Topological effects on intramolecular electron transfer via quantum interference. *Inorganic Chemistry,* 1997. **36**(22): pp. 5037–5049.
4. Sautet, P., and C. Joachim, Electronic interference produced by a benzene embedded in a polyacetylene chain. *Chemical Physics Letters*, 1988. **153**(6): pp. 511–516.
5. Joachim, C., J.K. Gimzewski, and A. Aviram, Electronics using hybrid-molecular and mono-molecular devices. *Nature*, 2000. **408**(6812): pp. 541–548.
6. Joachim, C., and M.A. Ratner, Molecular electronics: Some views on transport junctions and beyond. *Proceedings of the National Academy of Sciences*, 2005. **102**(25): pp. 8801–8808.

7. Loppacher, C., et al., Direct determination of the energy required to operate a single molecule switch. *Physical Review Letters*, 2003. **90**(6): p. 066107.
8. Moresco, F., et al., Conformational changes of single molecules induced by scanning tunneling microscopy manipulation: A route to molecular switching. *Physical Review Letters*, 2001. **86**(4): p. 672.
9. Cardamone, D.M., C.A. Stafford, and S. Mazumdar, Controlling quantum transport through a single molecule. *Nano Letters*, 2006. **6**(11): pp. 2422–2426.
10. Hettler, M.H., et al., Current collapse in tunneling transport through benzene. *Physical Review Letters*, 2003. **90**(7): p. 076805.
11. Hush, N.S., et al., Optimization and chemical control of porphyrin-based molecular wires and switches. *Annals of the New York Academy of Sciences*, 1998. **852**(1): pp. 1–21.
12. Ke, S.-H., W. Yang, and H.U. Baranger, Quantum-interference-controlled molecular electronics. *Nano Letters*, 2008. **8**(10): pp. 3257–3261.
13. Markussen, T., J. Schiotz, and K.S. Thygesen, Electrochemical control of quantum interference in anthraquinone-based molecular switches. *The Journal of Chemical Physics*, 2010. **132**(22): p. 224104.
14. Solomon, G.C., et al., Quantum interference in acyclic systems: Conductance of cross-conjugated molecules. *Journal of the American Chemical Society*, 2008. **130**(51): pp. 17301–17308.
15. Stadler, R., et al., Integrating logic functions inside a single molecule. *Nanotechnology*, 2004. **15**(4): pp. S115–S121.
16. Stafford, C.A., D.M. Cardamone, and S. Mazumdar, The quantum interference effect transistor. *Nanotechnology*, 2007. **18**(42): p. 424014.
17. Walter, D., D. Neuhauser, and R. Baer, Quantum interference in polycyclic hydrocarbon molecular wires. *Chemical Physics*, 2004. **299**(1): pp. 139–145.
18. Yaliraki, S.N., and M.A. Ratner, Interplay of topology and chemical stability on the electronic transport of molecular junctions. *Annals of the New York Academy of Sciences*, 2002. **960**(Molecular Electronics II): pp. 153–162.
19. Aradhya, S.V., et al., Dissecting contact mechanics from quantum interference in single-molecule junctions of stilbene derivatives. *Nano Letters*, 2012. **12**(3): pp. 1643–1647.
20. Fracasso, D., et al., Evidence for quantum interference in SAMs of arylethynylene thiolates in tunneling junctions with eutectic GaIn

(EGaIn) top-contacts. *Journal of the American Chemical Society*, 2011. **133**(24): pp. 9556–9563.

21. Kaliginedi, V., et al., Correlations between molecular structure and single-junction conductance: A case study with oligo(phenylene-ethynylene)-type wires. *Journal of the American Chemical Society*, 2012. **134**(11): pp. 5262–5275.
22. Hong, W., et al., An MCBJ case study: The influence of π-conjugation on the single-molecule conductance at a solid/liquid interface. *Beilstein Journal of Nanotechnology*, 2011. **2**: pp. 699–713.
23. Solomon, G.C., et al., Understanding quantum interference in coherent molecular conduction. *The Journal of Chemical Physics*, 2008. **129**(5): pp. 054701–054708.
24. Ernzerhof, M., et al., Electron transmission through aromatic molecules. *Journal of Chemical Theory and Computation*, 2006. **2**(5): pp. 1291–1297.
25. Solomon, G.C., et al., Exploring local currents in molecular junctions. *Nature Chemistry*, 2010. **2**(3): pp. 223–228.
26. Guedon, C.M., et al., Observation of quantum interference in molecular charge transport. *Nature Nanotechnology*, 2012. **7**(5): pp. 305–309.
27. Paulsson, M., and S. Datta, Thermoelectric effect in molecular electronics. *Physical Review B*, 2003. **67**(24): p. 241403.
28. Bergfield, J.P., M.A. Solis, and C.A. Stafford, Giant thermoelectric effect from transmission supernodes. *ACS Nano*, 2010. **4**(9): pp. 5314–5320.
29. Markussen, T., R. Stadler, and K.S. Thygesen, The relation between structure and quantum interference in single molecule junctions. *Nano Letters*, 2010. **10**(10): pp. 4260–4265.
30. Baer, R., and D. Neuhauser, Anti-coherence based molecular electronics: XOR-gate response. *Chemical Physics*, 2002. **281**(2–3): p. 353–362.
31. Hush, N.S., (private communication), 2007.
32. Richardson, D.E., and H. Taube, Electronic interactions in mixed-valence molecules as mediated by organic bridging groups. *Journal of the American Chemical Society*, 1983. **105**(1): pp. 40–51.
33. Siddarth, P., and R.A. Marcus, Calculation of electron-transfer matrix elements of bridged systems using a molecular fragment approach. *Journal of Physical Chemistry*, 1992. **96**(8): pp. 3213–3217.
34. Yoshizawa, K., T. Tada, and A. Staykov, Orbital views of the electron transport in molecular devices. *Journal of the American Chemical Society*, 2008. **130**(29): pp. 9406–9413.

35. Lee, H.W., Generic transmission zeros and in-phase resonances in time-reversal symmetric single channel transport. *Physical Review Letters*, 1999. **82**(11): p. 2358.
36. Andrews, D.Q., (private communication), 2008.
37. Ernzerhof, M., Simple orbital theory for the molecular electrician. *Journal of Chemical Physics*, 2011. **135**(1): p. 014104.

Chapter 12

Parallel Self-Assembly Strategies toward Multiple Single-Molecule Electronic Devices

Kasper Nørgaard and Titoo Jain
Nano-Science Center and Department of Chemistry, University of Copenhagen, Universitetsparken 5 Copenhagen, DK-2000 Copenhagen, Denmark
kn@nano.ku.dk

The construction of molecular devices from the "bottom-up" using self-assembly processes has the potential to overcome a number of shortcomings inherent to current-day lithographic techniques, and may provide a feasible route for the parallel fabrication of multiple molecular devices on a very large scale.

This chapter introduces possible fabrication schemes for molecular junctions through self-assembly processes as well as strategies for incorporating these molecular junctions into functional devices. Finally, the scope for upscaling these concepts into devices consisting of multiple molecular junctions will be discussed.

Handbook of Single-Molecule Electronics
Edited by Kasper Moth-Poulsen

ISBN 978-981-4463-38-6 (Hardcover), 978-981-4463-39-3 (eBook)
www.panstanford.com

12.1 Introduction

Conjugated organic molecules have become ever more important as the active component in electronic devices, both as low-tech, high-market-volume applications, such as organic light-emitting diodes, or as components in nanoscale devices based on a few (or even a single) molecules [1]. The development of nanoscale electronic devices is tied intimately to the parallel development of controlled fabrication schemes for materials and devices with a critical length scale in the nanometer regime [2].

One of the fundamental challenges in nanotechnology, and in the field of molecular electronics in particular, concerns the fabrication of electrode pairs separated by only a few nanometers that can accommodate a single molecule of interest to be optically or electrically characterized when residing in the gap between the electrodes [3]. Nanogap electrodes are applied in such diverse research areas as single-molecule electronics, spectroscopy, and plasmonics [4–6].

The aim of single-molecule electronics is to employ single molecules as electrical components, such as transistors, diodes, wires, or memory elements. This goal necessitates an understanding of how electron transport through molecules is governed by the chemical nature of the molecule to tailor its specific function through chemical synthesis. Candidate molecules are typically deposited from solution or by evaporation to bridge the nanogap between a pair of miniscule electrodes. The molecule–electrode contact is mediated by chemical anchoring groups that ensure a connection between the metal electrode and the molecule in question.

Despite being beyond the capabilities of present-day industrial fabrication techniques, nanogaps are successfully prepared in research laboratories around the world via "top-down" methods such as scanning-probe techniques [7, 8], mechanically controlled break junctions (MCBJ) [9], shadow mask evaporation [10], and electromigration [11]. Common limitations of these techniques include low throughput, extended time consumption, and the requirement for expensive and complicated experimental setups. An additional challenge regarding these top-down approaches is that molecules may bridge the nanogap differently with each experimental realization due to the random placement of the molecule, leading to variations at the metal–molecule interface, and potentially resulting in large differences in the observed device properties.

In contrast, the construction of nanogaps via "bottom-up" processes have the potential to be more cost effective as well as providing parallel fabrication schemes for the production of multiple devices on the molar scale (i.e., $\approx 10^{23}$) in just a few synthetic steps [12, 13]. In particular, self-assembly methods can potentially overcome a number of the shortcomings inherent to current lithographic techniques. An attractive scenario for device fabrication would, therefore, be a combination of top-down approaches (lithography) and bottom-up self-assembly, where the microstructured scaffold and interface to the outside world are produced by conventional lithographic techniques, while the nanostructured functional units are self-assembled at designated sites of the device.

12.1.1 *Self-Assembly of Nanomaterials*

For more than a century, synthetic organic chemists have perfected the rules that govern the fabrication and transformation of covalent molecular species. The level of understanding has presently reached a stage where even very complex organic target molecules can be synthesized (see Chapter 8). Unlike this established branch of chemistry, the field of molecular self-assembly is still in its early stages, as the ability to design and arrange individual molecular building blocks into predefined and well-ordered architectures is still relatively limited. These limitations are imposed by the spatial complexity and relative weakness of the intermolecular interactions that determine the supramolecular structure: electrostatic forces, hydrogen bonding, π–π stacking, van der Waals interactions, and the hydrophobic effect. The strength of these interactions is usually only a small fraction of that of a covalent bond (i.e., $\approx$ 350 kJ mol^{-1}), and supramolecular species are, therefore, generally more labile, adaptive, and flexible but less stable than their molecular constituents [14].

Self-assembled systems spontaneously form extensive ordered molecular arrays, consisting of relatively simple molecular building blocks held together by an appropriate balance between intermolecular noncovalent forces and the surrounding media. The self-assembly process thus relies on mutual recognition between complementary supramolecular building blocks through noncovalent interactions. Self-assembly is ubiquitous in nature; cell-membrane, double-stranded DNA helix, and the folding of polypeptides are prominent examples. While the self-assembly of artificial functional molecular systems still remains a challenge, supramolecular chemistry has often been proclaimed to hold the key to nanoscale engineering.

12.2 Self-Assembly on Surfaces

The most commonly studied systems for molecular electronics are based on self-assembled monolayers (SAMs), typically composed of various thiol-terminated molecules that are chemisorbed onto a gold surface (Fig. 12.1) [15, 16]. A general introduction to SAMs is given in Chapter 7.

Compared to single-molecule junctions, SAMs offer a more reproducible way for the fabrication of molecular electronic devices, more amenable to upscaling and integration into circuits. SAMs can be fabricated over relatively large areas with high reproducibility and would thus be ideal candidates for metal–molecule–metal sandwich junctions. However, direct evaporation of metallic top electrodes onto SAMs has proven to be very invasive and damaging

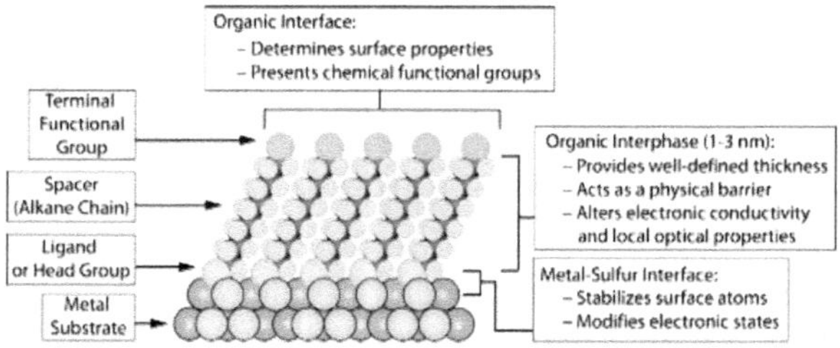

Figure 12.1 Schematic illustration of a self-assembled monolayer of alkanethiolates on a gold surface. Reprinted with permission from Ref. [16]. Copyright (2005) American Chemical Society.

to the molecular layer, where filamentary paths of metals or damages to the molecules can very easily cause electrical short circuits [17, 18].

To address this problem, a number of "soft" contacting methods have been developed, including the mercury-drop electrode [19], eutectic Gallium-Indium (EGaIn) [20], nanotransfer printing (nTP) [21], and surface-diffusion-mediated metal deposition [22].

A new method was recently developed in Copenhagen, where a very thin layer (5–7 nm) of reduced graphene oxide (rGO) was placed on top of the SAM, serving as a protective interlayer that shields the monolayer from the damaging effects of evaporating a top electrode [23]. Due to the high conductivity of rGO, this layer could even function as an electrode, obviating the need for an additional top electrode. Furthermore, since the rGO layer is highly transparent, it was shown that a monolayer device consisting of a molecular photoswitch could be converted between two states of different conductances by irradiating through the rGO electrode layer (Fig. 12.2) [24].

Using a similar approach, but employing a conducting polymer poly(3,4-ethylene-dioxythiophene) stabilized with polystyrene sulphonic acid (PEDOT:PSS) as the soft-contact barrier, a Dutch group

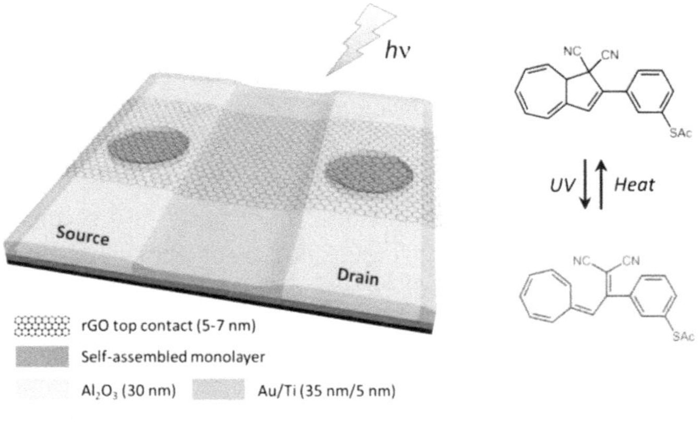

Figure 12.2 Schematic illustration of a molecular test-bed with rGO thin film as the transparent top electrode, and the molecular structure of the photoswitch. Charge transport measurements are made through adjacent micropore junctions in series. From Ref. [24].

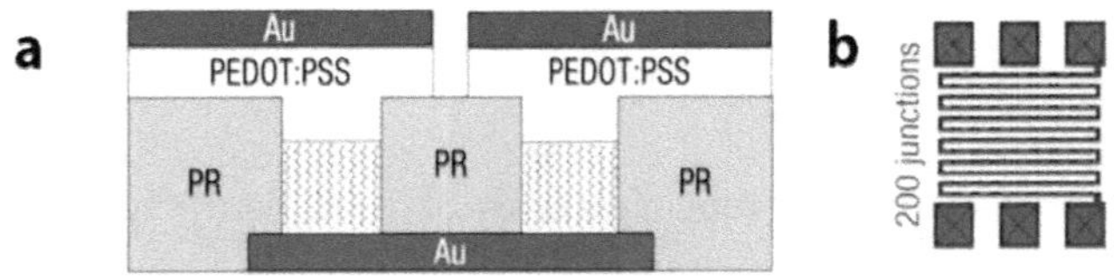

Figure 12.3 (a) Schematic cross section of two connected molecular junctions. (b) Layout of strings with up to 200 molecular junctions in series. Adapted with permission from Macmillan Publishers Ltd: Nature Nanotechnology [25], copyright 2008.

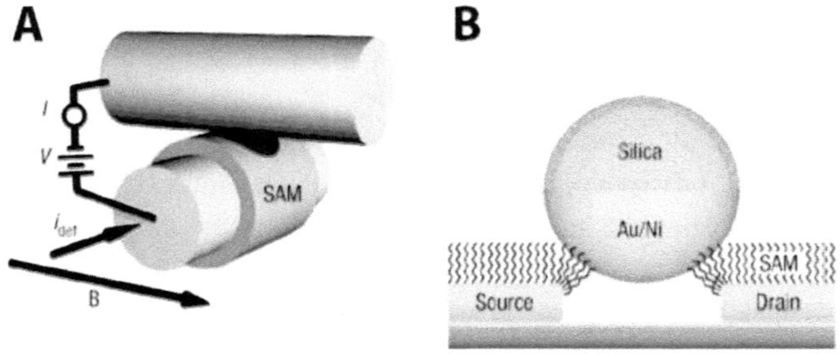

Figure 12.4 Schematic illustration of (a) crossed-wire tunneling junction, and (b) magnetic bead junction. Adapted with permission from Macmillan Publishers Ltd: Nature Materials [29], copyright 2005.

demonstrated that individual SAM junctions could be integrated in series of up to 200 with a yield of unity (Fig. 12.3) [25].

Kushmerick et al. [26] developed an elegant way to sandwich approximately 1000 molecules in a molecular junction. An SAM of the molecules was prepared on a 10 μm diameter Au wire. The SAM-covered wire was brought into the vicinity of a second wire in a crossed-wire geometry (Fig. 12.4a). A molecular junction was formed by applying a small DC current across the first wire in a magnetic field, causing the wire to deflect. Upon touching the other wire, the I-V characteristics of the molecular junction could be measured.

Amlani et al. [27] combined conventional photolithographic techniques with self-assembly aspects to form a metal–SAM–metal–SAM–metal junction (Fig. 12.4b). Two Au electrodes, with a separation of 40–100 nm, were covered with SAMs, and an Au nanoparticle ($d = 40$–100 nm) was trapped in the gap between the electrodes by applying an AC bias.

A conceptually similar approach was developed by Kushmerick et al. [28], who used Au-covered magnetic particles to undergo directed assembly into lithographically defined magnetic arrays functionalized with SAMs. The magnetic particles were made by depositing Ni and Au onto a silica microsphere.

12.2.1 *Self-Assembly at the Air–Water Interface*

Compared to SAMs on gold, the interface between air and water offers some advantages as a model system for studies of two-dimensional self-assembly processes, as it forms a well-defined, ideally flat, and homogeneous experimental situation [14].

Langmuir films are monomolecular layers of surface-active molecules (surfactants) floating at the interface between a gas and a liquid phase, typically between air and water. Such monolayers are thermodynamically stable due to intermolecular forces between the molecules in the layer as well as through a balance between hydrophilic and hydrophobic forces, which tend to respectively maximize and minimize the contact area between the water and specific parts of the molecules. Molecules that consist of both solvent-attracting and solvent-repelling moieties are generally termed amphiphilic.

The Langmuir technique is named after the 1932 Nobel laureate Irwin Langmuir, who was among the first to perform systematic studies on the physical–chemical properties of insoluble monolayers of organic surfactants on water surfaces. He introduced the term surface pressure, Π, a two-dimensional equivalent to pressure, defined as the difference between the surface tension of pure solvent, γ_0, and the experimentally measured surface tension of solvent-including surfactant, γ (i.e., $\Pi \equiv \gamma_0 - \gamma$).

When simple amphiphilic surfactants (e.g., long-chain fatty acids) are spread on the water surface of a trough, they form monolayers in which the hydrophilic carboxylic acid moieties are submersed in the water phase, resulting in an overall orientation of the molecules parallel to the surface normal. Using a set of movable barriers in the trough, the accessible surface area of the

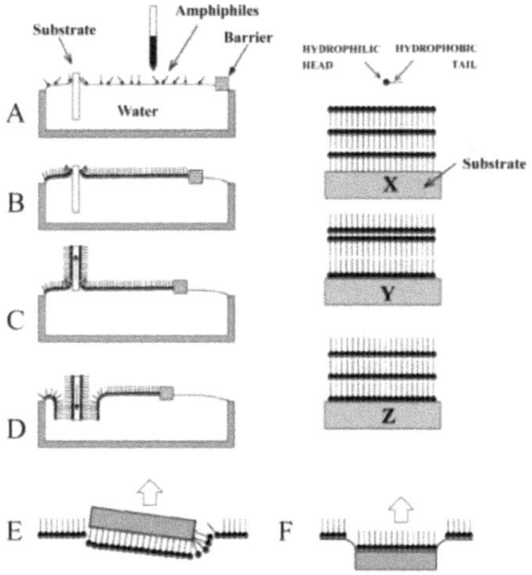

Figure 12.5 The Langmuir–Blodgett technique. Left: (A) deposition of amphiphiles on the air–water interface (with a solid substrate submerged); (B) compression of the monolayer by the barrier; (C) vertical transfer of the monolayer by pulling out the substrate; (D) multilayer build-up by repeated up and down strokes of the substrate; (E) monolayer transfer by horizontal dipping; and (F) horizontal lifting. Right: examples of different types of multilayer structures (X, Y, and Z), which can be built by the Langmuir–Blodgett technique. From Ref. [14]. Reproduced with permission from The Royal Society of Chemistry.

surfactants can be controlled, which in combination with surface pressure measurements allows the pressure–area relationship of a given surfactant to be determined. Langmuir's coworker Katherine Blodgett later discovered that consecutive monolayers of surfactants could be transferred onto a solid substrate by repeatedly dipping the substrate vertically through the water surface while maintaining a constant surface pressure. This method for multilayer fabrication has since been known as the Langmuir–Blodgett technique, which is illustrated schematically in Fig. 12.5. The produced multilayer films can have different molecular orientations of each layer, depending on the exact details of the dipping process.

Using the Langmuir–Blodgett technique, researchers from Purdue [30], Saclay [31], and later Copenhagen [32] and Basel [33] assembled alkanethiol-stabilized gold nanoparticles (AuNPs) into

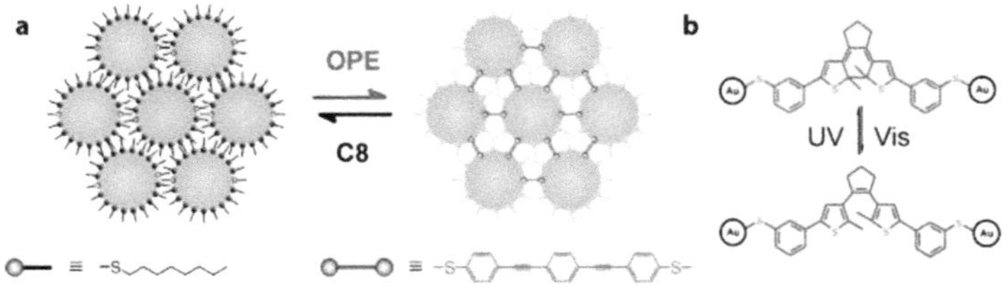

Figure 12.6 (a) The preparation of molecular junction networks using ligand- protected metallic nanoparticles. A subsequent in situ ligand-exchange reaction permits the reversible interlinking of neighboring nanoparticles [33]. Copyright (2006) Wiley-VCH Verlag GmbH and Co. (b) Reversible, light-controlled conductance switching of such molecular devices linked with photochromic diarylethene molecules was demonstrated. Adapted with permission from Ref. [34]. Copyright (2009) American Chemical Society.

two-dimensional hexagonal arrays at the air–water interface. The interparticle distance could be tuned by varying the length of the ligands [33]. After transfer to a solid support, the monolayer could be functionalized by adding double-ended organic wires [dithiolated oligo(phenylene ethynylene)s, OPEs, Fig. 12.6], which would replace some of the alkanethiols in the AuNP ligand shell by a place-exchange reaction. The two anchoring groups on the OPEs served to interlink neighboring AuNPs, thus forming a network of molecular junctions. In a later study [34], a diarylethene photoswitch was used in place of the OPE molecules (Fig. 12.6b), which resulted in devices that exhibited reversible, light-controlled conductance switching.

In an attempt to break up the hexagonal lattice of the AuNPs, the use of nanopatterning via phase-separation phenomena was investigated [35]. Alkanethiol-stabilized AuNPs were co-deposited with a phospholipid (dipalmitoylphosphatidylcholine, DPPC) at the air–water interface at various temperatures and surface compositions. TEM inspection revealed that the nanoparticles sintered into continuous quasi one-dimensional maze-like wire structures as shown in Fig. 12.7a. The width of the wires corresponded to the diameter of one AuNP (<3 nm), whereas the length extended over several tens of nanometers. The gaps between adjacent wire segments could be tuned by changing the length of the alkanethiolate ligands on the nanoparticles and thus varied between 1.5 nm and 3

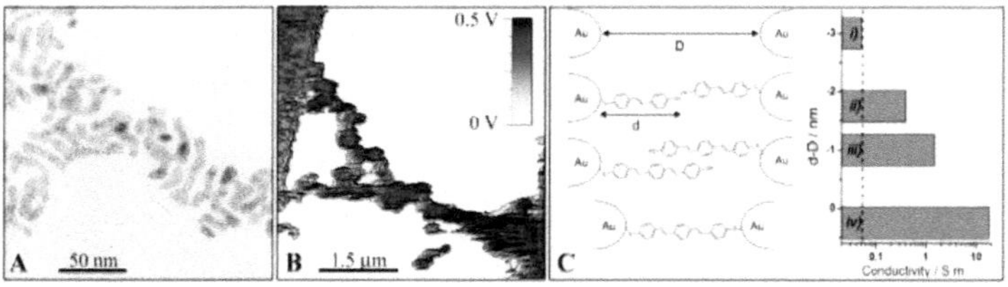

Figure 12.7 A network of OPV-linked gold nanowire segments, confined in a two-dimensional lipid matrix. (A) High-resolution TEM micrograph of gold nanowire segments. (B) Kelvin-force microscope image. A potential of 0.5 V is applied between two prefabricated gold electrodes, which are located at the top-left and bottom-right corners of the image. (C) Left: Schematic illustration of the average gap size (D) between neighboring gold-wire segments, compared to the length (d) of the OPV molecule for four different geometries. Right: Measured conductivities vs.d-D. The vertical dashed line marks the detection limit of the experimental setup. Adapted with permission from [36]. Copyright 2004 American Chemical Society.

nm. The individual wire segments could be interconnected using the molecular wire oligo-*p*-phenyenevinylene (OPV). When transferred to a substrate with pre-patterned macroscopic electrodes, a bias voltage could be applied across the network (Fig. 12.7b). The conductance of the network without the OPV molecules was lower than 0.05 Sm^{-1}. This value increased more than 350 times for the OPV-interlinked networks [36].

12.3 Solution Assembly of Molecular Nanogaps

The previous section introduced a number of approaches for fabricating permanent and scalable molecular electronic devices based on SAMs or Langmuir–Blodgett films. The self-assembled functional units could be inserted into pre-patterned electrode systems to generate hybrid self-assembled devices with the possibility to integrate multiple molecular junctions into the same chip. Although the dimensionality of the systems can be reduced, e.g., by utilizing phase-separation behavior, these devices address an ensemble of molecules, for which certain averaged values for single molecules can be deduced.

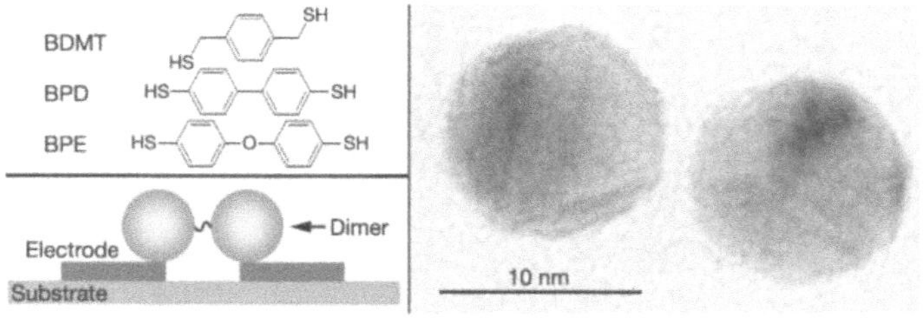

Figure 12.8 Self-assembled nanoparticle dimers, linked by a single molecule. The dimers are electrostatically trapped in a lithographically defined nanogap. Reprinted with permission from Macmillan Publishers Ltd: Nature [37], copyright (2005).

A complimentary strategy for fabricating multiples of self-assembled molecular devices that approach the single-molecule level is to synthesize the entire metal–molecule–metal junction in solution. This approach aims to increase the reproducibility of the individual junctions compared to the top-down approaches discussed above and, in this way, to afford a large number of identical copies of single-molecule devices in solution before depositing the structures onto a solid support for electrical characterization. The metallic "electrode" component of such self-assembled junctions could be provided by metallic nanoparticles or nanorods.

Dadosh et al. [37] explored solution-based dimerization of 10 nm AuNP with dithiol-endcapped conjugated molecules and used an electrostatic trapping technique to place them in lithographically defined electrode gaps, which permitted electrical characterization of the molecules (Fig. 12.8).

The use of gold nanorods (AuNRs) as metallic electrodes for self-assembled single-molecule devices has been pursued extensively in Copenhagen [12].

The synthesis of AuNRs follows a solution-based seed-mediated growth approach introduced by Murphy and Jana [38, 39]. Spherical AuNP seeds, a few nanometers in diameter, are subjected to an aqueous growth solution favoring anisotropic growth. The growth solution typically consists of the surfactant cetyltrimethylammonium bromide (CTAB), a gold salt ($HAuCl4$), and ascorbic acid as a reducing agent. Silver nitrate may be added to the growth solution to yield more monodisperse nanorods. The mechanism for AuNR

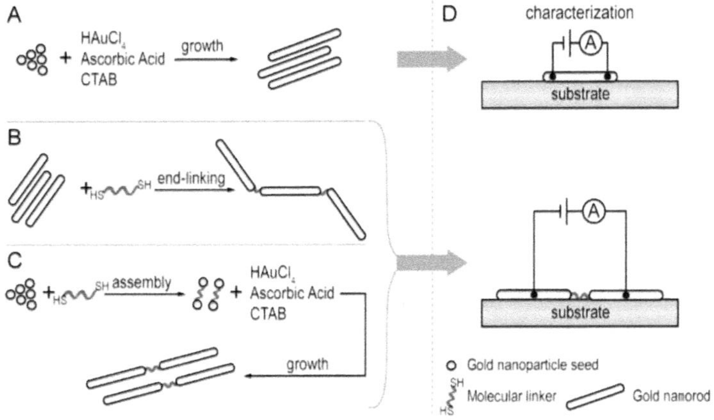

Figure 12.9 Schematic illustrations of AuNR synthesis and strategies for end-to-end assembly. (A) Gold nanoparticle seeds subjected to growth conditions favoring anisotropic growth of nanorods. (B) End-to-end linking of pre-grown AuNRs via reaction with a molecular linker. (C) Assembly of the seed precursors prior to growth resulting in end-linked structures. (D) Deposition of AuNR assemblies on substrates for optical and electrical characterization. Adapted from Ref. [12].

growth has been debated, but it is generally agreed upon that CTAB preferentially adsorbs to specific crystallographic facets of the developing seeds. CTAB has a larger affinity for the {110} and {100} side-facets of the evolving AuNRs, causing gold reduction to occur predominantly on the accessible {111} end-facets, which leads to anisotropic crystal growth (Fig. 12.9a). CTAB covers the resulting AuNRs in the form of a bilayer-oriented normal to the surface; the surfactant head group of the innerlayer interacts with the Au surface, and the outer layer connects through hydrophobic interactions between the lipophilic parts of the CTAB molecules leaving the head group of the outer layer exposed to the surrounding environment [40].

The dimensions of the AuNRs are tunable, depending on the exact reaction conditions, and can be prepared in sizes of ~50 × 10 nm (length × diameter) up to ~600 × 20 nm. The AuNRs possess unique optical properties due to their inherent anisotropy. Coherent oscillations of delocalized surface electrons (plasmons) can be induced upon interaction with light. When in resonance, these surface plasmons oscillate along both the long- and short-axis of the AuNRs giving rise to distinct absorption bands that can vary from the visible to the near-infrared part of the spectrum by tuning the size of the AuNRs. Plasmon excitation in nanorods leads to an amplified electromagnetic field due to charge up-concentration at the metal–dielectric interface. This amplification is in orders of magnitudes larger in nanogaps between dimers of metal NRs and can be exploited to study single molecules via the effect known as surface-enhanced Raman scattering (SERS). For detailed information on AuNR synthesis, growth mechanisms, and optical properties, the reader is referred to recent reviews [41–43].

Directed assembly of AuNRs has been realized using a variety of different strategies into one-, two-, or three-dimensional superstructures [44, 45]. Much research has focused on the assembly of low-aspect-ratio (<5) AuNRs by adding molecular linkers to pre-grown AuNRs (Fig. 12.9b). In the work of Thomas et al., both α,ω-alkanedithiols and aromatic dithiols were used to covalently assemble AuNRs into nanochains via Au–S bonds and study the effects of plasmonic coupling between AuNRs in close proximity [46]. The group also investigated the effect of SERS on the nanorod edges and in dimer junctions, and showed large SERS enhancement of molecules in the nanometer-sized junctions compared to molecules residing at the edges of the AuNRs [47].

Zhong et al. utilized bifunctional polyethylene glycol molecules to control the AuNR orientation by selectively binding to either the side- or the end-facets. In this way, AuNRs were selectively controlled into end-to-end, end-to-side, or side-to-side configurations with the end-to-end oriented AuNRs exhibiting the largest SERS enhancement compared to the other configurations [48].

In 2009, the group in Copenhagen developed a new approach for AuNR end-to-end linking [13]. Instead of assembling the AuNRs after growth (Fig. 12.9b), the group demonstrated that end-to-end-linked AuNRs could be grown directly from chemically linked precursor seeds (Fig. 12.9c). Linking citrate-stabilized AuNP seeds using water-soluble dithiol-functionalized polyethylene glycol (HS-PEG-SH), and then exposing the assembled seed dimers to growth conditions favoring anisotropic growth, would result in dimeric

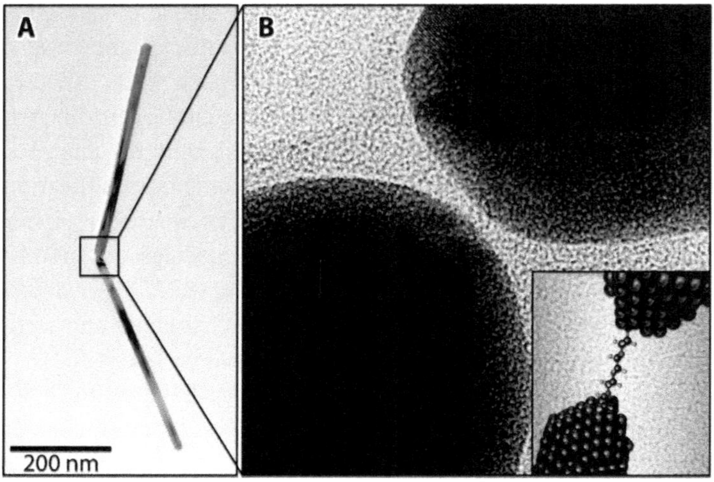

Figure 12.10 TEM images of (A) a gold nanorod dimer grown following the scheme in Fig. 12.9C, (B) the nanogap between the nanorods. Inset: Illustration of the linker molecule bridging the gap. Adapted with permission from Ref. [13]. Copyright (2009) American Chemical Society.

nanorods (Fig. 12.10) with a 1–3 nm gap, a size ideally suited for the insertion of molecular electronic components.

The PEG linker is electrically insulating (due to the nature of the polyethylene glycol backbone) and is, thus, relatively uninteresting as a component for molecular electronics. However, solution deposition of conducting molecular wires into the gap of surface-immobilized AuNR dimers is an alternative option. The AuNRs grown in the study were approximately 500 nm in length, making them well suited for contacting external macroscopic electrodes (Fig. 12.11).

Following the direct growth of AuNR dimers linked with a single molecule, the next step in producing molecular electronics would be to, instead, incorporate into the nanogap a π-conjugated molecule with interesting electrical properties. Thiol-endcapped OPEs and OPVs show exciting electrical properties [1] but are typically not soluble under the aqueous reaction conditions required for AuNR growth. Instead, the possibility to encapsulate these molecules in cyclodextrins (CDs), to make them water soluble and, hence, compatible with the presented strategy, has been investigated [49].

Figure 12.11 Dimeric AuNR structure connected with lithographically defined electrodes. Adapted from Ref. [12].

Fluorescence-titration experiments showed that OPV and CD form a water-soluble inclusion complex with a 1:3 stoichiometry (see Fig. 12.12).

As a water-soluble host, CDs serve the purpose of bringing the OPV molecule into aqueous solution, making it accessible for gold-seed attachment followed by growth into AuNRs (Fig. 12.12). In addition, the CD-host provides a structurally and dielectrically well-defined environment. Structurally, the OPV:CD inclusion complex will be preorganized in a way likely to limit the number of possible binding motifs between sulfur and gold. This confinement may provide a more reproducible molecule–metal interface compared to molecules randomly deposited into a prefabricated nanogap.

A complementary approach was explored by Wijaya and Hamad-Schifferly [50], who developed a customizable ligand-exchange procedure to replace the CTAB layer on the AuNRs with alkanethiols, rendering the nanorod hydrophobic and thus soluble in organic solvents. The group proceeded to functionalize the ligand-exchanged AuNRs with a variety of different molecules.

Considerable recent attention is given to interfacing biological systems with colloidal nanoparticles and electronics [51], and end linking of AuNRs can also be realized using biorecognition systems. Recently, a bioconjugation approach was developed to covalently end-link AuNRs of different aspect ratios via dicysteine-terminated oligopeptides of various lengths (Fig. 12.13), thereby controlling the gap sizes within a small range [52].

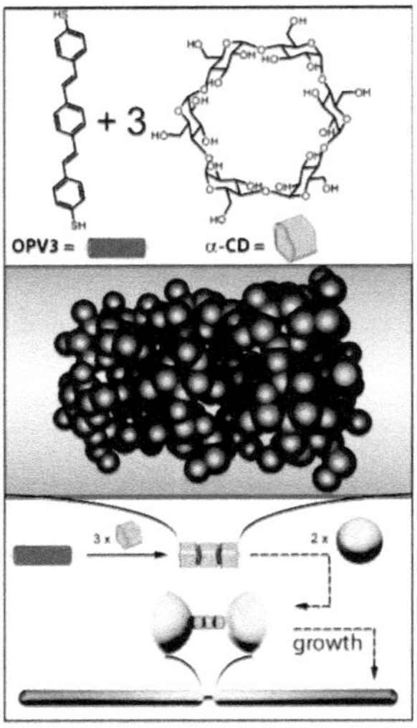

Figure 12.12 Strategy for incorporating π-conjugated molecular wires into the AuNR junctions prepared in aqueous solution. (A, B) Thiol-endcapped OPV3 forms a water-soluble inclusion complex with α-CDs with a 1:3 stoichiometry. (C) Strategy for attaching gold seeds to the inclusion complex via Au–S bonding and growing into AuNRs. Adapted with permission from Ref. [49]. Copyright (2010) American Chemical Society.

The development of an oligopepide-linking platform is motivated by the high modularity of peptides. This feature enables the possibility to integrate diverse functionality into molecular nanogap junctions. A propargyl-substituted glycine was incorporated into the oligopeptide backbone, which is uniquely identifiable using SERS (Fig. 12.13F). Introduction of this acetylene functionality does not disrupt the linking process, and it is an ideal chemical handle that may be utilized further to attach novel target molecules directly in the gaps via, e.g., copper-catalyzed click chemistry [53].

For device integration, AuNRs must be deposited or grown from solution onto suitable planar surfaces preferably with controlled positions and orientations of the individual AuNRs. Directional assembly of AuNRs can be obtained, e.g., by stretching AuNR-embedded poly(vinyl alcohol) polymer films [54], via electric field-assisted deposition between lithographically defined electrodes or electrospinning [55, 56]. Both positional and directional control

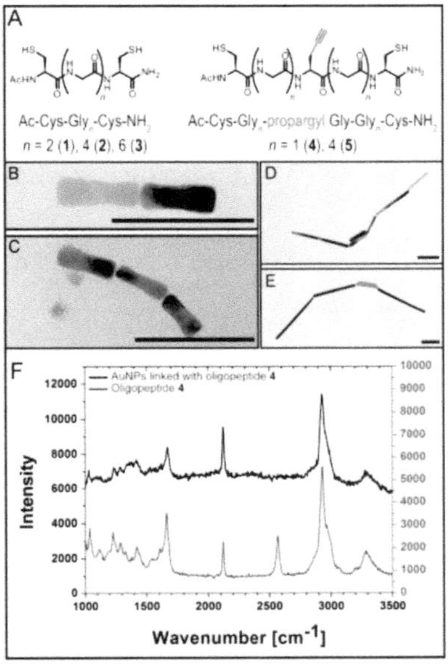

Figure 12.13 Controlled end-to-end assembly of AuNRs using dicysteine-terminated oligopeptides. (A) Chemical structures of (1–3) simple oligopeptides and (4–5) peptides with an incorporated acetylene moiety (red). (B, C) AuNRs with aspect ratios of 3.6 and (D, E) 18 assembled with oligopeptide 1. Scale bars are 100 nm. (F) Raman spectrum of oligopeptide 4 showing the resonance from the acetylene functionality at 2122 cm^{-1} and the S-H band from cysteine at 2566 cm^{-1}, and SERS spectrum of Au structures linked with oligopeptide 4 showing the acetylene mode. The loss of the S-H band indicates that the oligopeptide links the nanoparticles with Au–S bonds through its cysteine residues. Adapted from Ref. [52] with permission from Elsevier.

at the single AuNR level were recently demonstrated employing capillary forces [57].

All the above methods for placing AuNRs onto surfaces require the deposition of pre-synthesized AuNRs from solution. An interesting alternative to this approach is to grow AuNRs directly on substrates [58, 59].

Recently, a high degree of alignment was obtained by assembling seeds in developed guiding channels, tens of nanometers wide, in a spin-coated layer of poly(methyl methacrylate) (PMMA) on a silicon substrate. The walls of the PMMA channels restricted the growth direction of the evolving AuNRs, resulting in parallel aligned AuNRs on a large scale [60]. An additional handle for the positional control of the AuNRs was added by pre-patterning the substrate with lithographically defined electrodes, placed beneath and perpendicularly to the PMMA guiding channels (Fig. 12.14A–D).

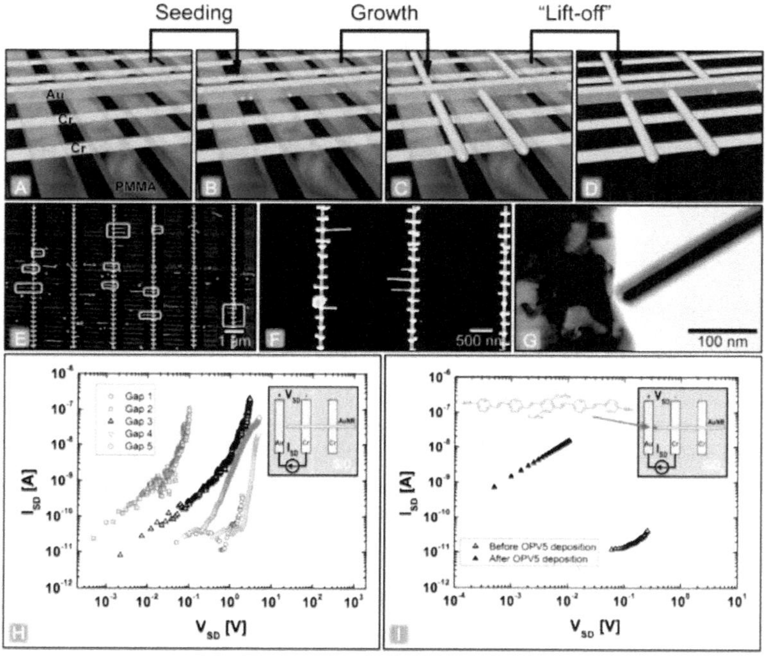

Figure 12.14 Aligned growth of AuNRs in PMMA channels and parallel fabrication of nanogaps. (A) Prefabricated chip containing five parallel electrodes (one Au wire surrounded by two Cr wires on each side) with perpendicular channels developed in PMMA. (B–D) Seeding of the chip by immersion into a seed solution, followed by growth and removal of the PMMA via "lift-off" in acetone. (E–F) AuNRs protruding from the main Au wire bridging two Cr electrodes (yellow markers). (G) TEM image of a typical ~1 nm gap formed between the central Au electrode and an AuNR. (H) Electrical characterization of five different gaps measured in the configuration given in the inset image. (I) I-V characteristics of a nanogap before (open markers) and after (solid markers) incubating the chip in a 1 mM solution of OPV5. The observed change in resistance of about four orders of magnitude suggests a successful insertion of OPV5 in the nanogap. Reprinted from Ref. [12].

The seeds were found to primarily assemble in the channels near the central Au wire, resulting in AuNRs protruding from the Au wire and bridging the two surrounding Cr electrode wires (Fig. 12.14E,F). A TEM study of the interface between the central Au electrode and the AuNR revealed the presence of nanogaps with an average size of 2.4 $\pm$ 1.5 nm (Fig. 12.14G), and this interface was further characterized through electrical measurements (Fig. 12.14H). The tunneling resistances of the nanogaps were typically measured to be in the order of 1–10 GΩ, which is expected for 1–2 nm wide gaps. In particular, the chemically prepared nanogaps were preliminary tested as a platform for molecular devices by depositing an OPV5 derivative from solution. The device displayed a change in resistance of almost four orders of magnitude (from 10 GΩ to 500 kΩ, Fig. 12.14I), which may be attributed to the OPV5 bridging the junction between the gold electrode and AuNR. The exact gap size may be tailored by attaching conjugated molecules to the central Au electrode prior to the addition of seed precursors and growth. Not only would the target molecule determine the gap size ensuring a perfect match between the molecule and gap length, but it would also circumvent random solution-based molecular deposition.

12.4 Conclusion

This chapter has described recent strategies in the preparation of soluble self-assembled nanogap electrodes, bridged by functional organic molecules, and their incorporation into solid-state systems. The successful realization of molecule-based electronics as a potential successor to modern-day microelectronics will require an exceptional level of understanding of the fundamental properties of individual and ensembles of molecules, and their integration into nanogap electrode systems.

The emerging insight obtained from single-molecule measurements, combined with the progress in the design and preparation of self-assembled nanoelectrodes, may eventually lead to the realization of actual electronic circuitry, fabricated through wet chemical synthesis from the "bottom-up."

Problems

1. The molecule octadecanoic acid [stearic acid, $CH_3(CH_2)_{16}COOH$] is an amphiphile and forms a monolayer on water. (a) Sketch the orientation of the molecule on the water surface. (b) Assuming a mean molecular area of 20 Å^2 and a density of 0.85 g/cm^3, what volume of stearic acid is required to cover a small pond with an area of 1000 m^2? (c) Estimate the length of a single stearic acid molecule.
2. Estimate the number of organic ligands in a dense self-assembled thiol ligand shell on a spherical nanoparticle with a diameter of 2 nm. Why is the real number of ligand molecules likely to be a bit higher? (The projected area of a thiol group is 21.4 Å^2.)
3. Examine the diarylethene molecule depicted in Fig. 12.6b. Which of the two forms of the molecule is expected to correspond to the most highly conducting molecular device and why?

References

1. Moth-Poulsen, K., and Bjørnholm, T. (2009). Molecular electronics with single molecules in solid-state devices. *Nat. Nanotech.*, **4**, pp. 551–556.
2. Heath, J.R. (1999). Nanoscale materials. *Acc. Chem. Res.*, **32**, p. 388.
3. Li, T., Hu, W., and Zhu D. (2010). Nanogap electrodes. *Adv. Mater.*, **22**, pp. 286–300.
4. Song, H., Reed, M.A., and Lee, T. (2011). Single molecule electronic devices. *Adv. Mater.*, **23**, pp. 1583–1608.
5. Mayor, M. (2009). Watching the gap close. *Angew. Chem., Int. Ed.*, **48**, pp. 5583–5585.
6. Pelton, M., Aizpurua, J., and Bryant, G. (2008). Metal-nanoparticle plasmonics. *Laser Photonics Rev.*, **2**, pp. 136–159.
7. Wei, Z., Li, T., Jennum, K., Santella, M., Bovet, N., Hu, W., Nielsen, M.B., Bjørnholm, T., Solomon, G.C., Laursen, B.W., and Nørgaard, K. (2012). Molecular junctions based on SAMs of cruciform oligo(phenylene ethynylene)s. *Langmuir*, **28**, pp. 4016–4023.

8. Xu, B.Q., and Tao, N.J. (2003). Measurement of single-molecule resistance by repeated formation of molecular junctions. *Science*, **301**, pp. 1221–1223.

9. Reichert, J., Ochs, R., Beckmann, D., Weber, H.B., Mayor, M., and von Lohneysen, H. (2002). Driving current through single organic molecules. *Phys. Rev. Lett.*, **88**, p. 176804.

10. Kubatkin, S., Danilov, A., Hjort, M., Cornil, J., Bredas, J.-L., Stuhr-Hansen, N., Hedegård, P., and Bjørnholm, T. (2003). Single-electron transistor of a single organic molecule with access to several redox states. *Nature*, **425**, pp. 698–701.

11. Zant, H.S.J., van der Kervennic, Y.-V., Poot, M., O'Neill, K., Groot, Z., de Thijssen, J.M., Heersche, H.B., Stuhr-Hansen, N., Bjørnholm, T., Vanmaekelbergh, D., Walree, C.A., and van Jenneskens, L.W. (2006). Molecular three-terminal devices: Fabrication and measurements. *Faraday Discuss.*, **131**, pp. 347–356.

12. Jain, T., Tang, Q., Bjørnholm, T., and Nørgaard, K. (2014). Wet chemical synthesis of soluble gold nanogaps. *Acc. Chem. Res.* **47**, pp. 2–11.

13. Jain, T., Westerlund, F., Johnson, E., Moth-Poulsen, K., and Bjørnholm, T. (2009). Self-assembled nanogaps via seed-mediated growth of end-to-end linked gold nanorods. *ACS Nano*, **3**, pp. 828–834.

14. Nørgaard, K., and Bjørnholm, T. (2005). Supramolecular chemistry on water: towards self-assembling molecular electronic circuitry. *Chem. Commun.*, pp. 1812–1823.

15. Porter, M.D., Bright, T.B., Allara, D.L., and Chidsey, C.E.D. (1987). Spontaneously organized molecular assemblies. 4. Structural characterization of normal-alkyl thiol monolayers on gold by optical ellopsometry, infrared-spectroscopy, and electrochemistry. *J. Am. Chem. Soc.*, **109**, pp. 3559–3568.

16. Love, J.C., Estroff, L.A., Kriebel, J.K., Nuzzo, R.G., and Whitesides, G.M. (2005). Self-assembled monolayers of thiolates on metals as a form of nanotechnology. *Chem. Rev.*, **105**, pp. 1103–1170.

17. de Boer, B., Frank, M.M., Chabal, Y.J., Jiang, W.R., Garfunkel, E., and Bao, Z. (2004). Metallic contact formation for molecular electronics: Interactions between vapor-deposited metals and self-assembled monolayers of conjugated mono- and dithiols. *Langmuir*, **20**, pp. 1539–1542.

18. Lau, C.N., Steward, D.R., Williams, R.S., and Bockrath, M. (2004). Direct observation of nanoscale switching centers in metal/molecule/metal structures. *Nano Lett.*, **4**, pp. 569–572.

19. Slowinski, K., Fong, H.K.Y., and Majda, M. (1999). Mercury-mercury tunneling junctions. 1. Electron tunneling across symmetric and asymmetric alkanethiolate bilayers. *J. Am. Chem. Soc.*, **121**, pp. 7257–7261.

20. Chiechi, R.C., Weiss, E.A., Dickey, M.D., and Whitesides, G.M. (2008). Eutectic gallium-indium (EGaIn): A moldable liquid metal for electrical characterization of self-assembled monolayers. *Angew. Chem., Int. Ed.*, **47**, pp. 142–144.

21. Loo, Y.L., Lang, D.V., Rogers, J.A., and Hsu, J.W.P. (2003). Electrical contacts to molecular layers by nanotransfer printing. *Nano Lett.*, **3**, pp. 913–917.

22. Bonifas, A.P., and McCreery R.L. (2010). 'Soft' Au, Pt and Cu contacts for molecular junctions through surface-diffusion-mediated deposition. *Nat. Nanotech.*, **5**, pp. 612–617.

23. Li, T., Hauptmann, J.R., Wei, Z., Petersen, S., Bovet, N., Vosch, T., Nygård, J., Hu, W.P., Liu, Y., Bjørnholm, T., Nørgaard, K., and Laursen, B.W. (2012). Solution-processed ultrathin chemically derived graphene films as soft top contacts for solid-state molecular electronic junctions. *Adv. Mater.*, **24**, pp. 1333–1339.

24. Li, T., Jevric, M., Hauptmann, J.R., Hviid, R., Wei, Z., Wang, R., Reeler, N.E.A., Thyrhaug, E., Petersen, S., Meyer, J.A.S., Bovet, N., Vosch, T., Nygård, J., Qiu, X.H., Hu, W.P., Liu, Y., Solomon, G.C., Kjaergaard, H.G., Bjørnholm, T., Nielsen, M.B., Laursen, B.W., and Nørgaard, K. (2013). Ultrathin reduced graphene oxide films as transparent top-contacts for light switchable solid-state molecular junctions. *Adv. Mater.*, **25**, pp. 4164–4170, DOI: 10.1002/ adma.201300607.

25. van Hal, P.A., Smits, E.C.P., Geuns, T.C.T., Akkerman, H.B., de Brito, B.C., Perissinotto, S., Lanzani, G., Kronemeijer, A.J., Geskin, V., Cornil, J., Blom, P.W.M., de Boer, B., and de Leeuw, D.M. (2008). Upscaling, integration and electrical characterization of molecular junctions. *Nat. Nanotech.*, **3**, pp. 749–754.

26. Kushmerick, J.G., Holt, D.B., Yang, J.C., Naciri, J., Moore, M.H., and Shashidhar, R. (2002). Metal-molecule contacts and charge transport across monomolecular layers: Measurement and theory. *Phys. Rev. Lett.*, **89**, p. 086802.

27. Amlani, I., Rawlett, A.M., Nagahara, L.A., and Tsui, R.K. (2002). An approach to transport measurements of electronic molecules. *Appl. Phys. Lett.*, **80**, pp. 2761–2763.

28. Long, D.P., Patterson, C.H., Moore, M.H., Seferos, D.S., Bazan, G.C., and Kushmerick, J.G. (2005). Magnetic directed assembly of molecular junctions. *Appl. Phys. Lett.*, **86**, p. 153105.

29. Blum, A.S., Kushmerick, J.G., Long, D.P., Patterson, C.H., Yang, J.C., Henderson J.C., Yao, Y.X., Tour, J.M., Shashidhar, R., and Ratna, B.R. (2005). Molecularly inherent voltage-controlled conductance switching. *Nat. Mater.*, **4**, pp. 167–172.

30. Andres, R.P., Bielefeld, J.D., Henderson, J.D., Janes, D.B., Kolagunta, V.R., Kubiak, C.P., Mahoney, W.J., and Osifchin, R.G. (1996). Self-assembly of two-dimensional superlattice of molecularly linked metal clusters. *Science*, **273**, pp. 1690–1693.

31. Bourgoin, J.P., Kergueris, C., Lefevre, E., and Palacin, S. (1998). Langmuir–Blodgett films of thiol-capped gold nanoclusters: Fabrication and electrical properties. *Thin Solid Films*, **327**, pp. 515–519.

32. Nørgaard, K., Weygand, M.J., Kjaer, K., Brust, M., and Bjørnholm, T. (2004). Adaptive chemistry of bifunctional gold nanoparticles at the air/water interface. A synchrotron x-ray study of giant amphiphiles. *Faraday Discuss.*, **125**, pp. 221–233.

33. Liao, J., Bernard, L., Langer, M., Schönenberger, C., and Calame, M. (2006). Reversible formation of molecular junctions in 2D nanoparticle arrays. *Adv. Mater.*, **18**, p. 2444.

34. van der Molen, S.J., Liao, J.H., Kudernac, T., Agustsson, J.S., Bernard, L., Calame, M., van Wees, B.J., Feringa, B.L., and Schönenberger, C. (2009). Light-controlled conductance switching of ordered metal-molecule-metal devices. *Nano Lett.*, **9**, pp. 76–80.

35. Hassenkam, T., Nørgaard, K., Iversen, L., Kiely, C.J., Brust, M., and Bjørnholm, T. (2002). Fabrication of 2D gold nanowires by self-assembly of gold nanoparticles on water surfaces in the presence of surfactants. *Adv. Mater.*, **14**, pp. 1126–1130.

36. Hassenkam, T., Moth-Poulsen, K., Stuhr-Hansen, N., Nørgaard, K., Kabir, M.S., and Bjørnholm, T. (2004). Self-assembly and conductive properties of molecularly linked gold nanowires. *Nano Lett.*, **4**, pp. 19–22.

37. Dadosh, T., Gordin, Y., Krahne, R., Khivrich, I., Mahalu, D., Frydman, V., Sperling, J., Yacoby, A., and Bar-Joseph, I. (2005). Measurements of the conductance of single conjugated molecules. *Nature*, **436**, pp. 677–680.

38. Jana, N.R., Gearheart, L., and Murphy, C.J. (2001). Wet chemical synthesis of high aspect ratio cylindrical gold nanorods. *J. Phys. Chem. B.*, **105**, pp. 4065–4067.

39. Murphy, C.J., and Jana, N.R. (2002). Controlling the aspect ratio of inorganic nanorods and nanowires. *Adv. Mater.*, **14**, pp. 80–82.
40. Nikoobakht, B., and El-Sayed, M.A. (2001). Evidence for bilayer assembly of cationic surfactants on the surface of gold nanorods. *Langmuir*, **17**, pp. 6368–6374.
41. Sharma, V., Park, K., and Srinivasarao, M. (2009). Colloidal dispersions of gold nanorods: Historical background, optical properties, seed-mediated synthesis, shape separation and self-assembly. *Mat. Sci. Eng. R.*, **65**, pp. 1–38.
42. Murphy, C.J., Thompson, L.B., Alkilany, A.M., Sisco, P.N., Boulos, S.P., Sivapalan, S.T., Yang, J.A., Chernak, D.J., and Huang, J. (2010). The many faces of gold nanorods. *J. Phys. Chem. Lett.*, **1**, pp. 2867–2875.
43. Xiao, J., and Qi, L. (2011). Surfactant-assisted, shape-controlled synthesis of gold nanocrystals. *Nanoscale*, **3**, pp. 1383–1396.
44. Guerrero-Martínez, A., Pérez-Juste, J., Carbó-Argibay, E., Tardajos, G., and Liz-Marzán, L.M. (2009). Gemini-surfactant-directed self-assembly of monodisperse gold nanorods into standing superlattices. *Angew. Chem., Int. Ed.*, **48**, pp. 9484–9488.
45. Sreeprasad, T.S., Samal, A.K., and Pradeep, T. (2008). One-, two-, and three-dimensional superstructures of gold nanorods induced by dimercaptosuccinic acid. *Langmuir*, **24**, pp. 4589–4599.
46. Pramod, P., and Thomas, K.G. (2008). Plasmon coupling in dimers of Au nanorods. *Adv. Mater.*, **20**, pp. 4300–4305.
47. Kumar, J., and Thomas, K.G. (2011). Surface-enhanced Raman spectroscopy: Investigations at the nanorod edges and dimer junctions. *J. Phys. Chem. Lett.*, **2**, pp. 610–615.
48. Zhong, L., Zhou, X., Bao, S., Shi, Y., Wang, Y., Hong, S., Huang, Y., Wang, X., Xie, Z., and Zhang, Q. (2011). Rational design and SERS properties of side-by-side, end-to-end and end-to-side assemblies of Au nanorods. *J. Mater. Chem.*, **21**, pp. 14448–14455.
49. Petersen, A.B., Thyrhaug, E., Jain, T., Kilsaa, K., Bols, M., Moth-Poulsen, K., Harrit, N., and Bjørnholm, T. (2010). First step in chemical preparation of metal nanogaps bridged by thiol end-capped molecular wires. *J. Phys. Chem. B*, **114**, pp. 11771–11777.
50. Wijaya, A., and Hamad-Schifferli, K. (2008). Ligand customization and DNA functionalization of gold nanorods via round-trip phase transfer ligand exchange. *Langmuir*, **24**, pp. 9966–9969.
51. Ofir, Y., Samanta, B., and Rotello, V.M. (2008). Polymer and biopolymer mediated self-assembly of gold nanoparticles. *Chem. Soc. Rev.*, **37**, pp. 1814–1825.

52. Jain, T., Roodbeen, R., Reeler, N.E.A., Vosch, T., Jensen, K.J., Bjørnholm, T., and Nørgaard, K. (2012). End-to-end assembly of gold nanorods via oligopeptide linking and surfactant control. *J. Colloid. Interf. Sci.*, **376**, pp. 83–90.

53. Chen, X., Braunschweig, A.B., Wiester, M.J., Yeganeh, S., Ratner, M.A., and Mirkin, C.A. (2009). Spectroscopic tracking of molecular transport junctions generated by using click chemistry. *Angew. Chem., Int. Ed.*, **48**, pp. 5178–5181.

54. Murphy, C.J., and Orendorff, C.J. (2005). Alignment of gold nanorods in polymer composites and on polymer surfaces. *Adv. Mater.*, **17**, pp. 2173–2177.

55. Ahmed, W., Kooij, E.S., van Silfhout, A., and Poelsema, B. (2009). Quantitative analysis of gold nanorod alignment after electric field-assisted deposition. *Nano Lett.*, **9**, pp. 3786–3794.

56. Roskov, K.E., Kozek, K.A., Wu, W.C., Chetri, R.K., Oldenburg, A.L., Spontak, R.J., and Tracy, J.B. (2011). Long-range alignment of gold nanorods in electrospun polymer nano/microfibers. *Langmuir*, **27**, pp. 13965–13969.

57. Kuemin, C., Nowack, L., Bozano, L., Spencer, N.D., and Wolf, H. (2012). Oriented assembly of gold nanorods on the single-particle level. *Adv. Funct. Mater.*, **22**, pp. 702–708.

58. Taub, N., Krichevski, O., and Markovich, G. (2003). Growth of gold nanorods on surfaces. *J. Phys. Chem. B*, **107**, pp. 11579–11582.

59. Wei, Z., Mieszawska, A.J., and Zomborini, F.P. (2004). Synthesis and manipulation of high aspect ratio gold nanorods grown directly on surfaces. *Langmuir*, **20**, pp. 4322–4326.

60. Jain, T., Lara-Avila, S., Kervennic, Y.-V., Moth-Poulsen, K., Nørgaard, K., Kubatkin, S., and Bjørnholm, T. (2012). Aligned growth of gold nanorods in PMMA channels: Parallel preparation of nanogaps. *ACS Nano*, **6**, pp. 3861–3867.

Chapter 13

Toward Circuit Design in Single-Molecule Electronics

Jaap Hoekstra
Department of Electrical Engineering, Mathematics, and Computer Science, Delft University of Technology, The Netherlands
j.hoekstra@tudelft.nl

Single-molecule devices may play an important role in future molecular electronics, like current solid–state devices in micro- and nanoelectronics, when the devices possess interesting electronic properties and can be modeled with circuit models that fit in the general circuit description used in today's electronic information processing.

One of the interesting electronic properties is the phenomenon of Coulomb blockade, which is observed when experimenting with single molecules trapped in a metal nanogap, if the coupling between the electrons in the electrodes and those in the molecule is weak.

This chapter clarifies the modeling of single molecules that are trapped between two or three electrodes and are controlled like in a solid-state device. The modeling concept is based on the assumption that the accumulation of charge on the single molecule can be represented by a capacitive circuit model.

Handbook of Single-Molecule Electronics
Edited by Kasper Moth-Poulsen

ISBN 978-981-4463-38-6 (Hardcover), 978-981-4463-39-3 (eBook)
www.panstanford.com

Among others, the Coulomb energy, the addition energy, and the energy scales involved are presented and discussed. Also how these energy quantities can be derived from measurements on three-terminal devices represented by so-called diamond plots is shown, indicating Coulomb blockade.

When discussing the model, special attention will be given to the possible values of the addition energy and when this amount of energy has to be delivered. An explanation is given why *e* times the critical voltage could be significantly below 1 eV, still an unsolved problem for molecules in metal nanogaps [1].

Single-molecule devices have been proposed in which the molecule is sandwiched between two metal wires, forming a conductive path, while a third lead is used to control the current through this path. In such a way, an electronic switch is formed. These switches might be used to design electronic circuits that include single molecules and might open the way to combine modern electronic technologies with single-molecule devices. The possible advantage of such an approach is that it opens the way for single-electron transport, or few-electron transport, through the molecule, resulting in very small currents through the switch—leading to low power dissipation.

To make modeling possible on the level of circuits, we have to translate the physical or electrochemical quantities—such as charge, Coulomb energy, electrostatic charging energy, addition energy, or energy levels—to the circuit quantities, charge, voltage difference, potential, current, and energy. The translation is done based on the impulse circuit model, which models single-electron transport through a tunnel barrier as an instantaneous charge pulse within a small time slot [2, 3]. In this model, the time associated with the transport of charge by tunneling or with the transport through the single molecule can be taken into account, which is necessary to make the description of current on the circuit level possible.

13.1 Tunneling, Coulomb Blockade, and Addition Energy

Consider an arrangement in which the single molecule is sandwiched between a source and a drain electrode, while a third electrode—the gate electrode—is applied to control the current through the molecule. Now, an important and distinguishing property of single molecules in solid-state devices is the strength of the coupling between the electrons in the electrodes (or leads) and the electrons in the molecule. If the coupling is weak, it is possible to describe all electrical properties of the system with well-defined energy levels and well-defined voltages across the molecule and a gate electrode, across the source electrode and the molecule, and across the drain electrode and the molecule.

The transport of electrons from the source to the molecule and from the molecule to the drain can be described as quantum mechanical tunneling; the interface between the leads and the molecule acts as tunneling barriers. According to their wave nature, electrons can "tunnel" through potential barriers, even if the total energy of an electron is smaller than the energy needed to overcome the barrier in a classical way. Electrons tunnel through the barriers and flow from source to drain; the result is a current. Weak coupling allows a description in terms of a two-step process in which an electron first tunnels from the source to the molecule, and then tunnels from the molecule to the drain.

Coulomb blockade is an electronic phenomenon observed in small metallic systems that include two tunnel junctions and a small metallic island in between; e.g., see Ref. [4]. For values of the voltages in the system smaller than certain critical values, the tunneling through the system is blocked. This phenomenon is generally associated with the quantization of charge—i.e., tunneling can be done only by whole electrons.

To overcome the Coulomb blockade, the so-called addition energy U_{add} has to be provided by energy sources in the electronic circuitry that excites the trapped molecule. The expression for the addition energy consists of two terms: the first term—the charging energy E_{ce}—associated with the energy necessary to bring a single electron on a classical "metallic island" in between the leads; and the second term—the quantum kinetic energy Δ—due to the fact that we are dealing with a molecule having separate energy levels instead of a metallic island. We note:

$$U_{add} = E_{ce} + \Delta \tag{13.1}$$

The charging energy is subject to a classical description in terms of capacitances, resistances, voltage and current sources, and single-electron charges. As such it does play an important role in modeling the single-molecule devices for application in single-molecule electronics. It is necessary to understand that any modeling effort of real devices involves a theoretical framework on a certain level. In general, in modeling single-molecule electronics, four different modeling levels can be distinguished at least.

At the lowest level, the molecule and the interaction with the gap are modeled by quantum physics or quantum chemistry. This is the level at which the tunneling of electrons through a potential barrier is described. At the level above this, phenomena are described by classical physics, such as thermodynamics. This is the level at which Coulomb blockade can be described as a condition that relies on the reduction in the free (electrostatic) energy of the system as a result of the tunneling event; e.g., see Ref. [5]. The distinction between these two levels is important: phenomena in one level cannot be described in the other. The "language" of modeling is different in the levels.

A similar problem exists when describing the analysis and synthesis of electronic circuits, including circuits with single-molecule devices. Circuits are modeled at the circuit level: phenomena are described in terms of the constitutional relations of the circuit components and the Kirchhoff current and voltage laws. These last two laws ensure energy conservation (Tellegen's theorem) before and after tunneling; any energy loss, if present, is supplied by sources during the tunneling event.

Before answering the question of how much energy is needed to let an electron tunnel onto a molecule, two simple questions has to be answered first. How much energy is needed to tunnel onto a metallic island? And, when is this amount of energy provided?

Figure 13.1a shows a small metallic island embedded in an insulating medium. We can view the system as a single capacitance C, in which the island forms one electrode of this capacitor and the environment forms the other electrode. (Both sides of the capacitor are not connected!)

The increase in energy when an electron is added to this island, to which already n electrons have been tunneled, is called the single–

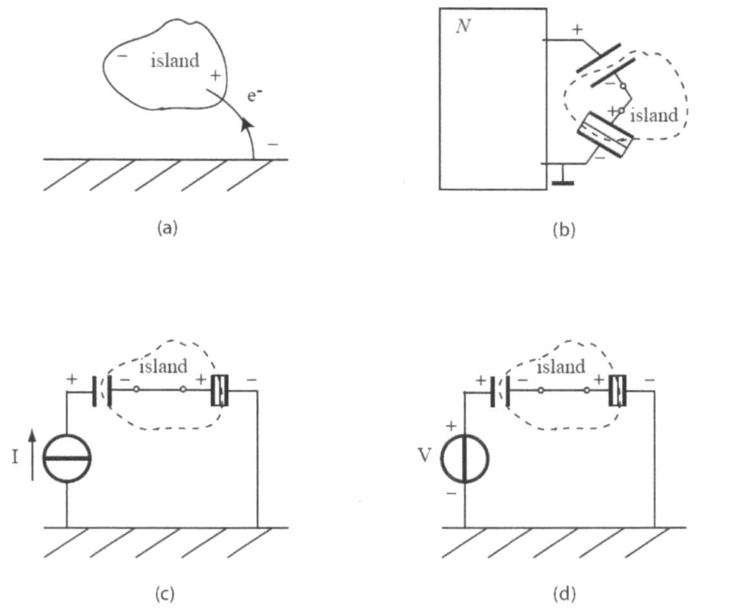

Figure 13.1 Tunneling of an electron toward an isolated island: (a) nonelectric energy source; (b) energy source in the electric network N; (c) energy provided by a current source; and (d) energy provided by an ideal voltage source.

electron charging energy $E_{ce}(n)$.

$$E_{ce}(n) = \frac{((n+1)e)^2}{2C} - \frac{(ne)^2}{2C} = \frac{e^2}{2C}(2n+1) \tag{13.2}$$

For initially uncharged islands, the expression for the single-electron charging energy $E_{ce}(n=0)$ is called the Coulomb energy E_C:

$$E_C = \frac{e^2}{2C} \tag{13.3}$$

Here C is the total capacitance seen by the island. Often the Coulomb energy is interpreted as the energy barrier felt by a single electron moving onto an electrically neutral island. That is, tunneling is forbidden until this barrier can be surmounted when an energy source is applied that provides enough energy *during* tunneling—a Coulomb blockade exists if enough energy is not supplied.

However, the question how much energy must be delivered to let an electron tunnel onto a molecular island and when this amount of energy has to be supplied is a bit more involved. To better understand the system, we represent the island by a capacitor and a tunnel(ing) junction and excite the island by an electric network (see Fig. 13.1b). This structure is a model to consider tunneling of single electrons through a tunnel barrier onto the (molecular) island. In the network, the excitation is done by an energy source. To clarify the properties of the system, we will discuss two cases: the energy source being an ideal current source or an ideal voltage source (Fig. 13.1c,d). We also have to distinguish between a zero or nonzero tunneling time.

13.2 Island Excited by an Ideal Current Source and Zero Tunneling Time

The current source is switched on at $t = 0$ and has a constant and finite value (it is a bounded current). Both the capacitor and the tunnel junction are initially uncharged. We start the analysis by finding the criteria for Coulomb blockade at the tunnel junction, assuming zero tunneling time.

The current through the circuit (Fig. 13.1c) will start charging both the junction capacitance and the capacitor. The voltage across the capacitor will increase continuously; the voltage across the tunnel junction will rise until a critical voltage is reached. Upon reaching this critical voltage, an electron will tunnel through the tunnel junction.

To find the value of the critical voltage, we use the property that the tunneling of a single electron is energy conserving (tunneling is elastic), which must be the case because in zero tunneling time, the current source (the source of energy) can neither deliver nor absorb any energy. If the energy has to be the same, the voltage across the tunnel junction before tunneling, $v_{TJ}(t_{0-})$, must be the same in absolute value (the energy of the capacitor is a quadratic function of the voltage across it) as the voltage across the tunnel junction after tunneling, $v_{TJ}(t_{0+})$. The electron tunnels from the negative to the positive side of the tunnel junction. At the same time, there cannot be any charge distribution in the circuit or in the island (the molecule); in zero time, a constant, finite current cannot displace any charge.

The critical voltage, v_{TJ}^{cr}, must be:

$$v_{TJ}^{cr} = \frac{e}{2C_{TJ}} \tag{13.4}$$

Here C_{TJ} is the capacitance of the tunnel junction only. Equation 13.4 predicts the critical voltage across the tunnel junction in case of a current source as energy supplier, and assuming a zero tunneling time.

Some remarks must be made. The current source charges the tunnel junction. As soon as the voltage across the tunnel junction reaches the critical voltage, a single electron will tunnel through the junction, and by doing so, the voltage across the junction will be reversed. The voltage being reversed, the source starts to discharge the junction and recharge it again (see Fig. 13.2). This oscillatory behavior in time is known as Coulomb oscillation and has been verified experimentally [6]. A second remark must be made on the inclusion of a tunneling time. Quantum mechanically, there is only a probability to find the electron on one side or the other side of the tunnel barrier; this probability changes with time. Classically, the electron is at one side or the other, and it is useful to define a tunneling time for the electron to pass the barrier. This tunneling time is generally taken to be in the order of 10^{-15} s. To include a tunneling time in the (electronic) circuit description, we need a more detailed circuit model of the tunneling of the electron. This model is

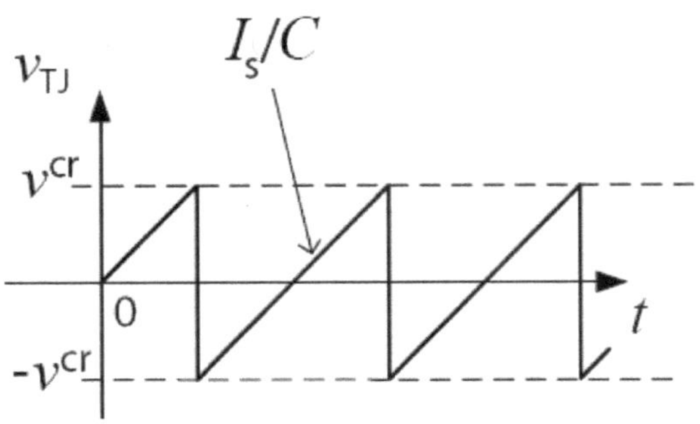

Figure 13.2 Coulomb oscillations at the tunnel junction in a circuit in which the (molecular) island is excited by a current source and the tunneling time is zero; for nonzero tunneling times, Coulomb oscillations can also appear, however at a lower critical voltage due to the redistribution of charge on the island.

treated in the next section. But it may be already clear that including a tunneling time will lead to the redistribution of charges on the island before the electron actually arrives at the other side of the tunnel junction. This redistribution leads to lower critical voltages. The last remark concerns the energy delivered by the source. It is clear that *during* tunneling, no energy is delivered by the source at all; in zero time, the current source cannot deliver any energy. All energy necessary has been delivered by the source *before* the actual tunnel event takes place; the junction capacitance must be charged.

Ideal current sources do not exist; it would require the source to be able to deliver any amount of energy. A real current source can be described by an ideal source in parallel with a resistor. This changes the behavior of the tunneling junction only slightly. The critical voltage remains the same, and the tunneling remains the same, but the voltage as a function of time will differ: the voltage during charging the capacitor from $v_{TJ} = 0$ until the critical voltage is reached follows $v_{TJ} = RI_s[1 - \exp(-t/RC)]$.

The analysis up to now has been done without taking into account the role of temperature. In general, due to temperature, some electrons may have enough energy to tunnel just before v_{TJ}^{cr} is reached and some dissipative mechanism during the tunnel event is necessary; however, at low temperatures, the above analyses are fine.

13.3 Island Excited by an Ideal Voltage Source and Zero Tunneling Time

We now consider the same structure but excited by an ideal voltage source. We are going to raise the voltage (ramp the voltage) until the critical voltage is reached. The circuit (Fig. 13.1d) consists of only an ideal voltage source and capacitances and will behave very differently because of this. First we drive some formula. Applying the Kirchhoff voltage law, we get

$$v(t) = v_C(t) + v_{TJ}(t) \tag{13.5}$$

Here $v(t)$ has the value of the voltage source V. Now we include the existence of an island charge (additional charge on the molecule):

$$q_i(t) = -q_C(t) + q_{TJ}(t) \tag{13.6}$$

Combining Eq. 13.5 and Eq. 13.6 and using the constitutive relation of the capacitor, $q(t) = Cv(t)$, we obtain expressions for the voltages across the capacitor and the tunnel junction:

$$v_C(t) = \frac{C_{TJ}}{C + C_{TJ}} v(t) - \frac{1}{C + C_{TJ}} q_i(t) \tag{13.7}$$

and

$$v_{TJ}(t) = \frac{C}{C + C_{TJ}} v(t) + \frac{1}{C + C_{TJ}} q_i(t) \tag{13.8}$$

We also obtain, using the constitutive relation of the capacitor, expressions for the charges on the capacitances:

$$q_C(t) = \frac{CC_{TJ}}{C + C_{TJ}} v(t) - \frac{C}{C + C_{TJ}} q_i(t) \tag{13.9}$$

And

$$q_{TJ}(t) = \frac{CC_{TJ}}{C + C_{TJ}} v(t) + \frac{C_{TJ}}{C + C_{TJ}} q_i(t) \tag{13.10}$$

Now we consider the *vi* relationship for each capacitance:

$$i(t) = C\frac{dv}{dt} \tag{13.11}$$

And find expressions for the current $i(t)$ through both capacitances:

$$i_C(t) = \frac{CC_{TJ}}{C + C_{TJ}} \frac{dv}{dt} - \frac{C}{C + C_{TJ}} \frac{dq_i}{dt} \tag{13.12}$$

And

$$i_{TJ}(t) = \frac{CC_{TJ}}{C + C_{TJ}} \frac{dv}{dt} + \frac{C_{TJ}}{C + C_{TJ}} \frac{dq_i}{dt} \tag{13.13}$$

Of course, due to the conservation of charge (that leads to the Kirchhoff current law), the current through both capacitances must be the same. Hence the island charge q_i must be piecewise constant. This has two consequences. First, the current through the circuit is zero before and after tunneling; this is conform classical descriptions of an ideal voltage source exciting two capacitances. Second, during tunneling (in zero time), the charge on the island and thus charges on both capacitances change. To describe this, we see the voltages on both capacitances changing stepwise before and after a tunnel event, what may lead to a current through the capacitances and through the source. This current can be described using a (Dirac) delta pulse: a pulse of infinite short duration transporting a finite amount of charge. The current described by a delta pulse is in circuit theory known as an unbounded current. We note that there is an immediate redistribution of charge in the circuit and on the island. The delta current can be calculated to be:

$$i = \frac{C}{C + C_{TJ}} e\delta(t) \tag{13.14}$$

We can calculate the energy delivered by the source necessary to tunnel onto the island if we know the critical voltage at the tunnel junction and thus the critical voltage of the source. To find the critical voltage, we again use the criterion that tunneling of a single electron is energy conserving, i.e., the voltage is reversed during tunneling:

$$v_{TJ}(t_{0+}) = -v_{TJ}(t_{0-}) \tag{13.15}$$

For a derivation of this condition, see Appendix B. We find the critical voltage of the source to be:

$$v_s^{cr} = \frac{e}{2C} \tag{13.16}$$

The energy delivered by the source w_s can be calculated, using the definition of the delta function (see Appendix A). We find:

$$w_s = \frac{e^2}{2(C + C_{TJ})} \tag{13.17}$$

This amount of energy equals the Coulomb energy.

Again, a few comments are in place. First, the description of a delta pulse current during tunneling is consistent with the previous description of the current source excited tunneling too. In that case, an electron tunneled toward the positive side of the junction capacitance and the tunneling can be modeled by

$$i = e\delta(t) \tag{13.18}$$

In this case, due to the well-defined (constant) current source, immediate redistribution of charges cannot take place. Second, we saw that the voltage source delivered energy during the tunnel event. Because the tunneling junction did not absorb any energy, all energy was stored on the capacitor, which increased its energy (there is more charge stored on this capacitor). Increasing the voltage above the critical voltage will cause more electrons to tunnel onto the island. The island behaves as an electron box. This behavior is also verified experimentally.

13.4 Single-Electron Tunneling Transistor

Now we change the structure and replace the capacitor in Fig. 13.1 by another tunnel barrier represented by a tunnel junction. Now the structure describes a molecular island separated from the outside circuitry by two tunnel barriers. To understand its behavior, we have to evaluate two possibilities: either the electron tunnels *from* the island through the tunnel junction TJ1, or the electron tunnels *onto* the island through TJ2. We start exciting the circuit with an ideal voltage source v_s. The ideal voltage source allows immediate redistribution of charges through the circuit, and also on the island.

First, we consider the tunneling through TJ2, while having an uncharged island. After tunneling, the island charge becomes $-e$. The critical voltage is found using the tunnel condition $v_{TJ2}(t_{0+}) = -v_{TJ2}(t_{0-})$:

$$\frac{C_{TJ2}}{C_{TJ1}+C_{TJ2}}v_s^{cr} - \frac{e}{C_{TJ1}+C_{TJ2}} = -\frac{C_{TJ2}}{C_{TJ1}+C_{TJ2}}v_s^{cr} \tag{13.19}$$

Giving

$$v_s^{cr} = \frac{e}{2C_{TJ2}} \tag{13.20}$$

And the critical voltage across the tunnel junction becomes:

$$v_{TJ1}^{cr} = \frac{C_{TJ2}}{C_{TJ1}+C_{TJ2}}v_s^{cr} = \frac{e}{2(C_{TJ1}+C_{TJ2})} \tag{13.21}$$

Second, if we consider the tunneling through the junction TJ1, while the island charge becomes zero again, we find that the critical voltage is always smaller than the voltage across this junction. So once an electron tunnels on the island, it immediately tunnels from the island again. The same holds for the next electron that can tunnel. The result is a current through the structure. When we do the same calculation for the first tunneling from the island, we find that the critical voltage for the source is:

$$v_s^{cr} = \frac{e}{2C_{TJ1}} \tag{13.22}$$

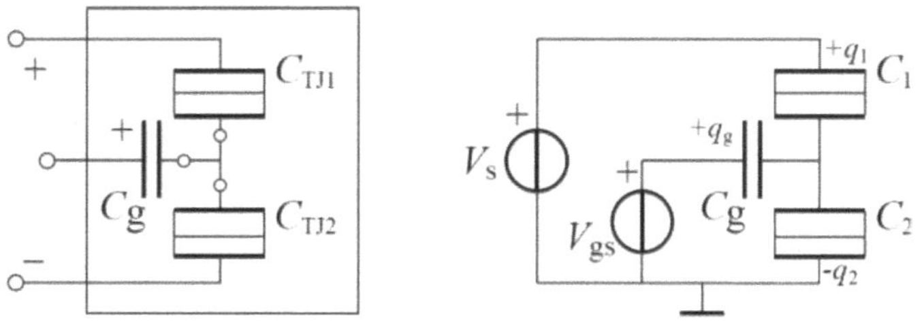

Figure 13.3 SET transistor subcircuit; SET transistor circuit excited by ideal voltage sources.

And we come to the conclusion that the structure will conduct a current as soon as the lowest critical value is met by the source.

We will now consider the SET transistor, a structure in which the island in a two-junction structure is connected to a normal capacitor (see Fig. 13.3). In the single-electron tunneling (SET) transistor, the charge on the island (or molecule) can be controlled by the gate-source voltage source. We anticipate that the resulting behavior is the behavior of the double-junction structure with a variable critical voltage, because the charge on the gate capacitor will be part of the island charge and as such will influence the charge distribution at both tunnel junctions.

We do the analysis in the case of ideal voltage sources. Writing down the two loop equations (KVL) and the equation expressing the island charge, we obtain three equations with three unknowns: v_{TJ1}, v_{TJ2}, and v_g.

$$\begin{aligned} v_s &= v_{TJ1} + v_{TJ2} \\ v_{gs} &= v_g + v_{TJ2} \\ q_i &= -q_1 + q_2 - q_s \end{aligned} \tag{13.23}$$

Solving using the constitutional relation of the capacitor, we obtain:

$$v_{TJ1} = \frac{C_g + C_2}{C_1 + C_2 + C_g} V_s - \frac{C_g}{C_1 + C_2 + C_g} V_{gs} - \frac{q_i}{C_1 + C_2 + C_g} \tag{13.24}$$

And

$$v_{TJ2} = \frac{C_1}{C_1 + C_2 + C_g} V_s + \frac{C_g}{C_1 + C_2 + C_g} V_{gs} + \frac{q_i}{C_1 + C_2 + C_g} \tag{13.25}$$

To find the conditions for tunneling, we start calculating $v_{TJ1;\text{after}} = -v_{TJ1;\text{before}}$ for a given value of the source V_{gs} to find the critical voltage of V_s, V_s^{cr}, for a given V_{gs}. (Having fixed V_{gs}, we increase V_s until one of the junctions starts to tunnel.) We define:

$$C_\Sigma = C_1 + C_2 + C_g \tag{13.26}$$

Assuming that tunneling through the junction TJ2 is not possible, consider tunneling through the junction TJ1 while the island charge is *ne*:

$$\begin{aligned} v_{TJ\,1;\text{before}} &= \frac{C_g + C_2}{C_\Sigma} V_s - \frac{C_g}{C_\Sigma} V_{gs} - \frac{ne}{C_\Sigma} \\ v_{TJ\,1;\text{after}} &= \frac{C_g + C_2}{C_\Sigma} V_s - \frac{C_g}{C_\Sigma} V_{gs} - \frac{(n+1)e}{C_\Sigma} \end{aligned} \tag{13.27}$$

For V_s^{cr}, using $v_{TJ\,1;\text{after}} = -v_{TJ\,1;\text{before}}$ we find:

$$v_s^{cr}(TJ\,1) = \frac{C_g}{C_g + C_2} V_{gs} + \frac{(2n+1)e}{2(C_g + C_2)} \tag{13.28}$$

Doing the same calculations for tunneling through TJ2 results in:

$$v_s^{cr}(TJ\,2) = -\frac{C_g}{C_1} V_{gs} + \frac{(1-2n)e}{2C_1} \tag{13.29}$$

The expressions for the critical voltage for the tunnel junction can be derived using Eq. 13.27:

$$\begin{aligned} v_{TJ\,1}^{cr} &= \frac{C_g + C_2}{C_\Sigma} v_s^{cr}(TJ\,1) - \frac{C_g}{C_\Sigma} V_{gs} - \frac{ne}{C_\Sigma} \\ &= \frac{e}{2C_\Sigma} \end{aligned} \tag{13.30}$$

We can do the same for the critical voltage $v_{TJ\,2}^{cr}$ and will find that the critical voltage is the same. We see that the critical voltage across the tunnel junctions is a property of the junctions and depends on the capacitances in the circuit only. We plot the lines indicating the critical voltage V_s^{cr} as a function of V_{gs} in Fig. 13.4 ($C_2 + C_g > C_1$); the shaded regions show the Coulomb blockade.

The model describes qualitatively the measurements on SET transistors, or single molecules, that show diamond patterns. The overall behavior shows periodicity along the V_g axis with period e/C_g. It shows that the behavior is the same for different values of additional or missing electrons.

The Coulomb blockade around the origin belongs to the blockade if the island charge is zero. The diamond on the left belongs to the

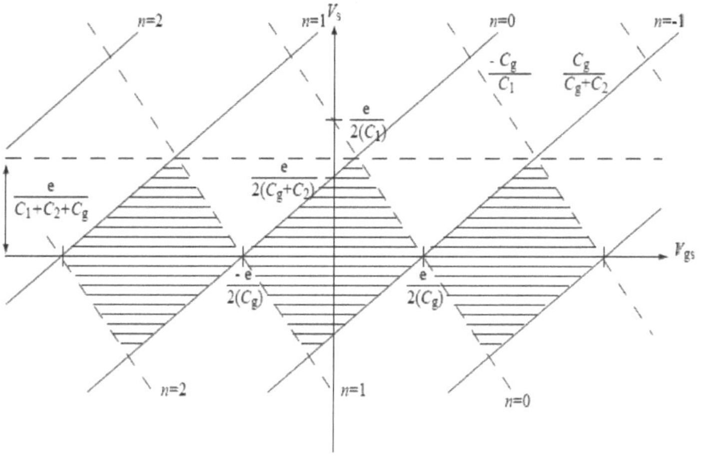

Figure 13.4 Coulomb blockades in SET transistor or single-molecule circuits showing diamond pattern; the straight lines indicate tunneling through TJ1, and the dotted lines indicate tunneling through TJ2. The shaded regions are the regions of Coulomb blockade. The v_{gs}-axis is horizontal.

blockade if the charge on the island is $(-e)$, $n = 1$. That charge on the island can increase and decrease is a consequence of the electron-box structure formed by the gate capacitor on one side and the tunnel junctions on the other side. One of the interpretations of the SET transistor characteristics is that if we take a fixed bias voltage (V_s) in absolute value smaller than $e/(2(C_g + C_2))$, the transistor can be switched into and out of the Coulomb blockade by switching the gate voltage.

Because the currents through the circuit in case of the voltage-biased SET transistor are unbounded delta pulses, an immediate redistribution of charge, on the island and through the remainder of the circuitry, will occur. This is the reason why the critical voltage deviates from $e/2C_{TJ}$, the critical voltage in case of a bounded current.

Experiments with this structure reveal that if we start increasing the source voltage above the critical voltage, a resistive behavior is observed. This resistive behavior can be described by many electrons that tunnel in a small time interval through the tunneling junctions. This point of view can be illustrated by an energy band diagram. Figure 13.5b shows the energy band diagram.

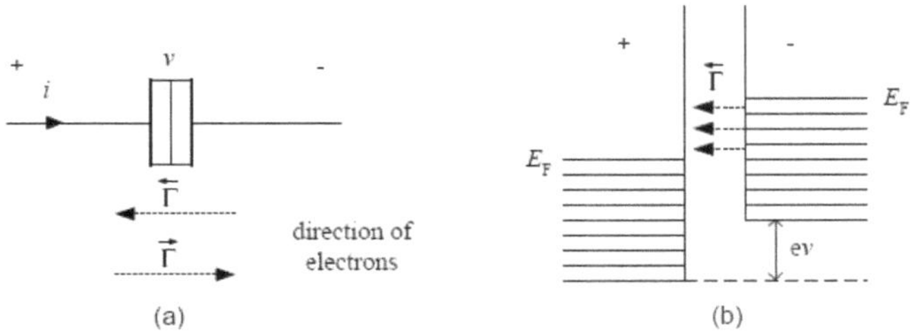

Figure 13.5 (a) Definition of the tunnel rates; and (b) energy band diagram showing empty and filled states when the voltage across the tunneling junction is fixed and above the critical voltage.

We simplify the discussion by considering a tunneling junction excited by an ideal voltage source. The fixed voltage across the junction ensures that levels filled with electrons (up to the Fermi level at the negative side of the junction) face empty energy levels at the opposite side. At a certain temperature, the actual current i through the junction is determined by the average number of electrons tunneling per unit time, from the negative to the positive side $\overleftarrow{\Gamma}$, minus the number of electrons that tunnel from the positive side to the negative side of the junction $\overrightarrow{\Gamma}$ in the same time span (this is possible when empty states are available below the Fermi level due to a nonzero temperature). Based on these definitions, the current through the junction is:

$$i = e(\overleftarrow{\Gamma} - \overrightarrow{\Gamma}) \tag{13.31}$$

At $T \approx 0\text{K}$, the average number of electrons will be proportional to the applied voltage: $\overleftarrow{\Gamma} \propto ev$. In general, the larger the potential across the tunnel junction, the more electrons will tunnel to the empty states at the other side. Knowing that we observe resistive behavior, we derive using $i = \Gamma e$ and $v = R_T i$:

$$\Gamma = \frac{v}{eR_T} = \frac{ev}{e^2 R_T} \tag{13.32}$$

Maxwell–Boltzmann statistics can be used to include the effect of a nonzero temperature:

$$\overleftarrow{\Gamma}(ev) = \frac{ev}{e^2 R_T\{1 - \exp(-\frac{ev}{k_B T})\}} \tag{13.33}$$

Tunneling in the other direction is:

$$\overrightarrow{\Gamma}(ev) = \exp(-\frac{ev}{k_B T}) \cdot \overleftarrow{\Gamma}(ev) \tag{13.34}$$

For $ev >> k_B T$, the tunnel current will be directly proportional to the voltage across the junction, and the tunneling junction can be described by a resistor.

13.5 Including Nonzero Tunneling Times

The currents through the circuit in case of the voltage-biased SET transistor, with only ideal voltage sources and capacitors, are unbounded delta pulses, and immediate redistribution of charge on the island and through the remainder of the circuitry will occur. For a bounded current, which will be the case if we have an ideal current source or resistive elements in the circuit exciting the junctions; a redistribution of charge is not possible if tunneling is immediate (i.e., the tunneling time is zero). Redistribution in this last case is possible, however, if we allow a nonzero tunneling time. During the time the electron passes the tunneling junction, charges in the circuit and on the island move due to the existence of a current during tunneling. As a consequence of this current, the critical voltage will decrease if we stick to the fact that energy is conserved during tunneling.

We understand the tunnel event as an electron sitting at both sides of the barrier at the same time, but the probability of finding the electron going from one side to the other is decreasing on one side of the barrier and increasing on the other side during tunneling. As a consequence of this description, charge is moved inside the tunneling junction. Now we consider that time is also involved. The time slot of interaction between the electron and the barrier is the tunneling time. During tunneling, the voltage/current source charges the junction capacitance. However, the leads do not always receive a current equivalent to the fundamental electron charge divided by the tunneling time. The very existence of Coulomb oscillations shows explicitly that during the tunnel event, enough current is not supplied (if there was, then the oscillations would not be there).

Seen from the outside of the tunneling junction, the charge on the junction is changed by both the tunneling electron and the charge delivered by the current during tunneling. The total charge on the tunnel junction, now, has three components: first, the amount of charge delivered at the junction before the tunnel event takes place; second, the amount of charge delivered at the junction during the tunnel event; and third, the change of charge by an amount of $(-e)$ Coulomb, e being the elementary charge, due to the actual tunnel event. To describe this last contribution, we realize that this change of charge is not associated with a current through the circuit (this argument holds only for bounded currents and not for unbounded currents). In these words, discrete (in time and charge) tunneling is combined with continuous charge transfer in the remainder of the circuit.

Again, we consider the tunnel junction excited by an ideal current source. The source will charge the junction capacitance until the tunnel condition is met. In terms of charge, the voltage on the junction just before tunneling is q/C_{TJ} and the voltage just after tunneling is $(q - e + \int_\tau i dt)/C_{TJ}$; within the parentheses, the first term is the charge on the junction before tunneling, the second term comes from the tunneling electron, and the third term is the contribution of the circuit during the tunneling time and makes it possible to take into account arbitrary circuit elements (including resistors) in the circuitry outside the junction. The only restriction here is that the current must be bounded.

Using the condition for tunneling, $v_{TJ;\text{after}} = -v_{TJ;\text{before}}$, we obtain

$$\frac{q - e + \int\limits_\tau i dt}{C_{TJ}} = -\frac{q}{C_{TJ}} \tag{13.35}$$

And using that the critical voltage is the voltage across the junction just before the tunnel event, we obtain:

$$v_{TJ}^{cr}(\tau) = \frac{e - \int\limits_\tau i dt}{2C_{TJ}}, \quad 0 \leq \int\limits_\tau i dt \leq e \tag{13.36}$$

For example, when we consider a single tunneling junction excited by a constant current I_s and showing Coulomb oscillations, we note that, for a given τ, the amplitude of the oscillations decreases depending on the value of the current, and find the critical voltage to be:

$$v_{TJ}^{cr} = \frac{e}{2C_{TJ}} - \frac{I_s \tau}{2C_{TJ}} \tag{13.37}$$

Figure 13.6 illustrates this situation.

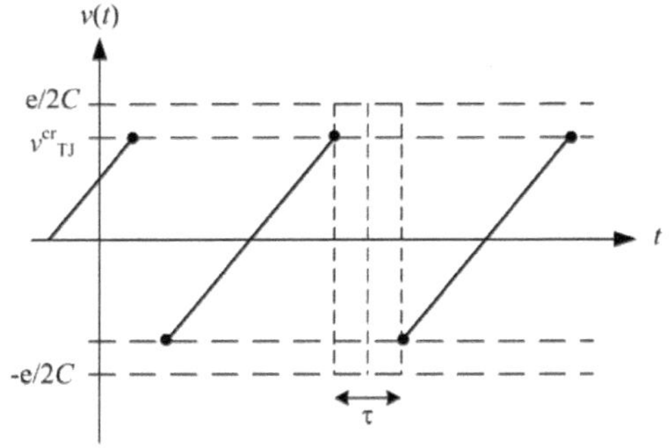

Figure 13.6 Decreased amplitude if nonzero tunneling times are considered.

The above formula—the critical voltage as a function of the current through the circuit just before the tunnel event, at a given tunneling time—makes it possible to use circuit simulators to obtain the correct value of the critical voltage and thus correctly simulate the circuits. For large values of the current source, the Coulomb oscillations are predicted to disappear (the critical voltage becomes zero). In the latter case, the amount of charge delivered at the junction during the time the electron is tunneling is equal or larger than a single-electron charge and consequently the condition for tunneling will be reached before the electron passed the junction; this will result in a continuous current of electrons.

Can the circuit model combined with the tunnel condition still model a tunneling electron and is energy conserved during the tunneling?

At the critical voltage, after time τ, the charge on the capacitor is reversed and so is the voltage across the junction. The equivalent circuit can be energy conserving only if the sum of the energy necessary for a possible temporal storage of charge, the energy gained by a discharge of charge, and the energy delivered or absorbed by the tunneling electron is zero during the transition (remember that we do not know the time evolution of the voltage across the junction during the transition). This also guarantees that the constant circuit current source did not deliver net energy during the total transition.

To find if it is possible to fulfill these requirements, we consider delta current pulses at various times within the interval t and $t + \tau$.

If the delta pulse appears immediately after the critical voltage is reached, then during the time τ, the source will deliver a charge of $I_S\tau$, but the voltage across the junction (and thus the source) will be negative during the whole interval; this is because the critical voltage for $\tau \neq 0$ is always smaller than $e/2$, while the charge contribution of the delta pulse equals e. The source will absorb energy and discharge the junction capacitance. The delta pulse itself will, however, deliver energy because the sum of the voltages just before and after is negative. An analogue reasoning for the delta pulse at the end of the interval shows that the source charged the junction capacitance and the delta pulse generates energy. It is not hard to see that at least in case of a constant current source, a net contribution of zero is obtained when the delta pulse appears just in the middle of the interval, and in which case the energy of the delta pulse is zero too. The circuit model represents a valid tunneling event. Note that in the physical description, tunneling was smeared out over the whole interval, and the circuit description is only a model. However, it is possible to describe a tunnel event and, more important, it is possible to define the expression for the critical voltage.

Because we did not take into account a possible stochastic behavior of the tunneling electron in this section, the tunnel interval τ started immediately after the critical voltage was reached. These tunnel events can also be illustrated as transitions in the capacitor's energy diagram, w_{se} being the energy stored. In Fig. 13.7, two transitions are shown. First, when $\tau = 0$, the electron is transported

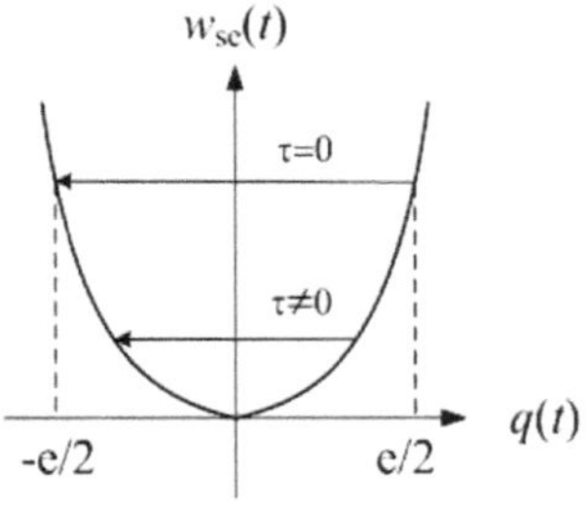

Figure 13.7 A tunnel junction charged by a bounded current at the moment the critical voltage is reached. The transitions, with and without assuming a nonzero tunneling time, conserve energy and change the polarity of the voltage across the tunneling junction.

to the other side, reversing the charge on the junction capacitance. Energy is conserved only when the charge before tunneling is $e/2$ Coulomb. Second, when $\tau \neq 0$, an electron is transported to the other side, but in the meantime the capacitor is charged by an amount $I_S\tau$. Consequently, the net charge difference is smaller than e Coulomb (the electron tunnels to the positive side). Still, energy is conserved before and after the tunnel event, because tunneling is considered to be elastic.

13.6 Circuit Perspectives

Using the above described circuit models for tunneling in island structures, it is possible, in principle, to design interesting electronic circuits; in principle because electronic design of such circuits is still at its infancy and many design issues are not yet resolved. However, it may be illustrative to show a possible example of logic design. We use the SET transistor in a circuit to perform logic operations. Remember that the SET transistor can be made by sandwiching the

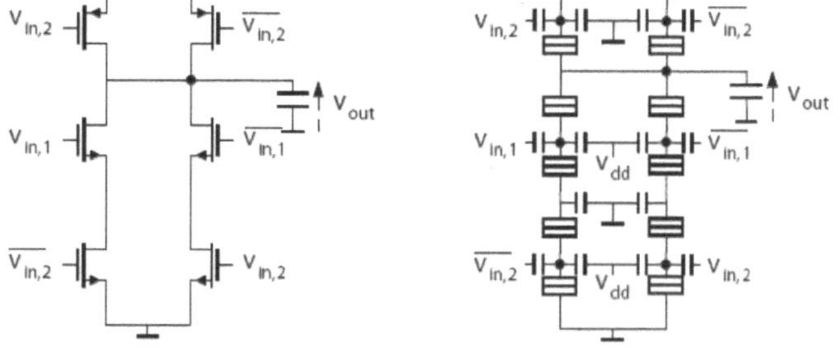

Figure 13.8 (a) Implementation of an EXOR function in standard CMOS; (b) implementation of an EXOR function in SET technology; also the inverse of both input signals is needed.

molecule between two leads and controlling the molecule by an additional capacitor.

The logic circuit [7], shown in Fig. 13.8, shows coupled SET transistors in such a way that depending on the input signal, only one of the SET transistors conducts, while the other is in the Coulomb blockade. Such a structure looks like a CMOS transistor pair. Because of this resemblance, logic gate can be designed based on known logic gate topologies in CMOS digital circuit design. The logic circuit shows the logic EXOR function.

13.7 Summary

This chapter described the modeling, by electronic circuit models, of a single molecule sandwiched between two metals wires in those cases where the interaction of the molecule and the wires can be described by a weak coupling. Experiments on those structures show diamond patterns if the current through the molecule is measured as a function of both the voltage across the molecule and a gate voltage. The diamond patterns are a consequence of the existence of a Coulomb blockade. The Coulomb blockade and the associated tunneling phenomenon can be modeled by electronic circuit models, what makes it possible to consider new circuits and integration with existing electronic technologies. The model description is based on modeling the tunnel event by an impulse current source transporting an electron charge during tunneling. The actual value of the critical voltage across the tunneling junction depends on the amount of charge that is redistributed during tunneling in the molecule and the remainder of the circuit. When the circuitry exciting the tunneling junction consists of a resistive circuitry, the critical voltage is determined by the junction capacitance of the tunneling junction only, unless a nonzero tunneling time is taken into account. In this last case, the current in the circuit determines (together with the tunneling time) the amount of charge that is redistributed through the molecule and thus the critical voltage. This critical voltage can be obtained by looking in a diamond pattern experiment. The chapter also discussed what the critical voltage would be in case of an unbounded current (Dirac delta pulse) that causes redistribution of charge. A small example illustrated the possibility to design circuits with single-molecular electronics.

Appendix A Delta Pulse Description

To describe circuits that include tunneling devices, we use the following definitions:

$$f(0^+) \triangleq \lim_{t\downarrow 0} f(t) \tag{13.38}$$

and

$$f(0^-) \triangleq \lim_{t\uparrow 0} f(t) \tag{13.39}$$

In most textbooks on circuit or system theory, models using initial conditions as above use operational calculus or Laplace transform to describe the models. For our purposes, the models are best described using (Heaviside's) operational calculus, making it possible to describe the models in the time domain. The following definitions are used for the operators p and $1/p$:

$$p \triangleq \frac{d}{dt} \tag{13.40}$$

and

$$\frac{1}{p} \triangleq \int_{-\infty}^{t} ()d\alpha \tag{13.41}$$

For a capacitor with a constant capacitance C holds:

$$i(t) = Cpv(t) \tag{13.42}$$

and

$$v(t) = \frac{1}{Cp} i(t) \tag{13.43}$$

Furthermore, the following notation is used:

- The unit-step function $\varepsilon(t)$ is defined by the relation

$$\begin{aligned} \varepsilon(t)1 &\triangleq t > 0 \\ \varepsilon(t)0 &\triangleq t < 0 \end{aligned} \tag{13.44}$$

At $t = 0$, the unit-step function is undefined but restricted between 0 and 1.

- The delta function or impulse function $\delta(t)$ is defined as:

$$\begin{aligned} &\delta(t)0 \triangleq t > 0 \\ &\int_{-\infty}^{\infty} \delta(\tau)d\tau = 1 \qquad t = 0 \\ &\delta(t)0 \triangleq t < 0 \end{aligned} \tag{13.45}$$

The $\varepsilon(t)$ and $\delta(t)$ are not regular analytic functions but generalized functions. They define each other:

$$\int_{-\infty}^{t} \delta(\tau)d\tau = \varepsilon(t) \quad \text{or} \quad \frac{1}{p}\delta(t) = \varepsilon(t) \tag{13.46}$$

And

$$\frac{d}{dt}\varepsilon(t) = \delta(t) \quad \text{or} \quad p\varepsilon(t) = \delta(t) \tag{13.47}$$

Appendix B Tunneling and the Delta Pulse Description

Tunneling is described by a delta pulse that transports the charge q while the voltage across the tunnel junction steps from v_{before} to v_{after}. Assuming a zero tunneling time, the stepping source can be represented by:

$$v(t) = v_{\text{before}}(t) + (v_{\text{after}}(t) - v_{\text{before}}(t))\varepsilon(t) \tag{13.48}$$

The energy involved in the tunnel event is in this case:

$$w_{TJ} = \int vi\,dt = q\int [v_{\text{before}} + (v_{\text{after}} - v_{\text{before}})\varepsilon(t)]\delta(t)\,dt \tag{13.49}$$

Using that all the voltages are constant just before and after the tunnel event and using the partial integration:

$$\int_{-\infty}^{\infty} \delta(t)\varepsilon(t)\,dt = \frac{1}{2}[\varepsilon^2]_{-\infty}^{\infty} = \frac{1}{2} \tag{13.50}$$

give us the important result

$$w_{TJ} = \frac{q}{2}(v_{\text{before}} + v_{\text{after}}) \tag{13.51}$$

This leads us immediately to the conclusion that for nondissipative (elastic) tunneling, the tunnel condition:

$$v_{\text{after}} = -v_{\text{before}} \tag{13.52}$$

must hold.

Problems

1. Equation 13.2 gives an expression for the single charging energy, i.e., the energy necessary to let an electron tunnel toward the molecule (island). The energy to let the first electron tunnel ($n = 0$) is

$$E_{ce}(n=0) = \frac{((n+1)e)^2}{2C} - \frac{(ne)^2}{2C} = \frac{e^2}{2C}(2n+1) = \frac{e^2}{2C},$$

   C is the total capacitance of the island. This energy can be delivered by a voltage source as in Fig. 13.1d. If we assume zero tunneling time, show that the energy delivered by the source is indeed $e^2/2C$, where C is the total capacitance.
2. Consider the case of current source, as in Fig. 13.1c. Why is the energy delivered by the current source *during* tunneling zero? What is the energy delivered by this current source *before* the electron tunnels?
3. Derive Eq. 13.29.
4. Show that the current modeling a single electron to tunnel, $i = e\delta(t)$, has indeed the dimension of a current.
5. Figure 13.7 shows how the energy is conserved during tunneling in the presented circuit model of single-electron tunneling. Compare this figure with the corresponding figure used in the so-called orthodox theory of single electronics. (This is another theory that tries to explain the experiments in Fig. 13.7, but from the point of view that tunneling will appear as soon as the tunnel event leads to less energy in the system, but without modeling a separate resistance.)

References

1. Moth-Poulsen, K., and Bjornholm, T. (2009). Molecular electronics with single molecules in solid-state devices, *Nature Nanotechnology*, pp. 551–556.
2. Hoekstra, J. (2010). *Introduction to Nanoelectronic Single-Electron Circuit Design* (Pan Stanford Publishing, Singapore).

3. Hoekstra, J. (2007). Towards a circuit theory for metallic single-electron tunneling devices, *International Journal of Circuit Theory and Applications, 35,* pp. 213–238.
4. Fulton, T., and Dolan, G. (1987). Observation of single-electron charging effects in small tunnel junctions, *Physical Review Letters,* 59, 1, pp. 109–112.
5. Likharev, L.K. (1999). Single-electron devices and their applications, *Proceedings of the IEEE,* 87, 4, pp. 606–632.
6. Bylander, J., Duty, T., and Delsing P. (2005). Current measurement by real-time counting of single electrons, *Nature,* 434, pp. 361–364.
7. Jeong, M.Y., Jeong, Y.H., Hwang, S.W., and Kim, D.M. (1997). Performance of single-electron transistor logic composed of multi-gate single-electron transistors, *Japanese Journal of Applied Physics,* 36, 11, pp. 6706–6710.

Index